Bosch
Fuel Injection
& Engine
Management

Theory of operation
Troubleshooting and service using common tools and equipment
High performance tuning
How to identify your Bosch system

by Charles O. Probst, SAE

Robert Bentley
Automotive Books and Service Manuals

Volkswagen Official Service Manuals

GTI, Golf, and Jetta Service Manual: 1985-1990 Gasoline, Diesel, and Turbo Diesel, including 16V models. *Robert Bentley.* ISBN 0-8376-0343-9

Corrado Official Factory Repair Manual: 1990. *Volkswagen United States.* ISBN 0-8376-0387-0

Passat Official Factory Repair Manual: 1990. *Volkswagen United States.* ISBN 0-8376-0377-3

Scirocco and Cabriolet Service Manual: 1985-1989, including 16V. *Robert Bentley.* ISBN 0-8376-0344-7

Volkswagen Fox Service Manual: 1987-1990, including Wagon and Sport. *Robert Bentley.* ISBN 0-8376-0340-4

Quantum Official Factory Repair Manual: 1982-1988 Gasoline and Turbo Diesel, including Wagon and Syncro. *Volkswagen United States.* ISBN 0-8376-0341-2

Vanagon Official Factory Repair Manual: 1980-1989 including Air-cooled and Water-cooled Gasoline Engines, Diesel Engine, Syncro, and Camper. *Volkswagen United States.* ISBN 0-8376-0345-5

Rabbit, Scirocco, Jetta Service Manual: 1980-1984 Gasoline Models, including Pickup Truck, Convertible, and GTI. *Robert Bentley.* ISBN 0-8376-0183-5

Rabbit, Jetta Service Manual: 1977-1984 Diesel Models, including Pickup Truck and Turbo Diesel. *Robert Bentley.* ISBN 0-8376-0184-3

Rabbit, Scirocco Service Manual: 1975-1979 Gasoline Models. *Robert Bentley.* ISBN 0-8376-0107-X

Dasher Service Manual: 1974-1981 including Diesel. *Robert Bentley.* ISBN 0-8376-0083-9

Super Beetle, Beetle and Karmann Ghia Official Service Manual Type 1: 1970-1979. *Volkswagen United States.* ISBN 0-8376-0096-0

Beetle and Karmann Ghia Official Service Manual Type 1: 1966-1969. *Volkswagen United States.* ISBN 0-8376-0416-8

Station Wagon/Bus Official Service Manual Type 2: 1968-1979. *Volkswagen United States.* ISBN 0-8376-0094-4

Fastback and Squareback Official Service Manual Type 3: 1968-1973. *Volkswagen United States.* ISBN 0-8376-0057-X

Audi Official Service Manuals

Audi 5000S, 5000CS Official Factory Repair Manual: 1984-1988 Gasoline, Turbo, and Turbo Diesel, including Wagon and Quattro. *Audi of America.* ISBN 0-8376-0370-6

Audi 100, 200 Official Factory Repair Manual: 1989, 1990. *Audi of America.* ISBN 0-8376-0372-2

Audi 80, 90 Official Factory Repair Manual: 1988-1990. *Audi of America.* ISBN 0-8376-0367-6

Audi 5000, 5000S Official Factory Repair Manual: 1977-1983 Gasoline and Turbo Gasoline, Diesel and Turbo Diesel. *Audi of America.* ISBN 0-8376-0352-8

Audi 4000S, 4000CS, and Coupe GT Official Factory Repair Manual: 1984-1987 including Quattro and Quattro Turbo. *Audi of America.* ISBN 0-8376-0373-0

Audi 4000, Coupe Official Factory Repair Manual: 1980-1983 Gasoline, Diesel, and Turbo Diesel. *Audi of America.* ISBN 0-8376-0349-8

Audi Fox Service Manual: 1973-1979. *Robert Bentley.* ISBN 0-8376-0097-9

Triumph Official Workshop Manuals

The Complete Official Triumph TR7: 1975-1981. Includes Driver's Handbook, Workshop Manual. *British Leyland Motors.* ISBN 0-8376-0116-9

Jaguar Official Workshop Manuals

The Complete Official Jaguar "E", all 6 Cylinder Models including Series 2 (4.2 E-type & 2 2). Includes Driver's Handbook, Workshop Manual, Special Tuning Manual. *British Leyland Motors.* ISBN 0-8376-0136-3

Automobile Books

The Design and Tuning of Competition Engines *Philip H. Smith, 6th edition revised by David N. Wenner* ISBN 0-8376-0076-6

New Directions in Suspension Design: Making the Fast Car Faster *Colin Campbell* ISBN 0-8376-0150-9

The Scientific Design of Exhaust and Intake Systems *Philip H. Smith and John C. Morrison* ISBN 0-8376-0309-9

The Technique of Motor Racing *Piero Taruffi with foreword by Juan Manuel Fangio* ISBN 0-8376-0228-9

Robert Bentley has published service manuals and automobile books since 1950. Please write Robert Bentley, Inc., Publishers, at 1000 Massachusetts Avenue, Cambridge, MA 02138 or call 1-800-423-4595 for a complete listing of current automotive literature, including additional Jaguar and Triumph titles and service manuals for Austin-Healey, MG, and Toyota and other foreign cars.

Bosch Fuel Injection & Engine Management

Theory of operation
Troubleshooting and service using common tools and equipment
High performance tuning
How to identify your Bosch system

by Charles O. Probst, SAE

Robert Bentley
Cambridge, Massachusetts

Published and Distributed by:

Robert Bentley, Inc., Publishers
1000 Massachusetts Avenue
Cambridge, Massachusetts 02138

Sole United Kingdom distribution by
Motor Racing Publications Limited
Unit 6, The Pilton Estate
46 Pitlake, Croydon CRO 3RY, England

Copies of this manual may be purchased from selected booksellers, automotive accessories and parts dealers, or directly from the publisher by mail. Robert Bentley Books are also available in bulk quantity for industrial or sales-promotion use. For details write to Special Sales Manager at the publisher's address.

CAUTION
It is assumed that the reader is familiar with basic automotive repair procedures. Before attempting any work on your car, read the cautions and warnings on page vii. Robert Bentley, Inc., the publisher, and the author have made every effort to ensure the accuracy of this manual. Neither can be held responsible for the result of any error in this manual.

NOTE TO USERS OF THIS MANUAL

The publisher encourages comments from the readers of this manual. These communications have been and will be carefully considered in the preparation of this and other manuals. Please write to Robert Bentley, Inc., Publishers, at the address on this page.

This manual was published by Robert Bentley, Inc., which has sole responsibility for its content. Robert Bosch Corporation has not reviewed and does not vouch for the accuracy of the technical information and procedures described in this manual.

The author and publisher would particularly like to thank and acknowledge Robert Bosch Corporation for the permission to use and reproduce many of their technical illustrations.

Many of the designations used by manufacturers and sellers to distinquish their products are claimed as trademarks. Where those designations appear in this book and Robert Bentley, Inc. was aware of a trademark claim, the designations have been printed in initial capital letters (e.g., Jetronic).

BOSCH and ⊕ are trademarks of Robert Bosch GmbH, Stuttgart, Federal Republic of Germany.

Library of Congress Catalog Card No. 89-61946
ISBN 0-8376-0300-5
Bentley Stock No. GFIB

92 91 90 89 10 9 8 7 6 5 4 3 2

The paper used in this publication is acid free and meets the requirements of American National Standard for Information Sciences—Permanence of Paper for Printed Library Materials. ∞

Since this page cannot legibly accommodate all the copyright notices, the art credits pages at the back of the book listing the source of the illustrations and photographs used, constitute an extension of the copyright page.

Front cover: Bosch LH-Jetronic air-mass sensor hot wire glows during development testing. For more information on hot wire air-mass sensors, see chapter 3 "Pulsed Injection—Theory," pages 23–25. Courtesy Bosch.

Back cover clockwise from top: *Audi 90 Quattro* IMSA-GTO race cars at Road America. Courtesy Audi Motorsport. *Volvo 740 GL Wagon* in SCCA Escort Endurance showroom stock competition at Sears Point. Courtesy Volvo Cars of North America. *BMW M3* Group "A" competition engine. Courtesy BMW of North America, Inc. *Porsche 911 Cabriolet*. Courtesy Porsche Cars of North America. Photograph reprinted with permission of Dr. Ing. h.c.F. Porsche AG. Porsche, the Porsche Crest, Carrera and Targa are registered trademarks of Dr. Ing. h.c.F. Porsche AG. *Volkswagen Golf GL*. Courtesy Volkswagen United States, Inc.

Manufactured in the United States of America

Contents:

Bosch Fuel Injection

Acknowledgments

Thanks to Robert Bosch GmbH, and special thanks to Robert Bosch Corporation, US, Automotive Service Department—especially to the Director of Service and the experts in Technical Service and Service Training. Many fuel-injection specialists have participated in the development of service information for Bosch systems, and they recognize the importance of its broad availability.

Some text and illustrations result from 10 years of cooperative development by the author and Bosch experts for use in gasoline fuel-injection audio-visual training programs. Other information comes from manufacturers, from SAE, and from the author's personal experiences. Thanks also to Volkswagen United States, Inc. and Audi of America, Inc. for their contributions.

I would also like to thank the editors who worked on getting this book to press, including Glen Grissom for many valuable comments in the book's early drafts, and at Robert Bentley, Inc., John Koenig, especially for his work on chapter 7, and John Kittredge for guiding it all through the final stages.

This book is an independent publication, not affiliated with Bosch, and has not been reviewed or approved by Robert Bosch Corporation.

Charles O. Probst
Palo Alto, CA July 1989

Please read these warnings and cautions before proceeding with maintenance and repair work.

WARNING –

● Never work under a lifted car unless it is solidly supported on stands intended for the purpose. Do not support a car on cinder blocks, hollow tiles, or other props that may crumble under continuous load. Do not work under a car that is supported solely by a jack.

● If you are going to work under a car on the ground, make sure that the ground is level. Block the wheels to keep the car from rolling. Disconnect the battery ground strap to prevent others from starting the car while you are under it.

● Never run the engine unless the work area is well ventilated. Carbon monoxide kills.

● Tie long hair behind your head. Do not wear a necktie, scarf, loose clothing, or necklace when you work near machine tools or running engines. If your hair, clothing, or jewelry were to get caught in the machinery, severe injury could result.

● Disconnect the battery ground strap whenever you work on the fuel system or electrical system. When you work around fuel, do not smoke or work near heaters or other fire hazards. Keep an approved fire extinguisher handy.

● Illuminate your work area adequately but safely. Use a portable safety light for working inside or under the car. Make sure its bulb is enclosed by a wire cage. The hot filament of an accidentally broken bulb can ignite spilled fuel or oil.

● Catch draining fuel, oil, or brake fluid in suitable containers. Do not use food or beverage containers that might mislead someone into drinking from them. Store flammable fluids away from fire hazards. Wipe up spills at once, but do not store the oily rags, which can ignite and burn spontaneously.

● Finger rings should be removed so that they cannot cause electrical shorts, get caught in running machinery, or be crushed by heavy parts.

● Keeps sparks, lighted matches, and open flame away from the top of the battery. If hydrogen gas escaping from the cap vents is ignited, it will ignite gas trapped in the cells and cause the battery to explode.

● Always observe good workshop practices. Wear goggles when you operate machine tools or work with battery acid. Gloves or other protective clothing should be worn whenever the job requires it.

CAUTION –

● If you lack the skills, tools, and equipment, or a suitable workshop for any procedure described in this manual, we suggest you leave such repairs to an authorized dealer or other qualified shop. We especially urge you to consult your authorized dealer before attempting any repairs on a car still covered by the new-car warranty.

● Before starting a job, make certain that you have all necessary tools and parts on hand. Makeshift tools, parts, and procedures will not make good repairs.

● Use pneumatic and electrical tools only to loosen threaded parts and fasteners. Never use such tools to tighten fasteners, especially on light alloy parts.

● Be mindful of the environment and ecology. Before you drain the crankcase, find out the proper way to dispose of the oil. Do not pour oil into the ground, down a drain, or into a stream, pond, or lake. Consult local ordinances that govern the disposal of wastes.

● Anti-theft radios can be rendered useless by disconnecting the battery. If power to the radio is interrupted, a protection circuit engages and disables the radio. For the radio to operate, a code must be entered into the radio after power is restored. If the wrong code is entered, the radio will lock up and can no longer be operated, even if the correct code is then entered. Make sure you know the correct code before disconnecting the battery. For more information, see your owner's manual.

Chapter 1

Bosch Fuel Injection— An Overview

1

Contents

1. INTRODUCTION

Here's my plan for you to get the most out of this book: This chapter introduces the idea of fuel injection, and tells you why virtually all cars being built today use fuel injection. I'll give you the broad picture of Bosch fuel-injection systems, and show how the different systems are separated into two basic types: pulsed and continuous injection. You'll see that what began as control of the air-fuel ratio has expanded to include the precise control of ignition timing and, often, idle rpm, leading to a new term, engine management.

In chapter 2, to help you understand the principles behind fuel injection, I'll look at the engine needs for each driving condition and also discuss the importance of fuel injection to meeting government standards for emission control and fuel economy. You'll see how the precise control of fuel metering, sometimes at a delivery rate of one drop of fuel in a few milliseconds of time, meets those needs while still providing good power.

In chapter 3, I'll describe how each different Bosch pulsed fuel injection system works, including D-Jetronic, L-Jetronic, LH-Jetronic, and Motronic. In chapter 5, I'll describe how each Bosch continuous fuel injection system works, including K-Jetronic, K-lambda, KE-Jetronic, and KE-Motronic. In chapter 4, I'll show you how to service pulsed systems, and in chapter 6, I'll show you how to service continuous systems. In many cases you can do this using only simple tools, a volt/ohmmeter (VOM) and a fuel-pressure gauge. Even if you don't want to do much of the work yourself, you'll still know enough about troubleshooting and repairs to deal with the technician who may use special tools to service your car.

Most of these fuel injection systems as installed by the manufacturer are capable of delivering extra fuel for engine modifications that increase performance. In chapter 7, for those who want more power from their fuel-injection system, I'll show you the different modifications for street-legal or off-road use, and discuss which ones work—and which ones don't. For you owners who can tweak a carburetor by ear with a screwdriver on Saturday afternoon, this book will help you accept the performance of fuel injection as it takes away your need for carburetor skills. Even so, you can do a lot more to your fuel-injected car than most people realize.

2. FUEL INJECTION IS HERE

Today's cars are changing under the hood. The tangles of vacuum hoses—as well as emission-control miseries—are being replaced by the orderly installation of fuel-injection systems. But before we talk about specific Bosch systems, you'll need to understand some basics about fuel injection.

Fig. 2-1. Electronic control of fuel injection and ignition (top) for engine management helps clean up the tangle of underhood hoses (bottom) which are necessary when vacuum circuits are used.

2.1   Air-Fuel Mixtures

Internal-combustion engines create power by burning fuel mixed with air. In gasoline-fueled engines, the proportions of air and fuel—the air-fuel ratios or "mixtures"—are of critical importance to the quality of combustion and, therefore, to engine power output and running characteristics. Since the amount of air required by the engine varies with rpm and load, the required amount of fuel varies too.

The overall purpose of the systems covered by this manual is to provide the engine with the best possible mixture—the optimum air-fuel ratio—for the constantly changing engine operating conditions.

2.2   What is Fuel Injection?

The throttle of a gasoline engine regulates only air flow into the engine. Since the proportions of air and fuel are critical, it is up to some other mechanism to meter the correct amount of fuel into the moving air. I'll call this other mechanism the fuel delivery system. The fuel delivery system—whatever type it may be—responds to throttle changes and adjusts to continuously supply the engine with a combustible mixture of air and fuel. Fuel injection is an accurate and sophisticated type of fuel delivery system.

2.3   Fuel Delivery

For modern gasoline automobiles, there are two basic types of fuel delivery systems in use today, carburetors and fuel injection. While these systems mix fuel and air, they achieve it in very different ways.

Carburetors

Carburetors take advantage of the venturi principle. Briefly, this principle says that as air flow increases, pressure decreases. Air flow through the carburetor throat, as determined by the throttle opening, creates a low pressure condition. This reduced pressure pulls fuel into the intake air stream where it is vaporized to form a combustible air-fuel mixture. A wider throttle opening causes more air flow which results in more fuel flow. A smaller throttle opening likewise reduces fuel flow. By this cause-and-effect relationship, fuel is "metered" in proportion to air flow.

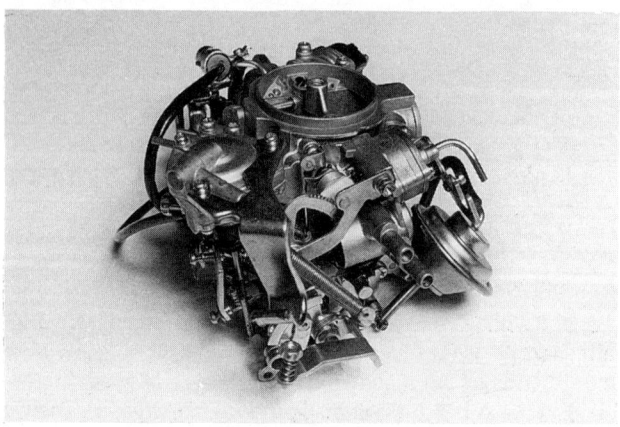

Fig. 2-2. Modern Complex feedback carburetors can be more expensive than fuel injection systems to install and service. Driveability cannot match most fuel-injected cars, and emission control is more difficult.

This relatively simple and crude fuel delivery technology has powered gasoline engines to acceptable levels of efficiency for many decades. In recent times, however, closer scrutiny of performance, fuel economy, and exhaust emissions has demanded even greater engine efficiency. This greater efficiency requires more precise control of fuel metering. Carburetors, although well developed after decades of use, are limited in their ability for precise fuel metering, especially under extreme operating conditions, even with their complex set of fuel circuits, jets, air bleeds, chokes and valves.

Fuel Injection

Fuel injection systems deliver fuel by forcing it into the incoming airstream. Fuel-injection systems actually measure the incoming air and pressurize the fuel to deliver it in precise amounts based directly on that measurement. Because fuel is delivered to the manifold under pressure, the quantity of fuel delivered can be more positively controlled. With this more positive control, fuel delivery can be more easily manipulated to meet the unique demands of extreme operating conditions. This results in greater efficiency over a wider range of operation.

3.   FUEL INJECTION TYPES

Bosch passenger-car fuel injection first appeared in 1927 as a diesel design using engine-driven pumps developing high pressure to deliver diesel fuel into the cylinders—direct injection. In 1937, these pumps were adapted to aircraft engines, with both diesel and gasoline fuel applications. In 1955, similar engine-driven gasoline injection pumps were developed for the Mercedes-Benz race cars and the 300SL production car, again injecting fuel at high pressure directly into the cylinders.

These mechanical systems used a precisely-timed, engine-driven pump to deliver measured quantities of fuel to each cylinder in time with the crankshaft and in proportion to throttle opening. The complex and expensive systems were an advance over the carburetors of their time, mainly in terms of performance, but production car applications were limited to exotic makes and special models. They were very difficult to properly set up, and proper maintenance was an expensive proposition.

The fuel-injection systems described in this book first appeared about 1967. They are completely different from engine-driven systems, delivering fuel at much lower pressures at the intake ports, and are generally electronically controlled. The early Bosch literature describes these systems as "non-driven". Bosch coined the term "Jetronic" to establish a common identity for their new designs in most of the 135 countries in which they operate.

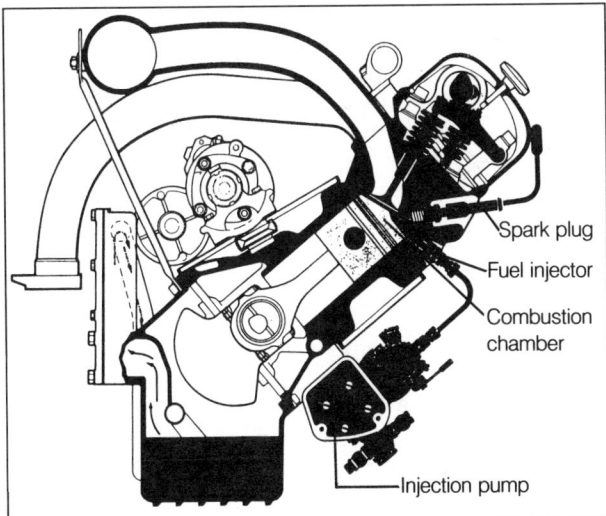

Fig. 3-1. Mercedes-Benz 300 SL cars use direct fuel injection. High-pressure engine-driven pump, hanging low under tilted block, injected gasoline directly into cylinder. Adapted from Bosch diesel injection systems.

Owners of fuel-injected cars experience better starting and driveability, especially when the engine is cold. Owners of the first fuel-injected BMWs proudly showed off their cars: even on a cold day, they would open the driver's window and, standing outside, reach in to turn the key, then boast about the smooth-running engine at cold idle.

For the manufacturer, fuel injection means better emission control and better fuel economy, both important in meeting increasingly stringent government regulations. Fuel injection offers the following advantages:

● reduces air-fuel ratio variability

● matches fuel delivery to specific operating requirements

● prevents stalling caused by fuel-bowl wash during cornering

● eliminates engine run-on (dieseling) when the key is turned off

Until a few years ago, because of looser emission limits in Europe, many European cars were built with fuel injection for delivery in the U.S., but had carburetors for delivery in Europe. When I spoke with a chief engineer of one of the largest European manufacturers, he expressed reluctance to change his engines from carburetors to the more expensive fuel injection, and suggested that we in the U.S. were a little paranoid about clean air. Now, Europeans recognize the importance of clean air, and fuel injection systems are spreading to more cars sold in Europe. In fact, many late-model German cars are sold in both the U.S. and Germany with the same fuel-injected engines using the same lead-free-fuel catalyst emission controls.

Multi-Point (or Port Injection)

All of the Bosch systems covered in this book are multi-point systems. They deliver fuel at the engine intake ports near the intake valves. This means that the intake manifold delivers only air, in contrast to carburetors or single-point fuel injection systems in which the intake manifold carries the air-fuel mixture. As a result, these systems offer the following advantages:

● greater power by avoiding venturi losses as in a carburetor, and by allowing the use of tuned intake runners for better torque characteristics

● improved driveability by reducing the throttle-change lag which occurs while the fuel travels from the throttle body to the intake ports

● increased fuel economy by avoiding condensation of fuel on interior walls of the intake manifold (manifold wetting)

● simplified turbocharger applications; the turbocharger compressor need only handle air

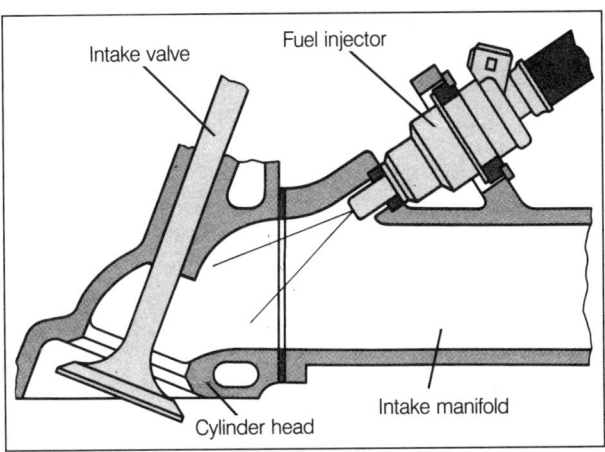

Fig. 3-2. Port injection delivers fuel to the manifold at the intake valve.

Fig. 3-3. On fuel-injected cars, tuned intake runners improve performance by reinforcing ram effect to pack in more air and increase torque.

Fuel Injection Types

The simplicity of a turbocharged Bosch system is evident in the Skip Barber Saab racing series cars which are powered by nearly stock production Saab Turbo engines equipped with Bosch LH-Jetronic fuel injection, as shown in Fig. 3-4.

Fig. 3-4. Turbochargers and fuel injection go well together, in both road cars and race cars.

3.1 Pulsed (Electronic) Systems

The pulsed systems are sometimes referred to as "Electronic Fuel Injection" (EFI), and these are the systems that most people think of when you say "fuel injection." There are several Bosch variations of pulsed systems, but their basic functions are the same.

In all of the pulsed systems, incoming air is measured by a sensor which puts out an electronic signal proportional to air flow. An electronic control unit (ECU), responding to the signals from the air-flow sensor and other sensors, meters fuel to the engine by way of electrically-operated solenoid valve injectors.

Fuel is injected in a series of short pulses, always under electronic control. In Bosch systems, the number of pulses is proportional to engine rpm. The length of time of each pulse is controlled electronically, so the injectors deliver more or less fuel per pulse depending on mixture requirements.

Since nearly all present day Bosch fuel injection systems feature some kind of electronic control, these systems can no longer be accurately identified by the EFI label. In this book, I'll use the term "pulsed" to refer to all systems with solenoid-valve injectors, opened in a series of pulses to meter the fuel. Engines equipped with pulsed systems can be quickly identified by the electric injectors connected to a common fuel supply rail, as shown in Fig. 3-5.

Fig. 3-5. For pulsed injection (EFI) look for electrically-operated injectors (**1**) connected to the fuel rail (**2**), as in this Porsche 944 Motronic.

3.2 Continuous Injection Systems

The continuous systems are sometimes referred to as mechanical or hydro-mechanical, because fuel metering is determined by the mechanical relationship between an air-flow sensor and a fuel distributor.

The first continuous systems were distinctly different from EFI systems, since there was no electronic control of basic fuel delivery. The continuous injection family has now grown and spawned more highly developed versions, and electronic control has been a part of almost all CIS fuel injection since 1980.

Fig. 3-6. For continuous injection (CIS), look for the bundle of fuel lines, usually braided like this, leading to simple injectors without electrical connectors, as in this Audi 5000 turbo.

In continuous systems, incoming air is measured by an air-flow sensor plate which is mechanically linked to the fuel distributor. Fuel is metered by this mechanical link in proportion to the incoming air flow, and delivered to the engine through pressure-actuated injectors.

Fuel is injected in continuous streams, all the time as long as the engine is running. This continuous fuel delivery gives the system its name, Continuous Injection System (CIS). The fuel distributor manipulates fuel pressure to control the volume of fuel delivered under various operating conditions.

4. ENGINE MANAGEMENT

Engine management is a term applied to systems that control more than fuel injection; in particular, they control ignition timing. Ignition timing has always been important to power and economy, even when Henry Ford's Model A provided the driver with controls for timing and fuel. Later, a form of automatic control was provided by flyweights in the distributor that advanced timing for increasing rpm, and by vacuum diaphragms that retarded timing for increasing engine load.

Fig. 4-1. In Ford Model A, driver controlled timing and fuel manually, by steering-column levers. Modern electronic engine-management systems automatically control fuel injection and ignition timing many times each second.

Beginning about 1982, "fuel injection" began to mean more than the control of fuel delivery. Ignition timing and fuel delivery are closely related. For example, careful adjustment of ignition timing at idle can increase fuel economy and reduce emissions. In Bosch "Motronic" engine-management systems, fuel delivery control and ignition timing control are combined in one control unit that processes all of the necessary engine information. Beginning in 1984, Motronic engine-management systems also controlled idle rpm.

4.1 Control of Ignition Timing

The correct timing of the spark-plug firing depends on many of the same variables that determine fuel metering, including engine speed, engine temperature, altitude and, in some cases, whether the engine is knocking. In engine-management systems, the ignition control uses these variables to compute the ignition timing point. In most cases, as in Motronic systems, the control unit refers to a timing map, a set of data points in the ECU memory that give the best timing point for all conditions.

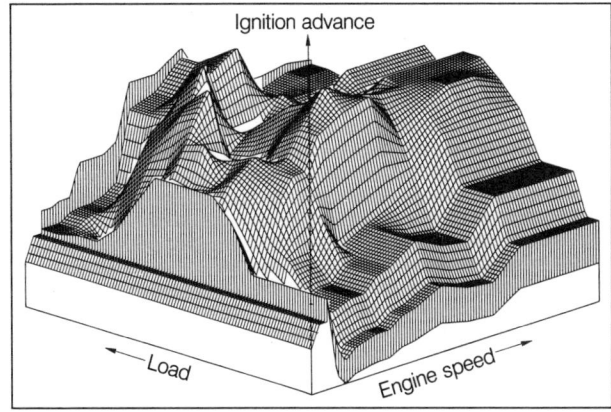

Fig. 4-2. Motronic ignition timing depends on ignition read-only memory (ROM) stored in the electronic control unit. Ignition timing can be set for best performance for any combination of load and rpm, based on extensive testing.

4.2 Control of Idle Speed

Control of idle speed contributes to fuel economy and reduced emissions. Using many of the same variables already input to the ECU, the control unit adjusts idle rpm by varying the amount of air bypassing the throttle valve, as well as varying the ignition timing.

4.3 Other Electronic Controls

Other engine-management functions may be included in the control unit functions. For example, some Motronic systems control the opening and closing of the fuel-vapor charcoal canister purge valve.

5. THE BOSCH JETRONIC FAMILY TREE

Bosch builds two basic types of fuel injection—pulsed and continuous—with several variations and improvements over the years. You'll want to know these differences in operation to understand the function of each distinct system, and to service each system. Fig. 5-1, a "family tree" diagram of all of the Bosch systems covered by this book, shows the progression of the two "branches."

5.1 Pulsed Fuel Injection Systems

D-Jetronic. The first Bosch Jetronic system. The D is short for Druck, the German word for pressure. Manifold pressure is measured to indicate engine load (how much air the engine is using). This pressure is an input signal to the control unit (ECU) for calculation of the correct amount of fuel delivery.

L-Jetronic. The L is short for Luft, the German word for air. Air flow into the engine is measured by an air-flow sensor with a movable vane to indicate engine load. Later systems have Lambda control (sometimes represented by the Greek letter λ) for more precise mixture control. For more on Lambda, see chapter 2. L-Jetronic was often called Air-Flow Controlled (AFC) injection, to further separate it from the pressure-controlled D-Jetronic.

LH-Jetronic. LH-Jetronic measures air mass (weight of air) with a hot-wire sensor instead of measuring air flow with an air vane (volume of air) as in L-Jetronic. Otherwise, L and LH systems are very similar. The H is short for Heiss, the German word for hot.

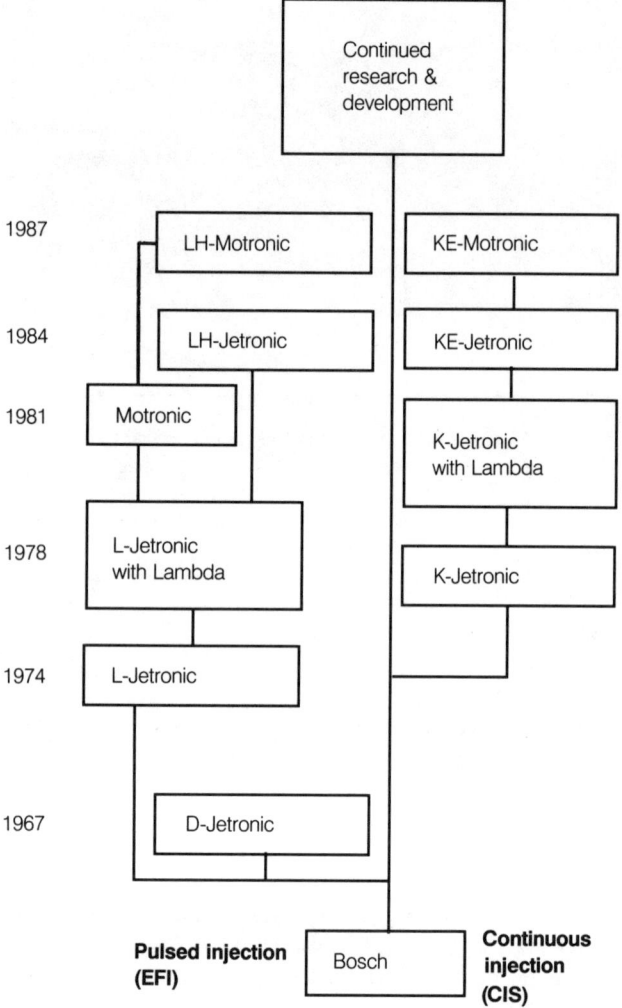

B224OVR.BCH

Fig. 5-1. The Bosch Jetronic family tree branches can be divided into the pulsed systems, often called EFI, and the continuous systems, often called CIS or mechanical. Some of the earlier definitions of these systems are less useful these days because all of the systems, including the "mechanical" ones, use electronics.

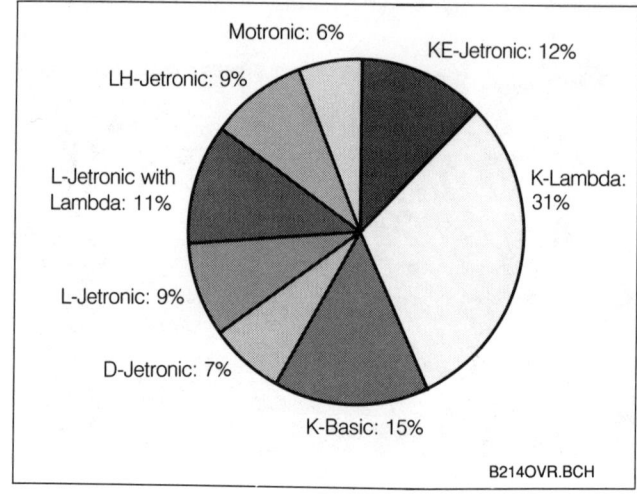

B214OVR.BCH

Fig. 5-2. Estimated Bosch-built fuel-injection systems in the U.S., through 1987. Out of about 6,000,000 cars, about 60% are continuous systems, and 40% are pulsed systems. About 7 out of 10 systems have lambda control.

5.2 Continuous Injection Systems

K-Jetronic. The first continuous system. Bosch called it K, for Kontinuerlich, the German word for continuous. Air flow is measured by a circular plate in the air-flow sensor. Until 1980, fuel delivery is strictly under mechanical control in direct relation to air flow; there are no electronics. I call this **K-basic** Volkswagen, Audi, and Mercedes call it CIS.

K-Jetronic with Lambda Control. This is an adaptation of K-basic mechanical control. The mixture is modulated with a limited electronic control in a feedback system to maintain a precise air-fuel ratio. As you'll see in chapter 2, Lambda (λ) refers to a certain air-fuel ratio. I call this version **K-lambda**.

KE-Jetronic. This combines K-Jetronic mechanical control with total electronic modulation of the mixture. It uses many of the same sensors used in L-Jetronic systems. Because it is based on K-Jetronic, it is still capable of fail-safe mechanical operation if the electronics fail. I call it **KE**; Volkswagen, Audi, and Mercedes call it CIS-E.

5.3 Engine-Management Systems

Bosch engine-management systems are often grouped under the single term "Motronic." This can be confusing, because while the original Bosch engine-management system was called Motronic to differentiate it from the Jetronic systems, there are now many different Motronic-type engine-management systems. So when you see a system called "Motronic", make sure you know which type it is.

Motronic. The first "engine-management" system. It combines L-Jetronic pulsed fuel injection with electronic ignition timing control in one control unit. In most engines, it also has electronic idle stabilization for a complete engine-management system. For clarity, I may sometimes refer to this system as **L-Motronic**. Beginning in about 1986, Motronic may also have additional functions:

- Knock regulation by ignition timing of individual cylinders.

- Adaptive circuitry – it adapts fuel delivery and ignition timing to actual conditions. It does this so well that, beginning in 1988, idle rpm and mixture are no longer adjustable.

- Diagnostic circuitry – the control unit recognizes system faults and stores fault information in its memory.

LH-Motronic. This is the same as Motronic, except that it uses a hot-wire air-mass sensor, hence the LH. All LH-Motronic systems have electronic idle stabilization.

KE-Motronic. Ignition-timing control added to KE-Jetronic fuel injection. It has the same additional functions as Motronic systems.

Monotronic. I mention it here, though it is not covered by this book. This is a single-point, throttle-body Motronic, lower-cost and simpler than Motronic, not used in cars currently imported to the U.S.

5.4 Bosch-Licensed Systems

Worldwide, almost all current fuel-injection systems are based on Bosch technology to a greater or lesser degree. Most Japanese fuel-injection systems are built under Bosch license, and many U.S. cars use Bosch components, built with Bosch principles under licensed production. The following systems are Bosch European-licensed systems installed in European-made cars sold in this country.

Volkswagen Digifant. This pulsed system was partly designed by Volkswagen, but operates pretty much as Bosch Motronic does. Its timing control map is less complicated than the Bosch Motronic map. It does not have a knock sensor.

Volkswagen Digifant II. A refined version of Digifant. Along with control improvements, it uses a knock sensor for more precise timing control.

Lucas. This system, used in Jaguar and Triumph cars, is a Bosch L-Jetronic system licensed for production by Lucas.

6. CARS WITH BOSCH FUEL INJECTION AND ENGINE-MANAGEMENT SYSTEMS

Bosch Jetronic fuel injection arrived in America in 1967 in the German-made Volkswagens. Emission-control legislation forced their adoption on the VW air-cooled engines, because their uneven cooling raises emissions. For more than ten years since then, fuel injection has been virtually standard in European cars delivered in the U.S. The Japanese are catching up; by 1989, over half their cars delivered in the U.S. are fuel-injected. The changeover to fuel injection is accelerating. Carburetors on new cars in the U.S. will soon be as rare as drum brakes on front wheels.

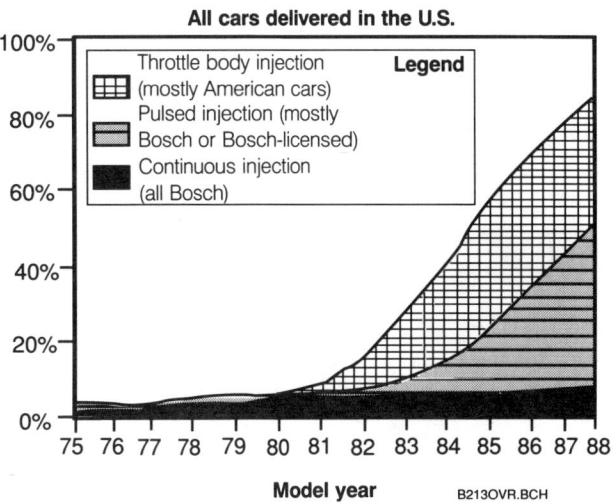

Fig. 6-1. Total car sales in the U.S. shows the replacement of carburetors by fuel-injection systems. The brief popularity of diesel fuel injection in 1979–1981 caused a "diesel dip" in gasoline injection production.

Cars with Bosch Fuel Injection and Engine-Management Systems

How can you tell if your car has a Bosch fuel-injection system? None of the European cars sold in the U.S. wear Jetronic or Motronic badges, but most have fuel-injection clues on the car badge. Lately, though, these clues are disappearing. Apparently, considering that almost all of these European cars sold in the U.S. have fuel injection, who needs to indicate that by badge? No longer does any car indicate disc brakes by spelling it out on the brake pedal, as was once the vogue.

Volkswagen, Audi. Early fuel-injected cars were sold at the same time as carbureted cars—look for the words "Fuel Injection" on the rear deck, whether the VW is assembled in the U.S. or is imported. Since 1977, all VWs have been fuel injected, with fuel injection badges. In GTI models, the I stands for injection. Audi, in the same family, also badged fuel injection; sometimes you'll see turbo instead, but the car is still fuel injected.

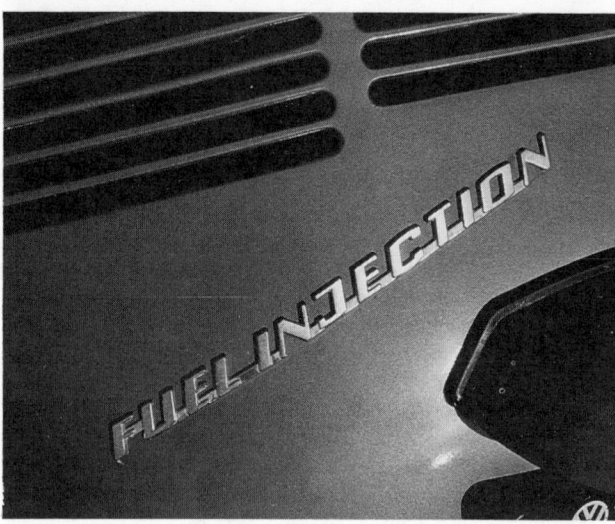

Fig. 6-2. VW was the first user of Bosch Jetronic. Emission control legislation forced adoption first in air-cooled engines. Because they cool unevenly, their emissions are hard to control.

Fig. 6-3. VW is the biggest user of Jetronic systems, both in cars imported from Germany and built in U.S. Most liquid-cooled VWs are fuel injected.

Cars with Bosch Fuel Injection
and Engine-Management Systems

Fig. 6-4. Sometimes badges boast about number of valves. Scirocco, as most 4-valve per cylinder engines, operates with fuel injection.

Fig. 6-5. Most Audis spell it out. Audi was one of the first users of K-Jetronic, in 1975.

Porsche, Saab, Volvo. These three began identifying fuel injection by adding E for Einspritz, the German word for injection. The E-for-injection might be lost if the badge included T for turbo, or SC for super coupe. By 1976, fuel injection had become normal, so we do not see Porsche 911E, 924E, or 928E even though all of these cars are fuel injected. Around 1980, the E on Volvo badges no longer indicates fuel injection, but indicates a level of luxury equipment.

Fig. 6-6. Saab is an early user, beginning in 1970. Early fuel-injected cars were badged E, for Einspritz. 99 Turbo was first to drop E.

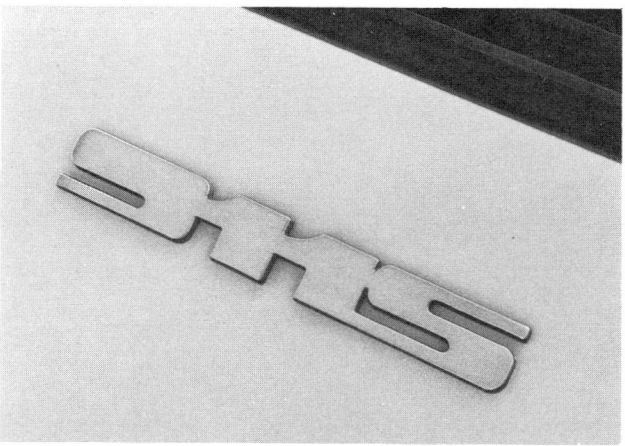

Fig. 6-7. Porsche has dropped the E. Porsche began using fuel injection with the 1970 914E.

Fig. 6-8. Volvo 164E hit the market with D-Jetronic in 1972. First use was on 1800E in 1970.

Fig. 6-9. Volvos and Saabs have been fuel-injected for so long that an indication of fuel injection on the badge has been replaced with a designation of the level of equipment.

Mercedes-Benz. They also began adding an E to identify the fuel-injected cars from those with carburetors. Most gasoline Mercedes show E. An exception is the two-seater SL, which is fuel-injected but omits the E to distinguish itself from the coupes, SEC, and from the long sedans, SEL. All Mercedes are fuel injected, the diesels are identified with D.

Fig. 6-10. Mercedes use KE-Jetronic. Switch from D-Jetronic to K-Jetronic was in 1976. D badges are diesels.

Cars with Bosch Fuel Injection
and Engine-Management Systems

Fig. 6-11. Omitting the E here helps differentiate this two-seater SL—Super Leicht (the German word for light)—from the SEC coupes.

Fig. 6-12. E for Einspritz is mixed with S for Super and L for Lang (the German word for long) on this Mercedes Sedan.

Fig. 6-13. BMW i means fuel injection. BMW began using L-Jetronic in 6-cylinder engines in 1975, K-Jetronic in 4-cylinder engines in 1977.

Fig. 6-14. On badge on fuel-injected BMW 325e, e stands for Eta, the Greek letter for efficiency.

BMW. BMW usually adds i for injection, beginning with the 6-cylinder 530i. BMW 2002ti (a carbureted car), stands for "touring international"; BMW 2002tii stands for "touring international, injection". Since 1979, BMW has used E for eta, the Greek letter for efficiency to indicate engines tuned for better torque and economy at low and mid-rpms; all BMWs are fuel injected but, in this case, E does not stand for Einspritz. What's es in 528es? "Eta, Sport".

Table a will enable you to identify what type of Bosch fuel injection system is in your specific make and model. It does not include U.S. hybrid systems that mix Bosch components with others, such as 1985+ Corvettes that use Bosch LH-Jetronic air-mass sensors and injectors. This list concentrates on European cars, including VWs made in the U.S.

Table a. Bosch Fuel Injection Systems Sold in the United States

Manufacturer/ Model	Injection system	Number installed (in round thousands)
ALFA		
6	L-Jetronic with Lambda	1,000
Graduate	L-Jetronic with Lambda	1,000
Graduate (1986 on)	Motronic	*
GTV 6	L-Jetronic with Lambda	13,000
Milano (1986)	Motronic	*
Milano (1987-1989)	LU-Jetronic	*
Spider (1981-1985)	L-Jetronic with Lambda	9,000
Spider (1985-1989)	Motronic	*
AMC		
Alliance (California)	L-Jetronic with Lambda	30,000
Alliance/Encore (California)	LU-Jetronic	15,000
AUDI		
100LS	K-Jetronic	27,000
Fox	K-Jetronic	88,000
4000 (1980)	K-Jetronic	10,000
4000 (1981 on) and +5	K-Jetronic with Lambda	43,000
4000 Quattro	KE-Jetronic	13,000+
4000 S (1984-1987)	KE-Jetronic	44,000+
5000 (1978-1979)	K-Jetronic	38,000
5000 (1980-1983)	K-Jetronic with Lambda	72,000
5000 S (1984-1985)	K-Jetronic with Lambda	82,000
5000 S (1986)	KE-Jetronic	3,000
5000 S (1987-1988)	KE3-Jetronic	19,000+
5000 Turbo (1980-1983, 1985)	K-Jetronic with Lambda	40,000
5000 Turbo (1984)	KE-Jetronic	6,000
5000 CS Turbo	K-Jetronic with Lambda	6,000+
5000 CS Turbo (1988)	KE3-Jetronic	*
5000 CS Quattro Turbo	K-Jetronic with Lambda	3,000+
5000 CS Quattro Turbo (1988)	KE3-Jetronic	*
80/90 W/AT	KE-Motronic	*
80/90 W/MT, 80/90 Quattro	KE3-Jetronic	*
100/100 Quattro	KE3-Jetronic	*
200/200 Quattro	KE-Jetronic	*
Coupe (1981-1984)	K-Jetronic with Lambda	11,000
Coupe (1985)	KE-Jetronic	4,000
Coupe GT (1986)	KE-Jetronic	3,000
Coupe GT (1987-1988)	KE3-Jetronic	2,000+
Quattro	K-Jetronic with Lambda	2,000

*information not available

Table a. Bosch Fuel Injection Systems Sold in the United States (cont'd)

Manufacturer/ Model	Injection system	Number installed (in round thousands)
BERTONE		
X1/9	L-Jetronic with Lambda	1,000
BMW		
3.0Si	L-Jetronic	19,000
318i	LU-Jetronic	62,000
320i (1977-1979)	K-Jetronic	46,000
320i (1980-1983)	K-Jetronic with Lambda	109,000
325/325E	Motronic	167,000+
528E (1979-1981)	L-Jetronic with Lambda	17,000
528E (1982 on)	Motronic	51,000+
530i	L-Jetronic	4,000
535i/535iS	Motronic	20,000+
630CSi	L-Jetronic	1,000
633CSi (1978-1979)	L-Jetronic	2,000
633CSi (1980-1981)	L-Jetronic with Lambda	2,000
633CSi (1982 on)	Motronic	10,000
635CSi	Motronic	3,000+
733i (1978-1979)	L-Jetronic	2,000
733i (1980-1981)	L-Jetronic with Lambda	3,000
733i (1982 on)	Motronic	17,000
735i	Motronic	20,000+
750	LH-Motronic	*
M3, M5, M6	Motronic	2,000+
DeLOREAN		
DeLorean	K-Jetronic with Lambda	2,000
EAGLE		
Medallion	LU-Jetronic	*
FERRARI		
308 GTB Si	K-Jetronic with Lambda	3,000
328GTB	K-Jetronic with Lambda	2,000+
Mondial/S Mondial	K-Jetronic with Lambda	2,000+
Testarossa	KE-Jetronic	1,000+
FIAT		
Brava	L-Jetronic with Lambda	16,000
Spider	L-Jetronic with Lambda	43,000
Strada	L-Jetronic with Lambda	21,000
X1/9	L-Jetronic with Lambda	15,000
JAGUAR		
XJ6/XJS/XJSV12 (1980-1986)	L-Jetronic with Lambda	95,000
XJ6/XJS/XJSV12 (1987-1989)	L-Jetronic with Lambda (Lucas)	23,000+
LANCIA		
Beta	L-Jetronic with Lambda	2,000
LOTUS		
Esprit Turbo	K-Jetronic with Lambda	2,000+
MASERATI		
Maserati	Motronic	1,000+

continued on next page

Cars with Bosch Fuel Injection and Engine-Management Systems

Table a. Bosch Fuel Injection Systems Sold in the United States (cont'd)

Manufacturer/ Model	Injection system	Number installed (in round thousands)
MERCEDES-BENZ		
190E (1984-1987)	KE-Jetronic	79,000
190E (1988 on)	KE3-Jetronic	*
260E	KE-Jetronic	6,000+
280E (1971-1972)	D-Jetronic	20,000
280E (1977-1979)	K-Jetronic	10,000
280E (1980-1981)	K-Jetronic with Lambda	3,000
300 (1972)	D-Jetronic	5,000
300E (1986, 1988 on)	KE3-Jetronic	28,000+
300E, SEL (1987)	KE-Jetronic	24,000
380E	K-Jetronic with Lambda	49,000+
420E (1985)	K-Jetronic with Lambda	*
420E (1986, 1988 on)	KE3-Jetronic	16,000+
420 SEL (1987)	KE-Jetronic	15,000
450/450E (1972-1975)	D-Jetronic	20,000
450E (1976-1979)	K-Jetronic	33,000
450E (1980)	K-Jetronic with Lambda	10,000
500E (1983)	K-Jetronic with Lambda	7,000
500 (1984-1985)	K-Jetronic with Lambda	10,000+
560 (1985)	K-Jetronic with Lambda	*
560 (1986, 1988 on)	KE3-Jetronic	15,000+
560 (1987)	KE-Jetronic	20,000
MERKUR		
XR4Ti, Scorpio, Sierra	D-Jetronic (Ford)	37,000+
OPEL		
Manta, 1900E	L-Jetronic	40,000
PEUGEOT		
505/505i	K-Jetronic with Lambda	60,000
505 Turbo, GL, STX	Motronic	9,000+
PORSCHE		
911 Carrera, Coupe, Cabriolet	Motronic	23,000+
911S (1974-1977)	K-Jetronic	15,000
911SC (1978-1979)	K-Jetronic	4,000
911SC (1980-1983)	K-Jetronic with Lambda	16,000
911T (1972)	K-Jetronic	2,000
911 Turbo Carerra (1976-1977)	K-Jetronic	2,000
911 Turbo (1978)	K-Jetronic	1,000
911 Turbo (1986 on)	K-Jetronic with Lambda	3,000+
912	L-Jetronic	1,000
914E (1970-1973)	D-Jetronic	45,000
914 (1974-1976)	L-Jetronic	27,000
924 (1976-1979)	K-Jetronic	17,000
924 (1980-1985)	K-Jetronic with Lambda	8,000
924 (1986)	Motronic	3,000
924S (1987 on)	Motronic	*
924 Turbo	K-Jetronic with Lambda	5,000+
928 (1978-1979)	K-Jetronic	7,000
928 (1980-1984)	L-Jetronic with Lambda	7,000
928S (1985-1986)	LH-Jetronic	6,000
928S (1987)	Motronic	2,000
928S4 (1988 on)	LH-Motronic	*
930T	K-Jetronic	1,000
944, S, Turbo	Motronic	50,000+
Carerra 4	Motronic	*

Table a. Bosch Fuel Injection Systems Sold in the United States (cont'd)

Manufacturer/ Model	Injection system	Number installed (in round thousands)
RENAULT		
Fuego/Fuego Turbo (1982-1984)	L-Jetronic with Lambda	25,000
Fuego (1985)	LU-Jetronic	5,000
Medallion	LU-Jetronic	14,000+
R17/R17 Gordini	L-Jetronic	6,000
R18i (1980-1984)	L-Jetronic with Lambda	30,000
R18i (1985)	LU-Jetronic	2,000
Sportwagon	LU-Jetronic	4,000
ROLLS-ROYCE		
Rolls-Royce (all)	K-Jetronic with Lambda	7,000+
SAAB		
99 EMS (1970-1974)	D-Jetronic	60,000
99 EMS (1975-1980)	K-Jetronic	42,000
99 Turbo	K-Jetronic with Lambda	4,000
900 (1979-1980)	K-Jetronic	11,000
900, 900S (1981 on)	K-Jetronic with Lambda	123,000+
900 Turbo (1979-1984)	K-Jetronic with Lambda	50,000
900 Turbo (1985 on)	LH2-Jetronic	33,000+
9000, S, Turbo	LH2-Jetronic	22,000+
STERLING		
825, 825S	Motronic (Lucas)	14,000+
TRIUMPH		
TR7	L-Jetronic with Lambda	25,000
TR8	L-Jetronic with Lambda	*
VOLVO		
1800E	D-Jetronic	25,000
140E (1971-1973)	D-Jetronic	65,000
140E (1974)	K-Jetronic	35,000
164E	D-Jetronic	65,000
240E (1975-1979)	K-Jetronic	120,000
240 (1986 on)	LH-Jetronic	103,000+
260E (1976-1978)	K-Jetronic	20,000
260E (1979)	K-Jetronic with Lambda	9,000
740, 740 Turbo	LH2-Jetronic	120,000+
760 (1983-1985)	K-Jetronic with Lambda	11,000
760 (1986 on)	LH2-Jetronic	10,000+
760 Turbo (1984-1985)	LH-Jetronic	13,000
760 Turbo (1986 on)	LH2-Jetronic	*
780	LH2-Jetronic	*
DL (1980)	K-Jetronic	39,000
DL (1981-1982)	K-Jetronic with Lambda	75,000
DL (1983-1985)	LH-Jetronic	132,000
GL (1980)	K-Jetronic	9,000
GL (1981)	K-Jetronic with Lambda	9,000
GL (1982-1985)	LH-Jetronic	63,000
GLE	K-Jetronic with Lambda	2,000
GL Turbo (1981-1984)	K-Jetronic with Lambda	38,000
GL Turbo (1985)	LH-Jetronic	2,000

continued on next page

*information not available

Cars with Bosch Fuel Injection and Engine-Management Systems

Table a. Bosch Fuel Injection Systems Sold in the United States (cont'd)

Manufacturer/ Model	Injection system	Number installed (in round thousands)
VOLKSWAGEN		
Type 3	D-Jetronic	85,000
Type 4 (1971-1973)	D-Jetronic	30,000 +
Type 4 (1974)	L-Jetronic	30,000
Beetle	L-Jetronic	173,000 +
Bus	L-Jetronic	235,000 +
Cabriolet	KE-Jetronic	28,000 +
Corrado	Digifant II	*
Dasher (1976-1980)	K-Jetronic	60,000
Dasher (1981)	K-Jetronic with Lambda	1,000
Fox	K-Jetronic/K-Jetronic with Lambda	40,000 +
Golf (1985-1987)	K-Jetronic with Lambda	100,000 +
Golf (1988 on)	Digifant II	*
Golf GTI (1985-1987)	KE-Jetronic	100,000 +
Golf GT (1987)	KE-Jetronic	*
Golf GT (1988 on)	Digifant II	*
Golf GTI 16V	KE-Jetronic	*
Jetta (1980)	K-Jetronic	9,000
Jetta (1981-1985)	K-Jetronic with Lambda	182,000
Jetta (1986-1987)	KE-Jetronic	60,000 +
Jetta (1988 on)	Digifant II	*
Jetta GLI (1985-1987)	KE-Jetronic	*
Jetta GLI (1988 on)	Digifant II	*
Quantum (1982-1984)	K-Jetronic with Lambda	43,000
Quantum (1985-1988)	KE-Jetronic	33,000 +
Rabbit (1977-1980)	K-Jetronic	453,000
Rabbit (1981 on)	K-Jetronic with Lambda	109,000 +
Rabbit GTI (1983-1984)	K-Jetronic with Lambda	25,000 +
Scirocco (1977-1980)	K-Jetronic	77,000
Scirocco (1981-1987)	K-Jetronic with Lambda	94,000
Scirocco 16V	KE-Jetronic	*
Vanagon (1980-1983)	L-Jetronic with Lambda	48,000
Vanagon (1984-1985)	Digijet (VW version of L-Jetronic with Lambda)	35,000
Vanagon (1986 on)	Digifant I	24,000 +

*information
not available

*Cars with Bosch Fuel Injection
and Engine-Management Systems*

Chapter 2

Engine Management Fundamentals

2

Contents

1. INTRODUCTION

What sets fuel-injection and engine-management systems apart from other fuel delivery systems is their ability to precisely control fuel metering and adjust it in response to changing operating conditions. The key elements in the success of Bosch fuel-injection and engine-management systems are (a) a comprehensive understanding of the engine's varying fuel delivery requirements over a broad range of operating conditions, (b) extensive research into how the factors which influence the engine's fuel needs can be measured and interpreted by control systems and, (c) highly developed methods for controlling fuel metering to optimize overall performance.

In this chapter, I'll review the basic factors governing air-fuel mixture and fuel delivery, I'll describe why various engine operating conditions demand different air-fuel mixtures, and I'll explain how the air-fuel ratio can be manipulated to improve driveability and control exhaust emissions with little sacrifice of power. In chapter 3 and chapter 5, you'll see how Bosch applies these principles to the design and operation of their fuel-injection systems.

2. BASIC FACTORS

First, the basics. We'll take a closer look at air-fuel ratios and the engine's basic demands for a combustible mixture. I'll describe additional factors affecting fuel delivery which are imposed by the demands of the car-buying public, and we'll examine pressure and pressure measurement—a subject that is fundamental to understanding how Bosch fuel-injection systems work.

2.1 Air-Fuel Ratios

An engine's throttle controls the amount of air the engine takes in. The main function of any fuel delivery system is to mix fuel with that incoming air in the proper ratio. Small variations in air-fuel ratio can have dramatic effects on power output, fuel consumption, and exhaust emissions.

The Basic Combustible Mixture

Laws of physics tell us that combustion of any substance requires a sufficient ratio of surface area to mass (small enough particles), and the correct amount of oxygen (in proportion to the amount of fuel). In internal combustion engines these conditions are met by atomizing the fuel into tiny droplets, and by metering fuel in correct proportion to the intake air. In round numbers, approximately 14 parts of air are required to support complete combustion of 1 part of fuel—in other words, an air-fuel ratio of about 14:1.

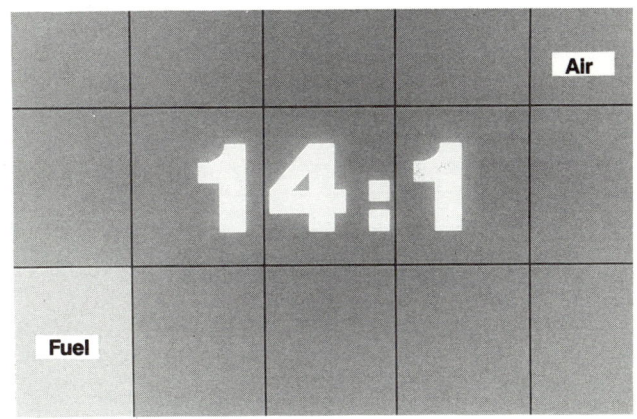

Fig. 2-1. By mass, it takes about 14 parts of air to support complete combustion of 1 part of fuel. Before today's concerns over fuel economy and emissions, an air-fuel ratio somewhere close to 14:1 was good enough.

Throughout this book, I'll follow generally accepted practice and discuss air-fuel ratios primarily in terms of mass. This is the simplest and best way to help you understand the basic factors governing fuel delivery and combustion. Many of the Bosch systems, however, measure air flow by volume. It is interesting to see that, by volume, the proportion of air to fuel is approximately 11,500:1.

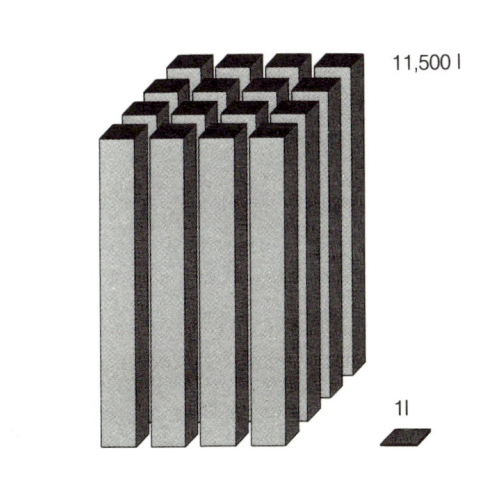

Fig. 2-2. By volume, it takes approximately 11,500 parts of air to support complete combustion of 1 part of fuel.

Notice that I am talking about combustion that is complete — combustion that makes the best, most thorough use of fuel. Air-fuel ratios which are higher or lower than approximately 14:1 will still burn, but such combustion produces unwanted by-products and other side effects. As you read further into this chapter, you'll see what those are, and why precise control of the air-fuel ratio has become so important.

Rich and Lean Mixtures

The terms "rich" and "lean" are used to describe mixtures which deviate from the theoretically perfect air-fuel ratio and burn less efficiently.

A rich mixture is one with a lower air-fuel ratio; there is insufficient air (oxygen) to support complete combustion of the fuel. Rich mixtures increase fuel consumption and emissions of hydrocarbons (HC) and carbon monoxide (CO) — the products of incompletely burned gasoline. They tend to reduce power, increase carbon deposits and, in the extreme case, foul spark plugs and dilute the engine's lubricating oil.

"Enrichment" is the process of metering more fuel for a given amount of air to produce a richer mixture.

Fig. 2-3. Rich mixtures contain more fuel than can be completely burned in the given amount of air.

A lean mixture is one with a higher air-fuel ratio; there is more air than necessary for complete combustion of the fuel. The fuel will burn completely, but more slowly and at a higher combustion temperature. Lean mixtures reduce power, elevate engine temperature, and increase emissions of oxides of nitrogen (NO_x) — a product of combustion at excessively high temperature. They also tend to cause driveability problems. In the extreme case, the high temperatures resulting from lean combustion will cause pre-ignition — violent, untimed combustion of the mixture which has the potential to cause serious engine damage.

Fig. 2-4. Lean mixtures contain more air than is necessary for combustion of the fuel. Inferior combustion reduces power and can cause engine damage.

Stoichiometric Ratio

When "gas was cheap and the air was dirty," carburetors were usually set up to deliver mixtures richer than 14:1, perhaps with an air-fuel ratio as low as 12:1. As the car aged and a little excess air leaked in around the gaskets of the carburetor or the intake manifold, the engine still got a good combustible mixture. Then, too, the carburetor was farther from the end cylinders than from the middle ones. A richer mixture was some compensation for unequal fuel distribution.

There was another reason for setting up carbureted engines to run a little rich. Air-fuel mixture was less precise, and some variations were to be expected. As shown in Fig. 2-5, variations in a rich mixture have only a small effect on power, while variations in a lean mixture affect power dramatically.

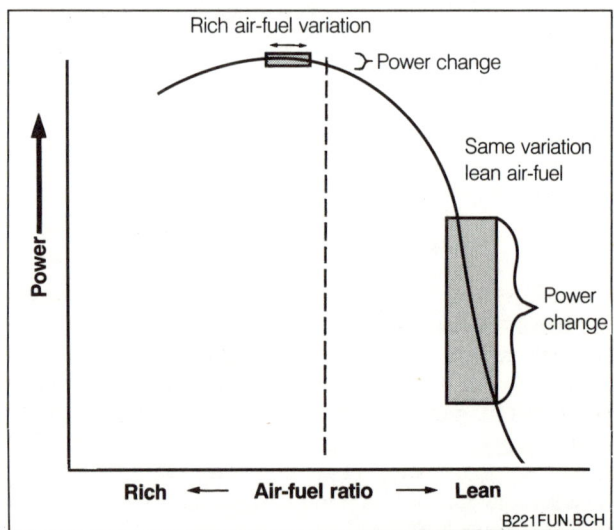

Fig. 2-5. Variation in a rich air-fuel ratio makes much less difference in power than the same variation in a lean air-fuel ratio.

For today's engines, with the increased emphasis on fuel economy and reduced emissions, the air-fuel ratio has to be controlled much more carefully. The "ideal" air-fuel ratio—the one which yields the most complete combustion and the best compromise between lean and rich mixtures—is 14.7:1. This is called the "stoichiometric" ratio. The mixture is neither rich nor lean.

The Excess Air Factor—Lambda

The stoichiometric ratio can also be described in terms of the air requirements of the engine. Bosch calls this the "excess air factor" and represents it using the Greek letter λ (lambda). At the stoichiometric ratio—when the amount of air equals the amount required for complete combustion of the fuel and there is no excess air—lambda (λ) = 1.

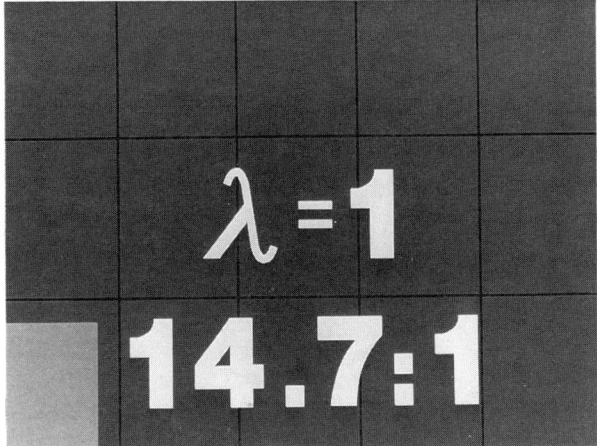

Fig. 2-6. The ideal or "stoichiometric" air-fuel ratio—when there is just enough air to burn all the fuel—is 14.7:1. This is also described as λ (lambda) = 1.

When there is excess air (air-fuel ratio leaner than stoichiometric), lambda is greater than one. When there is a shortage of air (air-fuel ratio richer than stoichiometric), lambda is less than one.

The concept of lambda (the excess air factor) was created specifically to support thinking about fuel delivery in terms of the air requirements of the engine. As you'll see later on, this concept plays a big part in controlling exhaust emissions.

2.2 Driveability and Emission Control

I've described the engine's basic requirement for a combustible mixture of air and fuel, how variations in mixture influence performance, and how older carburetor systems tended to run richer than the ideal air-fuel ratio in the interest of delivering smooth, reliable power.

While power is always a requirement, modern fuel delivery systems face additional demands. Increasing concern over the cost and availability of gasoline has resulted in greater demand for fuel economy. Environmental concerns and resulting legislation demand rigid control of harmful exhaust emissions. And, the car-buying public increasingly demands good driveability—quick-starting and smooth, trouble-free performance under any and all operating conditions. Each of these factors places different demands on the fuel delivery system, and there are trade-offs.

Adjusting the system for maximum power also means increasing fuel consumption. Minimizing fuel consumption means sacrificing power and driveability. Choosing either maximum power or minimum fuel consumption means increased exhaust emissions. The modern fuel delivery system must be able to maintain strict control of air-fuel ratio in order to achieve the best compromise and meet these conflicting demands in the most acceptable way. In general, this means slight sacrifices of power and fuel economy in exchange for optimum emissions control.

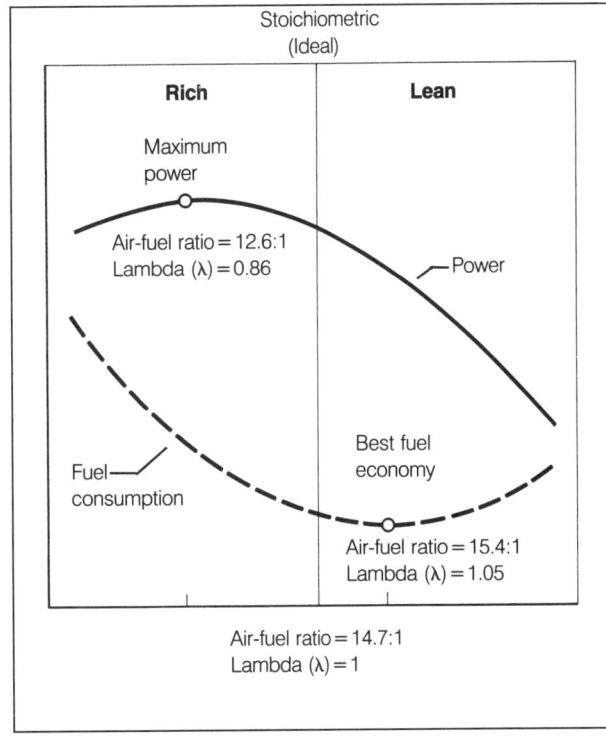

Fig. 2-7. The air-fuel ratio which delivers maximum power is slightly richer than stoichiometric; the one which delivers minimum fuel consumption is slightly more lean than stoichiometric. The stoichiometric ratio is a compromise which sacrifices very little of either.

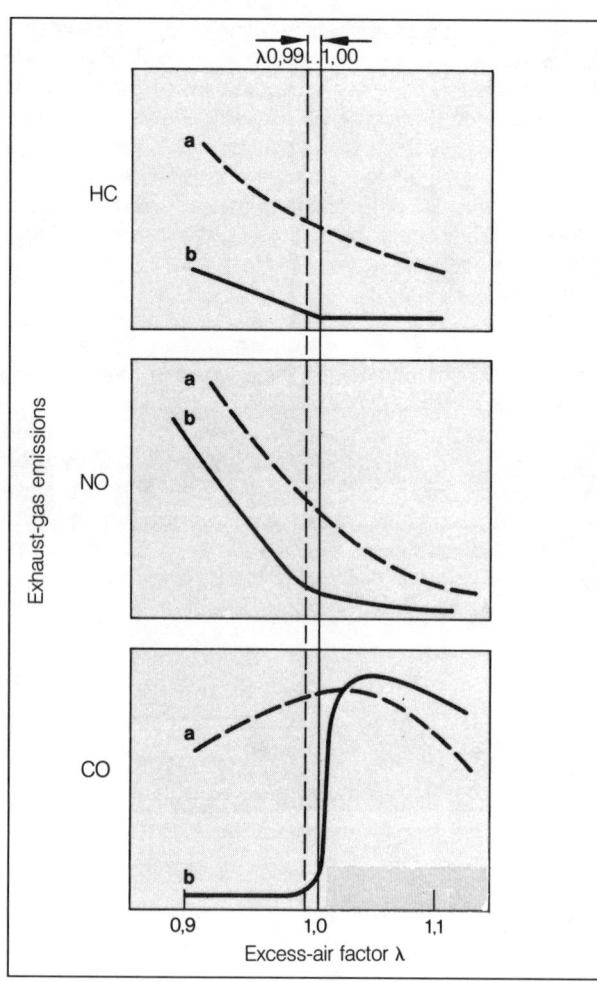

Fig. 2-8. Stoichiometric air-fuel ratio (λ = 1) yields the best compromise for control of hydrocarbon (HC), carbon monoxide (CO), and oxides of nitrogen (NOₓ) emissions, before the catalytic converter (a) or after it (b).

It is fuel injection's ability to maintain the air-fuel ratio within close tolerances that makes it superior to carburetor systems. For the manufacturer, fuel-injection means better emission control and better fuel economy, both important in meeting increasingly stringent government regulation. For the owner, fuel-injection means achieving fuel economy and emission control while preserving driveability and maximum power.

Fuel Economy - CAFE

Along with the general demand for fuel economy, each manufacturer must consider another factor: mandated federal standards for rated fuel economy — the Corporate Average Fuel Economy (CAFE) standards. The industry trend in rated miles-per-gallon (mpg) turned upward beginning in 1975, as catalytic converters replaced engine de-tuning as a means of emission

control. In addition, government legislation established an average mpg standard to apply to the total fleet of cars each manufacturer delivers each year. Further, the target mpg standard rose each year, starting at 18 mpg in 1978, and rising to 27.5 mpg in 1985, but cut back in 1986 to 26 mpg. Fuel injection's precise control of fuel delivery minimizes fuel consumption while still providing good driveability.

2.3 Air Flow, Fuel Delivery, and Engine Load

You know that air is drawn into the engine with each intake stroke of each piston. The piston moving down on its intake stroke increases cylinder volume and lowers pressure in the cylinder. With the intake valve open, air at higher pressure rushes in from the intake manifold to fill the cylinder. The amount of fuel necessary to create a stoichiometric mixture depends on how much air rushes in.

In simplest terms, intake air flow occurs because normal atmospheric pressure is higher than the pressure in the cylinders. Air rushes in during the intake stroke, trying to equalize the pressure. In gasoline engines, the throttle valve restricts intake air flow. As you open the throttle, the opening to atmospheric pressure raises manifold pressure. So, in practice, the amount of air that rushes into the cylinder on the intake stroke depends on the difference between the pressure in the intake manifold and the lower pressure in the cylinder. Pressure in the intake manifold depends on throttle opening.

The greatest intake air flow occurs when the throttle valve is fully open. The throttle valve causes almost no restriction, and full atmospheric pressure is admitted to the intake manifold. This creates the greatest possible difference between manifold pressure and cylinder pressure, and the greatest intake air flow.

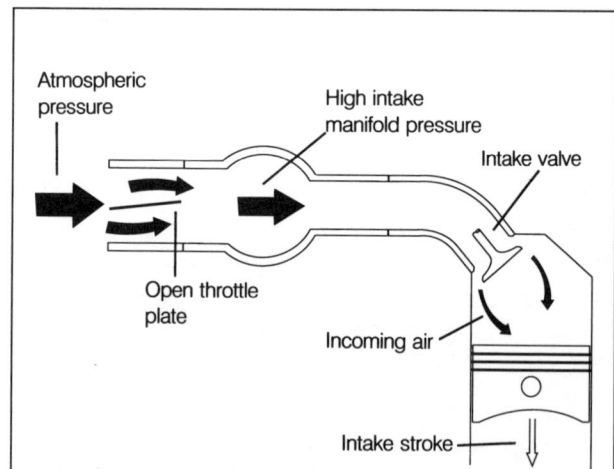

Fig. 2-9. At full throttle, nearly unrestricted atmospheric pressure raises manifold pressure. Large pressure differential between manifold and cylinders increases air flow.

The least intake air flow occurs when the throttle is nearly closed. The restriction of the throttle valve limits the effect of atmospheric pressure. There is little difference between manifold pressure and the low pressure in the cylinders, and air flow is low.

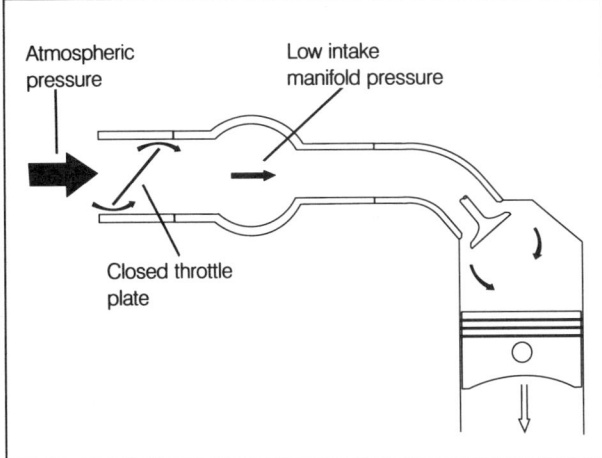

Fig. 2-10. With the throttle closed, atmospheric pressure has little effect on manifold pressure. Low pressure differential between manifold and cylinders results in little air flow.

Fuel delivery requirements depend more than anything else on how much work you are asking the engine to do—on how much of a "load" you are placing on it. To accelerate, you step down harder on the accelerator. This opens the throttle valve, increasing manifold pressure. The greater pressure difference between the manifold and the cylinders increases intake air flow, and therefore fuel flow, to increase power and accelerate the car.

Driving down a level road, you can cruise along comfortably and maintain a desired speed with a relatively small throttle opening. When you come to a hill, it is necessary to press farther down on the accelerator to maintain the same speed, even though engine rpm is unchanged. The hill has demanded more work from the engine—created a higher load—and the engine has demanded more air and fuel to match that load.

Regardless of engine speed, the air flow and fuel delivery demands of the engine depend on the load being placed upon it. That load, and the resulting throttle opening, directly affect manifold pressure. Manifold pressure in turn affects air flow and thus fuel requirements.

2.4 Fuel Pressure

Both carburetors and fuel-injection systems rely on pressure or pressure differential to dispense fuel into the intake air stream. The atomized fuel vaporizes and mixes with the air to produce a combustible mixture.

In a carburetor system, the fuel system supplies fuel to the carburetor float bowl. The bowl is vented, and the fuel in the bowl is at atmospheric pressure. Intake air passing through the carburetor's venturi creates low pressure in the venturi—lower than atmospheric pressure. This pressure differential—between atmospheric pressure in the carburetor bowl and reduced pressure in the venturi chamber—causes fuel to flow from the bowl through the discharge nozzle, and into the intake air stream.

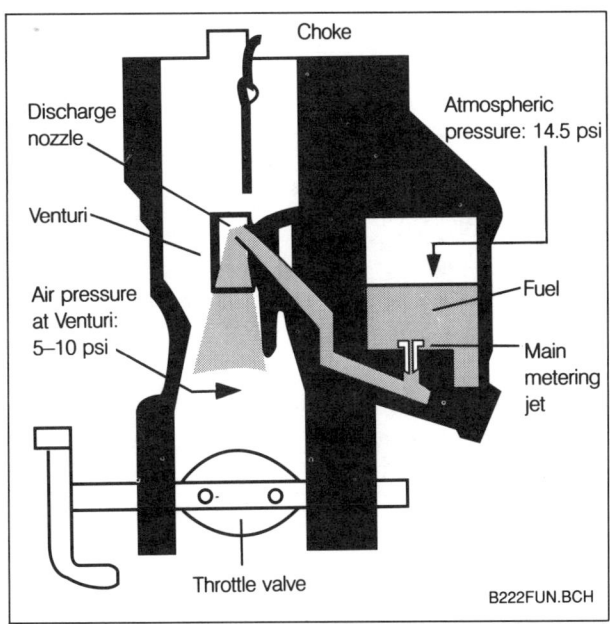

Fig. 2-11. The fuel-metering part of a carburetor operates with small differences in pressure: fuel in the bowl at atmospheric pressure vs. reduced pressure at the discharge nozzle in the venturi.

The pressure differential in the fuel-metering system of a carburetor is usually pretty small—only a few pounds-per-square-inch (psi). In contrast, fuel-injection systems introduce the fuel at higher pressures, anywhere from 30 to 90 psi depending on the system and specific operating conditions. Fuel delivery at these higher pressures ensures that the fuel is better atomized, and it is vaporized more completely in the airstream. Higher fuel pressure also makes possible more precise control of fuel delivery.

2.5 Pressure Measurement

As I get into the specific functional details of each of the various fuel-injection systems, you'll see that many functions and relationships are defined in terms of pressure. They may be fuel pressure values in the fuel system, manifold pressure in the air intake system, or atmospheric pressure. I may be talking about a differential pressure—the difference between two opposing pressure values somewhere in the system.

In any case, pressure values are always expressed in one or more of the following units or terms. All are correct. With the appropriate conversion factors, all are interchangeable. Fortunately, the math required to convert between the different units is simple.

- **Pounds-per-Square Inch (psi)**. The English system unit of pressure defined as force (pounds) divided by area (square inches). Atmospheric pressure at sea level is approximately 14.5 psi.

- **Bar**. A metric term derived from barometer or barometric pressure – atmospheric pressure. One bar is approximately equal to standard atmospheric pressure at sea level, so 1 bar = 14.5 psi. Typical fuel pressure in a pulsed-injection system might be 2.5 bar, or about 36 psi. European manufacturers' specifications and pressure gauges usually refer to pressures in bar.

- **kiloPascal (kPa)**. The proper metric unit for pressure. 1 bar = 100 kPa, so atmospheric pressure at sea level is 100kPa. Used mainly in the U.S.

- **Inches of Mercury (in. Hg)**. Originally refers to measurement of pressure using a mercury manometer. (Hg is the chemical symbol for mercury). This is a term used to specify manifold vacuum; 29.92 in. Hg is the difference between standard atmospheric pressure at sea level and absolute vacuum.

Table **a** lists the various units of pressure and appropriate conversion factors.

Table a. Units of Pressure Conversion

Unit	Atmospheric pressure (sea level)	to convert to:			
		psi	bar	kPa	in.Hg
		multiply by:			
psi	14.5	—	0.069	6.9	2.06
bar	1	14.5	—	100	29.92
kPa	100	0.145	0.01	—	0.299
in.Hg	29.92	0.485	0.033	3.34	—

If you watch the local TV weatherperson, you'll see how barometric pressure readings (in. Hg) are used to describe atmospheric pressure changes. Listen to the references to "highs" and "lows". Changing atmospheric pressure changes the density of the air. Denser intake air can slightly alter the air-fuel ratio and may affect how your engine operates. Some Bosch fuel-injection and engine-management systems have features which allow the system to compensate for variations in air density.

Gauge Pressure vs. Absolute Pressure vs. Vacuum

I've described engine intake air flow and load in terms of manifold pressure, and I've discussed the units of measure used to describe pressure. Now it is important that you understand exactly what you are measuring.

For many years, people have traditionally thought about engine air flow and load in terms of vacuum – the "vacuum" created in the intake manifold by the pistons' intake strokes. Using atmospheric pressure as a baseline, as zero, the lower manifold pressure is expressed as a negative value – vacuum.

By the 1980s, the automotive industry had moved away from thinking in terms of vacuum. Widespread use of fuel injection began to demand more accurate measurement of manifold "vacuum", and turbocharging – where manifold pressure may exceed atmospheric pressure – began to blur the distinction between vacuum and pressure. A simpler and more useful approach is to think in terms of manifold absolute pressure (MAP). MAP refers to the pressure in the intake manifold which is positive compared to zero absolute pressure, rather than negative compared to atmospheric pressure.

If that seems hard to understand, think of atmospheric pressure at sea level. We tend to think of this as zero pressure and, in fact, an open fuel-pressure gauge will read 0 at sea level. But remember, atmospheric pressure at sea level is actually 1 bar (14.5 psi). Almost all pressure gauges use atmospheric pressure as their reference. Any measurement made with the gauge reads pressure only with respect to atmospheric pressure. As indicated by Fig. 2-12, a gauge which reads 0 pressure at sea level (atmospheric pressure) is actually measuring 1 bar on the absolute pressure scale.

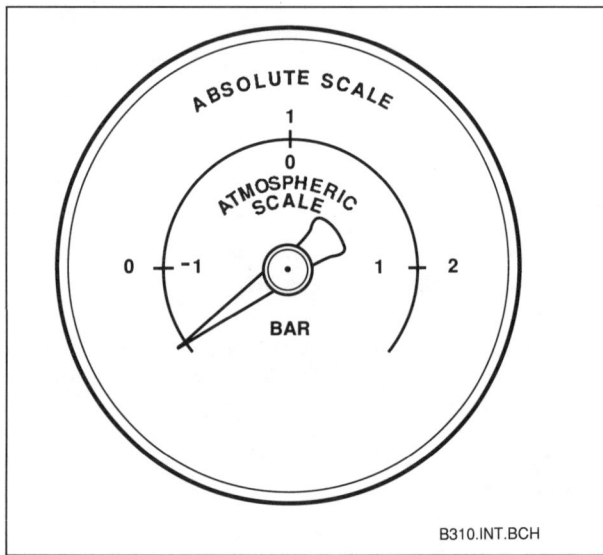

Fig. 2-12. Calibration of pressure gauge influences how you interpret pressure readings.

Using absolute pressure as the reference point, the piston on its intake stroke is creating a very low pressure in the cylinder—approaching zero absolute pressure. The pressure in the intake manifold—manifold absolute pressure (MAP)—is higher and always a positive number. Higher still is atmospheric pressure outside the engine, its influence on manifold pressure controlled by the throttle. Boost from a turbocharger or supercharger is pressure above atmospheric pressure. Many European boost gauges read MAP. They are calibrated from zero absolute pressure so, with the engine off, the gauge reads 1 bar—normal atmospheric pressure.

I've been driving with a "vacuum" gauge connected to the intake manifold of each of my last 9 cars. The gauge comes calibrated in terms of vacuum, so at wide open throttle (WOT) it reads close to zero. I have also marked my gauges to read MAP, positive manifold pressure relative to absolute pressure at or near sea level. That means that at WOT (near zero vacuum), my gauge reads about 9 on the homemade MAP scale, as shown in Fig. 2-13. I can read it two ways: multiply by 10 to read pressure in kPa (90 kPa), or divide by 10 to read pressure in bar (0.9 bar). The important point is that the numbers increase as the load and power increase. Fig. 2-14 shows approximate range of gauge values for various driving conditions and throttle positions. By coincidence, 65 mph cruise reads about 65 kPa or 0.65 bar.

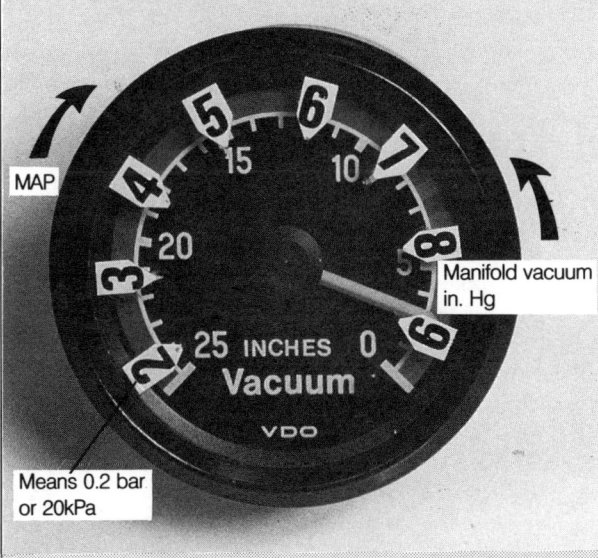

Fig. 2-13. Manifold vacuum gauge calibrated to read vacuum (in. Hg) can also be calibrated to read manifold absolute pressure (MAP). With fuel-injection or forced-induction systems, MAP makes it easier to relate engine load and throttle opening to manifold pressure.

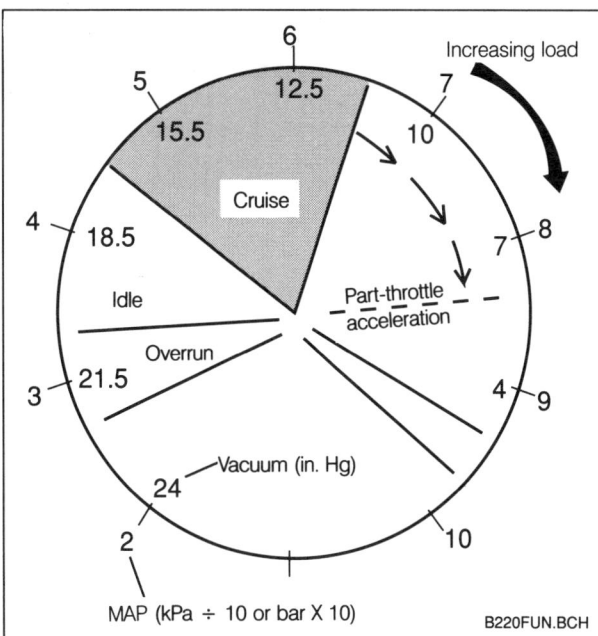

Fig. 2-14. Gauge schematic shows how vacuum and MAP indications relate to throttle opening.

On a turbocharged car, boost will read above 10 (100 kPa) on the MAP scale. MAP changes continuously from about 20kPa during overrun (coasting with closed throttle) to perhaps 160kPa at maximum boost. When we discuss the importance of manifold pressure to fuel injection, you will find it an advantage to think in terms of the positive MAP values (from absolute zero) rather than vacuum.

3. OPERATING CONDITIONS AND DRIVEABILITY

If I were talking about the requirements of a stationary industrial engine, I'd expect it to operate under basically fixed conditions: constant rpm, constant load, nearly constant temperature, limited stops and starts, no acceleration, and no heavy-footed driver. Such an engine would operate quite nicely at a fixed air-fuel ratio. It could be easily tuned to maximize fuel economy, and would require only the simplest of fuel systems.

Cars, however, are a different story. We expect them to perform under the widest possible variety of operating conditions. And we have given "performance" a new definition it is not only impressive horsepower and torque, but also maximum fuel economy, and controlled exhaust emissions. As if these performance demands were not enough, we also expect the car to meet these demands effortlessly, at any time, under any conditions, and at the turn of a key. We expect what has come to be called "driveability"—the ability of the car to provide smooth, trouble-free performance under virtually any conditions while delivering power, fuel economy, and controlled emissions.

Driveability is a term which evolved out of the early days of strict emission control and the '70s energy-crisis concerns over fuel economy. The less-developed technology of the time dictated an approach to both problems which often resulted in the engine running too lean. Running too lean robbed power and contributed to rough idle, poor throttle response, stumbling and stalling, and overall poor running. Fuel injection and more complex engine-management systems offer the precise control and flexibility necessary to meet modern performance requirements.

To meet these demands under different operating conditions, the engine has different air-fuel requirements. I'll discuss these operating conditions, the effect they have on basic air-fuel requirements, and how the capabilities of fuel-injection and engine-management systems are used to meet these requirements. I'm describing what these systems in general can do, not necessarily what a basic Jetronic fuel-injection system will do, particularly those systems on cars manufactured before 1982.

3.1 Normal (Warm) Cruise

Normal cruising at light load with the engine fully warmed up is the baseline operating condition. The basic fuel system is designed to meet the engine's need for the proper air-fuel ratio under these simple cruising conditions. The design must be flexible enough to handle fuel delivery at different speeds and loads, but this basic fuel metering is the fuel system's simplest task. Though the driver may add power on a hill or to pass another car, or cut back power to slow down, fuel management is still relatively simple. All other parts of the fuel system which compensate for different operating conditions do so by making adjustments to this main fuel-metering function.

When you are cruising down the interstate, if the road is level the engine is operating under relatively constant normal conditions. The fuel quality may vary from one tank fill to the next; weather and outside temperature may change; it may be dry or it may be damp. The ideal air-fuel ratio will be different for each of these conditions. A fuel-injection system can adjust to these changing conditions with little challenge, maintaining air-fuel delivery near the ideal (stoichiometric) ratio of 14.7:1 to satisfy the most important considerations of fuel economy and low exhaust emissions.

3.2 Starting

As an owner, you expect the engine to start instantly; whether it's at sub-zero temperatures or parked in the desert, too hot to touch; whether it's been sitting for five minutes at the store or for five weeks in the garage.

As far as fuel system requirements are concerned, starting makes different demands depending on temperature. I'll discuss cold start, warm start, and hot start.

Cold Start

This means the engine is cold; that is, it probably hasn't run for at least 12 hours. In most cases, the temperature needle is at the low end of the gauge. The engine might be as cold as −20°F (−30°C), sometimes called cold-cold; or as hot as 115°F (45°C), sometimes called warm-cold. In either case, the engine is still cold compared to its normal operating temperature of about 195°F (90°C).

Fig. 3-1. Engine-cold temperature varies according to outside temperature. Cold start means the engine has not run for several hours.

To understand the distinctions I've made in cold-start temperature, consider the Environmental Protection Agency (EPA) definition of cold start. For EPA testing, cold is room temperature, about 68°F (20°C); all cars are tested on dynamometers inside the lab. The engine must cold-soak at 68°F in the lab next to the dynamometer for 12 hours before the test. Cars are pushed onto the dynamometer for the cold-start test. As it happens, 68°F is close to actual cold start temperatures in Southern California and much of southern U.S. Engine temperature affects the driveability of your engine, which may be much colder than an EPA cold engine. For scientific comparison testing, and for the regulatory aspects of EPA testing for emissions and fuel economy, uniformity is the most important factor.

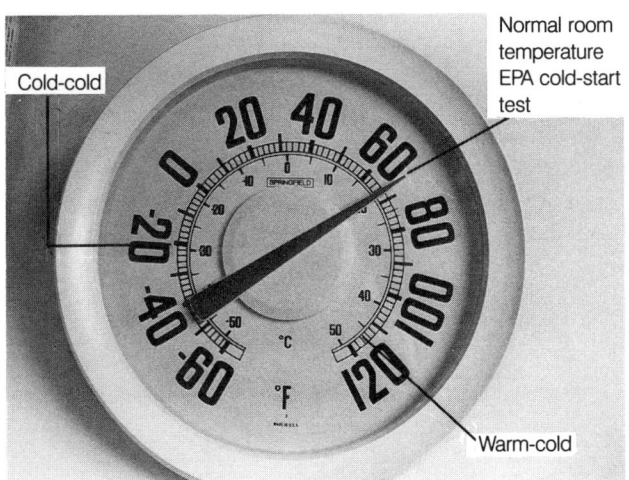

Normal room temperature
EPA cold-start test

Cold-cold

Warm-cold

Fig. 3-2. Fuel requirement for engine starting varies according to outside air temperature: cold-cold or warm-cold. These temperatures are both cold compared to normal engine operating temperatures.

In most parts of the country, you will at times be faced with cold-cold starting conditions; the most challenging of all. Gasoline is less likely to vaporize when it is cold. Even if it is adequately vaporized, some fuel condenses on the cold parts of the engine before it can be burned. The engine requires extra fuel for starting so that, in spite of vaporization and condensation problems, the engine still receives a combustible air-fuel mixture.

What constitutes a combustible mixture depends on air temperature, the volatility of the gasoline, altitude, barometric pressure and humidity. While a carburetor relies on a relatively crude choke mechanism to increase fuel delivery for cold starting, fuel injection compensates for many of these factors. Temperature is the most important.

Cold-start enrichment quickly becomes a problem if the engine does not start right away; due to a marginal battery, ignition components in poor condition, or whatever. Enrichment during cranking must be cut back in a matter of seconds; if it goes on too long, the air-fuel mixture will be too rich to ignite. The spark plugs may become fuel-fouled, particularly when they are cold, and the engine will not start.

Cold starting also needs help in terms of intake air flow. A closed throttle and slow cranking speeds do not allow enough air for starting. Drivers of carbureted cars develop intricate starting procedures: depress the accelerator to the floor, release, then hold the throttle about ⅓ open (for example). The first depression is necessary to set the carburetor choke and fast-idle cams; the second is to open the throttle and make sure the engine is getting enough air. Fuel-injection systems can control the air by bypassing the closed throttle, admitting more air when cold without any effort or attention from the driver.

Warm Start

Warm starting is very much like warm running. The air-fuel demands are simple. A warm engine helps promote fuel vaporization, and fuel is much less likely to condense out of the air stream onto warm engine parts

Depending on the temperature, the engine may benefit from some small portion of the usual cold-start compensations. A carbureted car may benefit from a light push on the throttle, using the accelerator pump to deliver a little extra fuel. A fuel-injection system sensing temperature only a little below normal may just slightly enrich the mixture and slightly increase intake air to improve warm starting.

Hot Start

This means the engine is hot; perhaps after being driven for some time on a hot day, and then allowed to sit for a short time, unable to significantly cool. Under these conditions, sometimes called "hot soak" under-hood temperatures may exceed 250°F (120°C).

The most important aspect of hot soak and subsequent hot starting is the possibility of overheating the fuel. For a carbureted engine this may mean boiling the fuel in the fuel bowl and losing it as fuel vapor into the atmosphere. Hot starting becomes more difficult, since it takes time for the fuel system to refill the bowl so that fuel can reach the discharge nozzle.

In any engine, the excess heat may raise fuel temperatures and cause "vapor lock" — vaporized fuel in the lines. Unlike fuel, the resulting vapor in the lines is compressible, so the fuel pump cannot necessarily overcome the problem and deliver fresh fuel. In fuel-injection systems, the number of fuel lines to individual injectors and the routing of lines near the engine makes heat-soak a more acute problem. To ensure quick hot restart, the system must maintain pressure in the fuel lines and the injectors while the engine cools to prevent vapor lock.

Some fuel-injection systems combat the fuel vaporization problem by adding small doses of extra fuel, similar to cold start enrichment. The extra fuel while cranking slowly helps the engine start; then the increased air flow and fuel demands of the running engine can overcome the vapor lock.

Ignition Timing and Starting

For the more sophisticated engine-management systems, adjusting the ignition timing can help start a cold engine. If the engine is turning over a little slowly because of thick, cold oil in the crankcase, or reduced battery voltage at low temperatures, the best ignition timing is near Top Dead Center (TDC). More advanced timing, even as little as 10 degrees before TDC, allows the slow-turning engine to fire while the piston is still rising; the reverse torque may damage the starter.

Operating Conditions and Driveability

On a warm day, ignition timing of a cold engine can be advanced because the engine spins a little faster. But if the temperature were high (hot start), retarded ignition timing can prevent reverse torque and starter damage caused by igniting the air-fuel mixture before the piston reaches TDC. Also, the engine may knock during starting if intake air temperature is too high. In short, it is desirable to retard ignition timing for conditions of low cranking speed, and on the basis of high intake-air temperature. More advanced ignition timing is permissible under less challenging, warm starting conditions.

3.3 Post-Start and Warm-Up

Post-start refers to the period immediately after the engine fires. The engine requires less enrichment than for starting, but it is still cold. Some enrichment is required to compensate for poor fuel vaporization and for condensation; and to improve torque and throttle response, even though acceleration at this point might not be good driving practice. Engine torque must be sufficient to carry the load of a cold automatic transmission shifted from Neutral to Drive; many cold engines have a tendency to stall when so shifted.

The amount of post-start enrichment required depends on temperature and time; a colder engine requires more enrichment for a longer period of time. On the other hand, exhaust emission control demands that enrichment be cut back properly. An engine can fail its EPA emission test in the first thirty seconds after "cold" start if the post-start enrichment causes excessive exhaust emissions.

Post-start ignition timing can be advanced to improve the engine running and incidentally to reduce the requirement for enrichment during this period. Post-start can end in a few seconds or may last as long as fifteen or twenty seconds if the temperature is extremely cold. But, strangely, ignition retard can be used to heat the exhaust, and therefore the lambda sensor and catalytic converter, which work as intended only at high temperatures. That means emission control can be effective quickly after a start.

Warm-up refers to that period when the engine is in transition from its cold, post-start temperature, to its normal operating temperature. As the engine begins to warm up, it begins to require less and less enrichment. Since warm-up is a smooth, gradual process, the reduction of enrichment needs to be gradual too. If the engine is really cold, warm-up can last for several minutes. Since the car is probably being driven during the warm-up period, enrichment which is appropriate for the conditions at any given time is essential to ensure driveability.

On a carbureted engine, the choke mechanism is sensitive to temperature. As the engine warms the choke warms, gradually reducing warm-up enrichment. Fuel-injection and engine-management systems also use temperature as the main input for control of warm-up enrichment. Engine coolant temperature is the most common input. Some systems also respond to changes in intake air temperature.

For systems that have the capability to control ignition timing, advancing the timing during warm-up under part-throttle load can aid driveability. Retarding timing during warm-up under closed-throttle deceleration will reduce hydrocarbon (HC) emissions which risk being excessive anyway due to the warm-up enrichment.

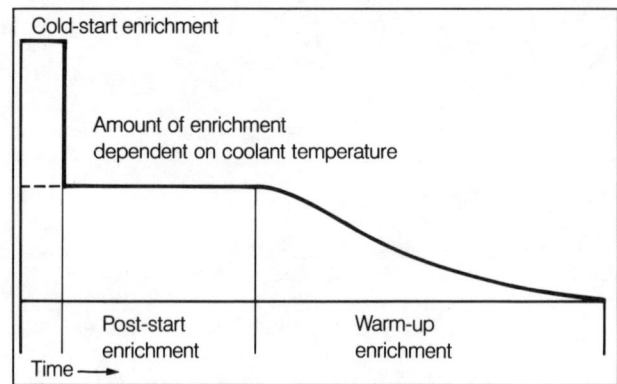

Fig. 3-3. Until the engine is fully warm, various degrees of enrichment are required. Starting requires maximum enrichment, depending on engine temperature. Post-start enrichment keeps the engine running for the first 30 seconds or so. Warm-up enrichment slowly drops off as the engine warms to normal operating temperature.

3.4 Idle

Our cars spend a lot of time idling, so a smooth idle and good off-idle throttle response are important aspects of driveability. First, there are some unique requirements for idle control during post-start and warm-up. Once the engine is warm, there are other factors which influence idle speed, and fuel injection's capabilities make possible improved idle characteristics.

Cold and Warm-up Idle

As I've already suggested under **3.2 Starting**, a cold engine has different air-fuel requirements than a warm engine. At idle, more air and more fuel are needed just to overcome the higher running friction of a cold-engine. A slightly higher idle speed is necessary to keep the cold engine from stalling, and to support good off-idle throttle response.

Fuel-injection systems, using both temperature inputs and direct rpm input, can precisely control idle speed by controlling intake air flow. With this precise control, cold idle is higher, but driveability is maintained. One of the joys of driving cars with fuel-injected engines is the freedom from a too-fast idle that can jar the car when an automatic transmission is shifted into Drive or Reverse.

Warm Idle

The main requirements of a warm engine running at idle speed are smoothness, and smooth response once the throttle is opened. Some engines require a richer mixture at idle for smooth running, and to ensure good off-idle response. With ignition timing control capability, it is also possible to operate with a leaner mixture and achieve a smooth idle by changing ignition timing.

In general, the engine should idle at the lowest speed at which the engine will still run smoothly enough to satisfy the driver. Reducing idle speed to a minimum reduces noise and fuel consumption.

The biggest obstacle to low idle speeds is the variation in load on the engine at idle. At idle, small changes have big consequences. Friction loads change with temperature; the power required to operate the charging system varies with electrical load (headlights, for example); air conditioning compressors switch on and off; on cars with automatic transmissions, shifting into Drive or Reverse at idle increases the load.

Fuel injection's capabilities include idle stabilization (also called idle-speed control). By monitoring and constantly correcting idle speed, it is not necessary to maintain a higher-than-minimum idle to handle variations in engine performance, changing of loads, and similar causes of stalling from idle.

Advanced idle-speed stabilization systems satisfy some sophisticated requirements, particularly on cars with air conditioning (A/C) and automatic transmission:

1. In Neutral, A/C off: maintains the minimum idle speed that is required for engine smoothness

2. In Neutral, A/C on:
   - for small compressors, increases idle speed to run the compressor faster (increase A/C performance)
   - for large compressors, maintains idle speed by compensating for added compressor load
   - through A/C relay, delays compressor-clutch engagement for ½ second to prevent temporary rpm drop

3. In Drive:
   - for smooth-idling engines, reduces idle speed to minimize stoplight creep
   - for rough-idling engines, increases idle speed to minimize idle/load-induced vibration

4. In Drive, when A/C switches on: controls idle speed increase to avoid inadvertent surge when driver does not expect it

3.5 Acceleration

Any time you press the accelerator, opening the throttle increases manifold pressure which in turn increases intake air flow. The fuel system, responding to this increase in air flow, meters additional fuel to maintain the correct air-fuel mixture.

Accelerating – opening the throttle very quickly – places additional demands on the system. The sudden transition from closed throttle to open throttle produces sudden spikes in manifold pressure. The fuel system responds quickly to these changes, but not instantly. Some additional fuel delivery capability is required to compensate for these changes in manifold pressure – to prevent stalling and produce the quick throttle response we have come to expect and demand.

For quick, smooth throttle response under acceleration, the basic system of air flow sensing and fuel metering is too slow. The fuel system must be able to instantly compensate for rapid increases in manifold pressure with additional fuel. In many carburetors, additional fuel is supplied by the accelerator pump. It mechanically squirts additional fuel into the intake air stream whenever the throttle is opened suddenly. As you will see in later chapters, fuel-injection systems employ various methods of acceleration enrichment.

In **2. Basic Factors** I explained that fuel injection systems are designed to maintain the air-fuel ratio in the narrow range around stoichiometric, but that maximum power output is achieved with an air-fuel ratio that is slightly richer. In addition, the engine may need a richer mixture to reduce the tendency to knock. Many fuel-injection systems recognize full-throttle acceleration as a special case, and provide additional acceleration enrichment keyed to the full-open throttle position.

During acceleration, systems which can control ignition timing can deliver the best balance between advanced timing for maximum torque, and retarded timing for knock control.

3.6 Deceleration and Coasting

In the quest for minimum fuel consumption, engineers are quick to take advantage of opportunities to reduce fuel waste. One such opportunity is coasting cut-off, also known as overrun or deceleration cut-off.

The basic idea of deceleration fuel cut-off is this: the engine takes in some air and the fuel system delivers some fuel even when the throttle is completely closed (at idle, for example). On the road, when the car is decelerating or coasting with the throttle closed, this fuel is wasted. It is passing through the engine and being expended, but doing no useful work. Cutting off fuel during these conditions can save fuel, even in city driving. It also reduces exhaust emissions. To prevent stalling, fuel flow must be re-established at some programmed engine speed (above idle).

Fuel cut-off behavior may be less satisfactory when the engine is cold, so normally the rpm limits for cut-off and for re-establishing fuel injection are higher for a cold engine. Coasting cut-off must also be restricted so it will not operate during shifting of manual transmissions, and so it will not interfere with cruise control.

Engine-management systems which can control ignition timing can retard the timing during the transition to smooth the response to the cut-off and resumption of fuel injection.

> Deceleration fuel cut-off was a feature of the early D-Jetronic systems supplied on some Volkswagens, but the air-cooled engine cooled irregularly during the cut-off so emissions were actually increased; the feature was discontinued.
>
> Water-cooled engines equipped with Jetronic systems have greater temperature stability, and can use deceleration cut-off successfully to save fuel and reduce emissions. Fuel cut-off is less successful in carbureted engines and those with throttle-body injection because the fuel is delivered so far from the cylinders. As fuel flow is resumed, the lag between restoring fuel delivery and resuming combustion spoils driveability with jerky response.

3.7 Altitude

At altitudes above sea level, the air is less dense; for a given volume of air, there is less oxygen to support combustion of fuel. In carburetors, and in fuel-injection systems which measure intake air by volume, less dense air changes the effective air-fuel ratio; at higher altitude, the mixture is richer.

Some fuel-injection systems can measure and respond to changes in air density, changing the air-fuel ratio as required to maintain the ideal (stoichiometric) ratio.

3.8 Engine Shut-Off

Engine shut-off is another chance to prevent wasting fuel. Carburetors dispense fuel into the intake air stream as long as the engine is turning over, whether or not the ignition is on. In contrast, turning off the ignition of a fuel-injected car also cuts power to the electric fuel pump. Fuel flow is immediately cut-off to prevent dieseling, or run-on.

3.9 RPM Limitation

Precise control of fuel delivery allows the fuel system to be used to limit engine rpm and prevent damage from over-revving. Before the days of catalytic converters, rpm was limited by simply cutting out the ignition when the rpm got too high.

Cutting the ignition, however, still allows unburned fuel into the catalytic converter and causes it to overheat; sometimes dangerously. Since the mid-1970s and the advent of catalytic converters, rpm limitation has been accomplished by the fuel-injection system cutting back on fuel delivery. The goal is to keep the engine running at its limitation without dumping unburned fuel into the catalytic converter.

3.10 Ignition Timing

For some of the specific operating conditions described above, I've described how ignition timing adjustments can be used to enhance performance and satisfy driveability requirements.

In terms of all four major aspects of performance—power, fuel economy, emission control, and driveability—ignition timing is a significant factor; perhaps as significant as air-fuel mixture. Combustion in the cylinders takes a certain amount of time. In terms of how we usually think about time, combustion is very rapid, but it is not instantaneous.

As the piston is compressing the mixture on the compression stroke, the exact point at which the spark plug fires to ignite the mixture—ignition timing—has profound effects on the quality of combustion and the amount of power that is produced. Because of its influence on the combustion process, ignition timing also affects combustion temperature which can also significantly affect exhaust emissions.

Modern fuel-injection systems have evolved into engine-management systems—systems which manage much more than fuel delivery. Chief among the extra capabilities of an engine-management system is simultaneous control of ignition timing and fuel delivery. Controlling both factors opens up new possibilities for power and driveability improvements while maintaining tight control of exhaust emissions and fuel economy.

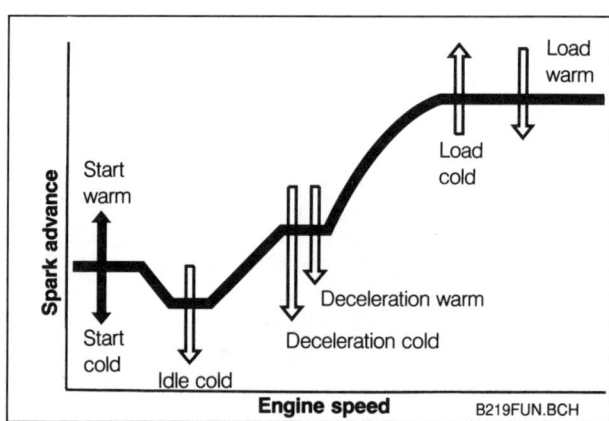

Fig. 3-4. Typical ignition timing advance curve can be modified for best performance, from starting to acceleration.

4. EMISSION CONTROL

It is practically impossible to separate the development of today's fuel-injection and engine-management systems from the increasing demand for control of harmful exhaust emissions. Exhaust emission control could not have been accomplished as successfully without fuel injection and, likewise, fuel injection may not have been so successful and widely used were it not for the challenges of meeting emission control regulations.

With changing legislation and tougher regulatory standards, fuel-injection systems have undergone significant changes so that engines can meet the emission standards while providing the driveability demanded by owners. The superiority of these systems is also demonstrated by the fact that, even as U.S. fuel economy standards have tightened, fuel injection has added both economy and power to smaller engines.

It is interesting to note that many fuel-injected cars are able to eliminate some types of emission control that interfere with driveability on carbureted cars, including exhaust gas recirculation (EGR) and air pumps. If you know something about emissions and the limits placed on them by legislation, you'll understand more about how fuel-injection and engine-management systems work.

4.1 Combustion By-Products

Combustion of the air-fuel mixture in the engine cylinders creates gaseous by-products which make up the exhaust. Some of these are harmless, and some are known to be harmful. Three gases: hydrocarbons (HC); carbon monoxide (CO); and oxides of nitrogen (NO_x) are the most harmful ones. Emission of these gases in vehicle exhaust is regulated by the Federal Clean Air Act. The same gases are often subjected to tighter limits by state regulation in California.

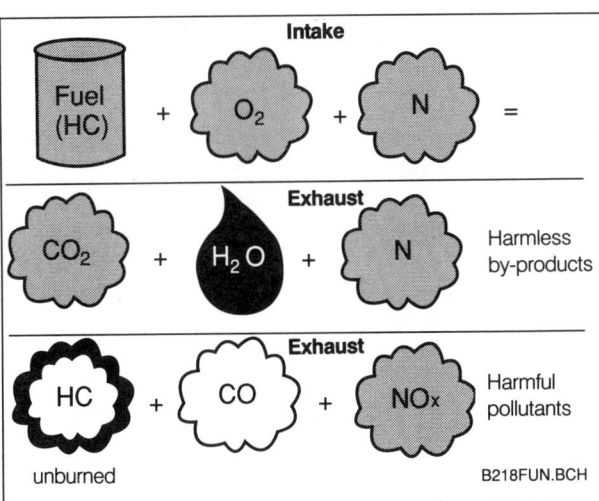

Intake

Fuel (HC) + O_2 + N =

Exhaust

CO_2 + H_2O + N   Harmless by-products

Exhaust

HC + CO + NO_x Harmful pollutants

unburned   B218FUN.BCH

Fig. 4-1. The emission equation: Engine takes in fuel, oxygen, and nitrogen (top). Combustion produces harmless by-products: carbon dioxide, water, and nitrogen (middle); and harmful pollutants: unburned hydrocarbons, carbon monoxide, and oxides of nitrogen (bottom).

Hydrocarbons (HC). Gasoline is a compound composed of hydrogen and carbon; in the combustion process, these elements combine with oxygen to form the by-products water (H_2O) and carbon dioxide (CO_2); HC in the exhaust is unburned gasoline, the result of incomplete combustion.

Carbon Monoxide (CO). CO, a poisonous gas, is another result of incomplete combustion; when gasoline burns completely, the result is CO_2.

Oxides of Nitrogen (NO_x). (NO_x) refers to several kinds of nitrogen oxide which result from nitrogen and oxygen during combustion. Nitrogen and oxygen are normal parts of air, but they exist in air as separate elements. And, as long as the combustion temperature stays below about 2000°F (1100°C), the nitrogen and oxygen do not combine. The nitrogen passes out the exhaust pipe just as it came in, separate and harmless. Combustion temperatures only slightly higher, however, cause these two elements to combine chemically into various forms of nitrogen oxide (NO_x), a key element of smog.

Remember, these exhaust gases are normally colorless and invisible. A clear tailpipe is not the sure sign of a clean-burning engine; it may be pumping out invisible pollutants.

Until recently, carbon dioxide (CO_2) was considered a harmless emission. But now we must consider the "greenhouse" effect. Recent studies show that CO_2 is accumulating in the upper atomsphere, trapping global heat much as glass traps heat in a greenhouse. The probable results are rises in global temperatures, successive heat waves, and iceberg melting, which could raise ocean levels to flood seaside properties worldwide.

Any burning of fossil fuels such as oil, gasoline, and coal produces CO_2. Automobiles are a significant source. What can we do in driving to reduce CO_2? Avoid unnecessary idling, for one thing. This highlights the need to adjust traffic patterns to reduce traffic jams. What else could the "greenhouse" effect mean to car owners? Smaller engines, because CO_2 increses with displacement. Also, it could mean restrictions on driving, and increased pressure from Congress for higher CAFE (more rated mpg) from every maker.

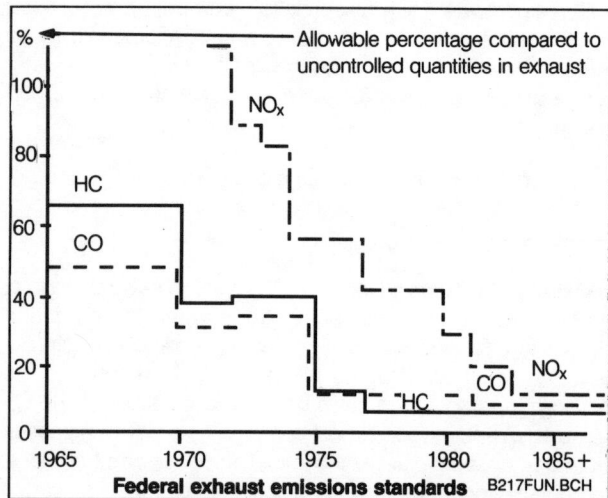

Fig. 4-2. Evolution of U.S. government-mandated emission control standards. Jetronic fuel injection came in to meet the early exhaust emission standards. As limits tightened after 1980, virtually all European cars sold in the U.S. were equipped with Bosch fuel injection.

4.2 Effects of Air-Fuel Ratios on Pollutants

Although exhaust pollutants are not normally visible, some visual clues are a tip-off to poor air-fuel mixture control. Black smoke coming from the exhaust pipe is usually a sign of unburned gasoline; a rich mixture. But look around you on the street; black smoke is something rarely seen on fuel-injected cars.

Just as variations in the air-fuel ratio affect power output and fuel consumption, they also affect exhaust emissions. As I described in **2. Basic Factors**, the air-fuel ratio is a key to complete combustion of the fuel. It also affects combustion temperature which in turn affects the formation of pollutants. Too rich (too little air), and the fuel will not be completely burned. The unburned fuel will enter the atmosphere as CO and HC. Too lean (too much air), and elevated combustion temperature increases NO_x. Fig. 4-3 shows how relative levels of each of the three major pollutants in the engine exhaust are affected by variations in the air-fuel ratio.

While Fig. 4-3 illustrates the effects of air-fuel ratio on the formation of harmful emissions, it also illustrates the fundamental problem in the control of these three pollutants: adjusting air-fuel ratio to ensure complete combustion and minimize the production of HC and CO has the undesirable side-effect of increasing NO_x. Likewise, richening the mixture to minimize combustion temperature and the formation of NO_x brings unacceptable increases in HC and CO.

It becomes clear then that manipulation of the air-fuel ratio is simply not enough to adequately control all three of the most harmful pollutants, especially in the face of increasingly tough emission control regulations. Control of air-fuel ratio to optimize emissions is a first step, to be followed by treatment of the exhaust after it leaves the engine. Maintaining precise control of the air-fuel ratio by electronic control of fuel metering is important not only to the composition of the combustion by-products but also, as you'll see, to the operation of three-way catalytic converters.

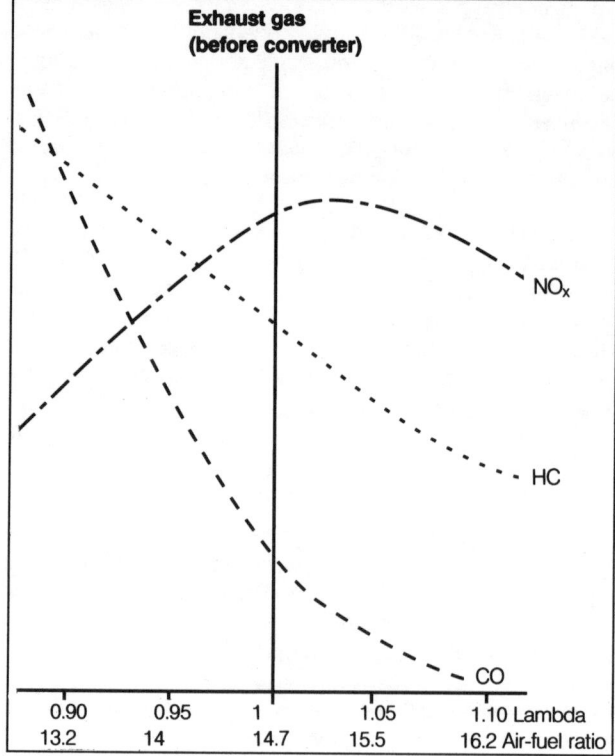

Fig. 4-3. Pollutant levels in engine exhaust change drastically as air-fuel ratio changes. Notice the trade-off between control of HC and CO, and control of NO_x.

When I flew piston-engine aircraft, I adjusted for a rich air-fuel mixture during high-power climbs to keep the engine cool. Then, during cruise, I leaned the mixture. If the needle on the temperature gauge started to rise above the green arc, I knew I had leaned the mixture too much. If you install an exhaust gas temperature (EGT) gauge on your car engine, you can make similar observations (though the mixture is not cockpit-adjustable). I added an EGT gauge when I installed an aftermarket turbocharger.

4.3 Exhaust Gas Recirculation (EGR)

Exhaust Gas Recirculation (EGR) is a technique for reducing the formation of oxides of nitrogen (NO_x). A small amount of exhaust gas is rerouted back into the combustion chambers, diluting the combustible mixture and lowering combustion temperature. You'll remember that excessive combustion temperature is the cause of NO_x formation.

EGR is necessary to properly control emissions in almost all carbureted engines; fuel-injected engines can often do away with EGR by maintaining proper control of the air-fuel ratio and treating the exhaust with a three-way catalyst.

4.4 Exhaust Gas Aftertreatment

As I discussed above under **3.2 Effects of Air-Fuel Ratios on Pollutants**, precise control of the air-fuel ratio alone is not enough to provide adequate reduction and control of engine exhaust pollutants. While this precise control is essential, the necessary reductions require additional treatment of the exhaust after it leaves the engine—aftertreatment.

A variety of methods of exhaust gas aftertreatment have been employed as the allowable levels of pollutants have been forced lower. All rely to some degree on particular exhaust characteristics to function properly, so the success of all depends on precise control of the air-fuel ratio.

AIR, Air Injection Reaction

One of the early approaches to reduce emissions was air injection. An air-pump, commonly reffered to as a "smog pump", delivers air into the exhaust manifold. Adding air tends to burn HC and CO as they exhaust from the cylinders, reducing emissions but increasing underhood heat. You'll find air injection on most carbureted engines and even on many U.S. fuel-injected engines. But Bosch fuel injection controls fuel delivery with enough precision that air injection is seldom necessary; it is used only on the largest Bosch-equipped engines.

Catalytic Converters

Catalytic converters are installed as part of the exhaust system, located between the exhaust manifold and the tailpipe. The interior surfaces of a catalytic converter are coated with special materials—catalysts—that promote additional chemical reactions with the pollutants in the exhaust gas and convert them into less harmful substances.

Oxidation catalysts make use of excess air supplied by an air pump to oxidize CO and HC—add oxygen—and convert them to CO_2 and H_2O. Reduction catalysts work without the addition of excess air to reduce NO_x. The combination of an oxidation catalyst and a reduction catalyst in one housing—a dual-bed catalyst—produces a complex series of chemical reactions which reduce all three pollutants. A disadvantage of dual-bed catalysts is that they rely on a slightly rich air-fuel ratio which increases fuel consumption.

To work most efficiently, the converter must be as hot as possible. For this reason, the best placement is in the exhaust system as near to the engine as possible. In addition to operating at high temperature, the reactions themselves produce heat. Most catalytic converters require heat shields to prevent combustion of something under the car. Even so, drivers are cautioned to avoid parking a hot car near anything combustible, such as tall grass or dry leaves.

Fig. 4-4. Catalytic converter treatment of exhaust generates heat, so converter has heat shield to protect car and anything combustible under the car.

Three-way Catalytic Converters

The most advanced—and by now the most widely used—catalyst is the three-way catalyst, so-called because its chemical reactions operate to lower all three controlled pollutants to levels which could not be achieved previously.

Remember that, in complete combustion, $HC + O_2 + N = CO_2 + H_2O + N$; incomplete combustion produces CO instead of CO_2; and high temperature combustion, as from a lean mixture, combines $N + O_2$ to form NO_x.

In the three-way catalytic converter, we want to (1) add oxygen to oxidize the HC and CO to make H_2O and CO_2 and, (2) take away oxygen to reduce NO_x, separate it into N and O_2.

You might think it's as simple as taking the oxygen away from the NO$_x$ and giving it to the CO. Stated simply, that is what happens in some three-way catalytic converters.

The catalytic material in the converter helps these chemical reactions take place. In order for reduction (taking away oxygen from NO$_x$) to match oxidation (adding oxygen to CO and HC), the proportion of the gases in the engine exhaust must be controlled very closely; that means the intake air-fuel ratio must always be in the narrow range near stoichiometric—lambda (λ) = 1, or an air-fuel ratio of 14.7 parts of air to one part of fuel. In some three-way converters, oxygen is pumped in after the reduction to further enhance oxidation.

Fig. 4-5 illustrates the degree of emission control afforded by a three-way catalyst on an engine running very near the stoichiometric ratio. You can see that if the air-fuel strays from λ = 1, the proportion of exhaust gases (HC, CO, & NO$_x$) exiting the converter changes. As the air-fuel ratio becomes leaner, hotter combustion temperature causes increased production of NO$_x$. A rich mixture will produce an excess of HC and CO.

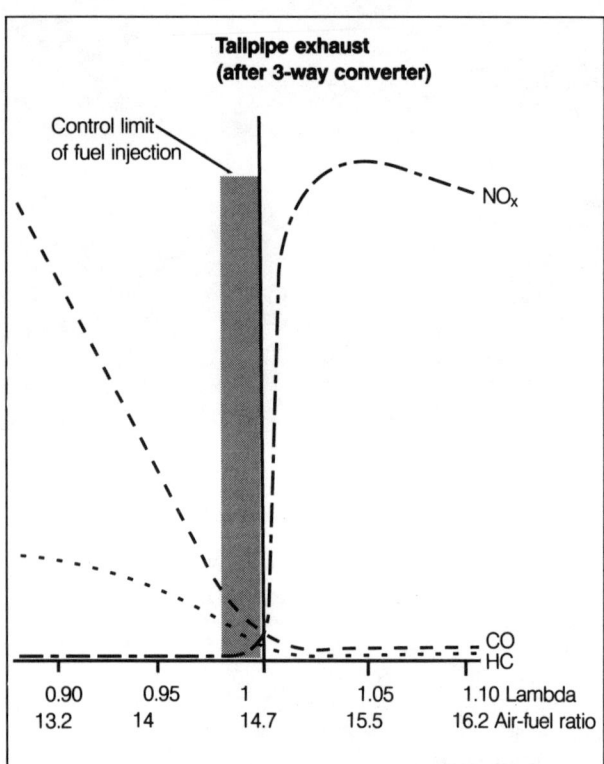

Fig. 4-5. Three-way catalytic converter treats three exhaust gases. HC and CO are oxidized, NO$_x$ is reduced so tailpipe gases meet emission limits. Engine can be tuned to stoichiometric air-fuel ratio for better drive-ability and economy. Compare with untreated exhaust (Fig. 4-3 above).

When the air-fuel ratio is maintained at λ = 1, the ideal air-fuel ratio, the emission of all three pollutants is reduced to very low levels. Precise control, however, is very important to the successful operation of three-way converters. Any signifi-

Emission Control

cant deviation from λ = 1 upsets the balance of the chemical reactions in the converter and the level of one or more pollutants increases dramatically. Development of three-way catalytic converters has been accompanied by development of more sophisticated systems for the fine control of air-fuel ratio.

5. CONTROL SYSTEMS

By now, you are aware that control plays an overwhelmingly important part in maintaining the acceptable balance of power, fuel economy, emission control, and driveability. The modern fuel-injection system, by responding to measured inputs and precisely metering the appropriate amount of fuel for the conditions, offers unparalleled control.

Fuel injection's basic control systems are one-way or "open-loop" controls. They take the information about operating conditions received from various sensors, and then use that information to determine—either by mechanical means or using pre-programmed electronics—exactly how much fuel it should dispense to achieve the desired air-fuel mixture. Accuracy of the fuel metering and the resultant air-fuel mixture depends entirely on how well the system—whatever type it may be—can predict the engine's needs based on its "knowledge" of operating conditions.

A major advance in fuel-injection control systems has been the advent of "closed-loop" controls—systems which not only try to predict the engine's needs based on operating conditions, but also measure the results of their fuel metering, using that information as an input to achieve ever more precise control.

5.1 Closed-loop Control Systems

In a closed-loop or feedback control system, information about whatever is being controlled is continuously fed back to the system as an input. The operation of a thermostat in an automatic heating system is an example of closed-loop control. As the temperature falls, the thermostat senses lower temperature and signals the furnace to add heat. As soon as the temperature rises above the setting, the thermostat senses the results of its own control action—heat produced by the furnace—and signals the furnace to cut back the heat.

An open-loop system may, for example, sense low temperature and simply turn on the heat for a predetermined amount of time; however, the closed-loop control is automatic, temperature stays relatively constant, and energy consumption is probably reduced. All in all, the result is better, more precise control.

Lambda Sensor

As you saw above in Fig. 4-5, the three-way catalytic converter operates best when the air-fuel ratio is near stoichiometric, when lambda (the excess air factor) is within a very narrow range around λ = 1. Fuel-injection systems, while very good at controlling the air-fuel ratio, cannot hold the air-fuel mixture within the required range. The necessary precision requires the additional feedback available from a closed-loop control system.

The source of this feedback is a lambda sensor (also called an exhaust-gas sensor, or oxygen sensor), installed in the exhaust system. The sensor generates a low-voltage signal; the signal's strength is based on the amount of unused oxygen remaining in the exhaust stream — an indirect measurement of the air-fuel ratio. The lambda sensor signal provides feedback to the fuel-injection ECU, indicating by actual results whether the air-fuel ratio needs to be corrected. The system can then adjust its fuel metering so that $\lambda = 1$, and the exhaust remains as clean as possible.

Fig. 5-1. Lambda sensor, usually installed in exhaust manifold, is about the size of a spark plug. Mounting location near engine is required to maintain proper sensor temperature.

Because of tightening exhaust emissions regulations and the need for three-way catalysts, you'll find a lambda sensor on virtually every car made since 1981, domestic or import, fuel-injected or carbureted. For more information on the lambda sensor and its closed-loop control, see chapter 3.

Changing Engine Conditions

You can see that closed-loop lambda sensor system provides automatic control of air-fuel ratio, continuously adjusting the amount of fuel injected in response to the amount of oxygen in the exhaust. The lambda sensor system can also adjust fuel metering to compensate for changing engine conditions over time. If general wear or a leaking valve causes a change in combustion, the feedback system can compensate within its limits and still provide the best possible air-fuel mixture. It has been described as the equivalent of having a skilled technician riding under the hood, continually tuning the mixture for the best operation under the current conditions.

The ability to finely adjust fuel metering to match conditions is what makes lambda control so important, but lambda-sensor closed-loop control is limited to small adjustments within its range of operation. It is often referred to as fine-tuning; in fact, as you'll see in chapters 4 and 6, one of the measures of properly set mixture control is consistent output of CO — the same whether the lambda sensor is working closed-loop, or whether its circuit is disconnected and running open-loop. This means that the basic mixture adjustment is correct and the lambda sensor is operating in the center of its range of adjustment, giving it maximum possible correction range.

2

Chapter 3

Pulsed Injection – Theory

Contents

2   PULSED INJECTION – THEORY

TABLES

1. GENERAL DESCRIPTION

You'll find a number of different Bosch fuel-injection systems which are all based on electronically-timed, pulsed injection. See **Table a** on the next page. These systems are sometimes referred to by the term EFI (Electronic Fuel Injection), or AFC (Air-Flow Controlled) injection. The pulsed systems I'll discuss in this chapter include:

- L-Jetronic
- LU-Jetronic
- LH-Jetronic
- Motronic
- LH-Motronic
- D-Jetronic
- Digifant II

I use the term "pulsed" instead of "electronic" to describe this branch of the Bosch family tree because, since 1980 the other branch, the continuous injection systems, also use some electronic controls to adjust fuel metering.

This chapter begins with a general description of the main components and operating principles of all Bosch pulsed injection systems. It then describes pulsed systems in detail, starting with L-Jetronic. For more basic information on fuel injection, see chapter 1 and chapter 2. To determine what system is installed on your car, see **1.2 Applications**, and the detailed applications table at the end of chapter 1.

1.1 Pulsed Injection Systems

All of the systems in this chapter are pulsed injection systems. They meter fuel to the engine by electronically controlling the amount of time that the fuel injectors are open. In contrast to the continuous systems where the injectors are open and flow fuel from the moment the engine starts, pulsed injectors open and close in time with the engine to deliver the fuel. The main components of pulsed systems are the air-flow meter, the electronic control unit, and the fuel injectors. See Fig. 1-1.

In pulsed injection, all air entering the engine first flows through an air-flow meter. The air-flow meter measures this air, which indicates engine load, and converts that measurement into an electrical signal to the control unit. The control unit uses the air flow and engine rpm inputs to compute the amount of fuel necessary to give a good air-fuel mixture, and then electrically opens the injectors at each cylinder intake port to inject the appropriate amount of fuel into the intake air stream. The control unit times the injections to the rotation of the crankshaft. The main fuel pump supplies fuel under pressure to the injectors.

Fig. 1-2. Motronic pulsed injection system on Porsche 944. Note air-flow sensor (**1**) and fuel injectors (**2**). Components may differ depending on system.

Bosch pulsed systems also use many additional sensors that monitor engine operating conditions. The control unit monitors the signals from these sensors and increases or decreases injector opening time—increasing or decreasing the amount of fuel delivered—to give the best air-fuel mixture for the various conditions.

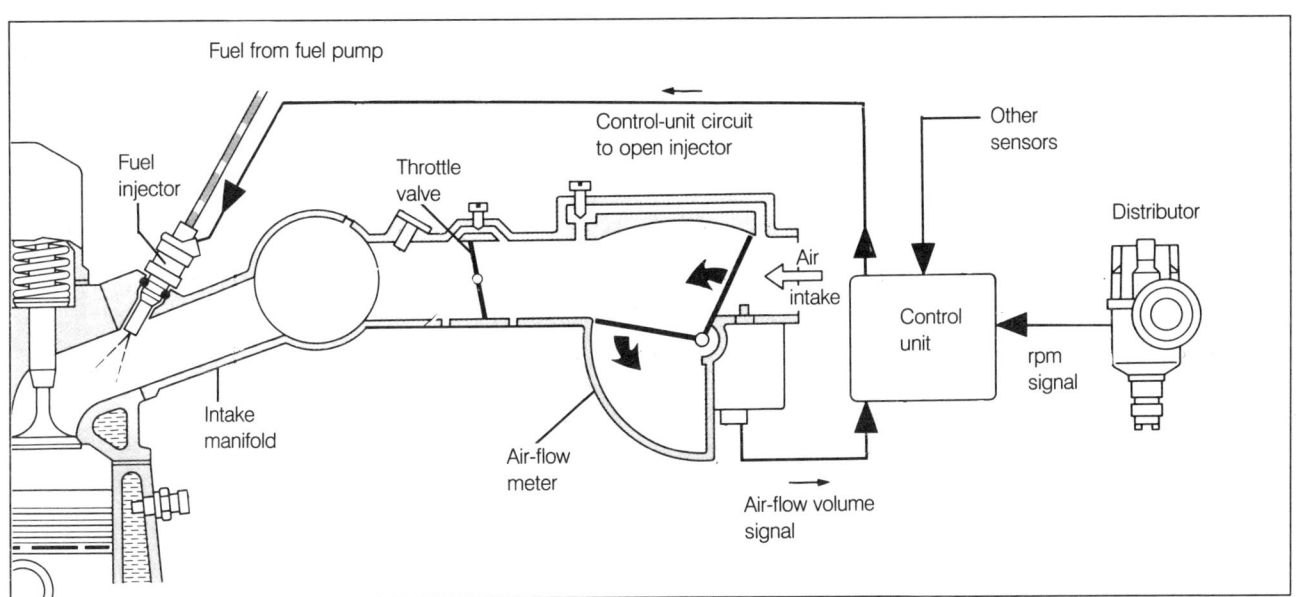

Fig. 1-1. Schematic of pulsed injection operation. Air-flow meter measures air entering engine. Control unit computes proportional amount of fuel required and opens injectors.

Table a. Bosch Pulsed Fuel Injection Systems

Name	First used	Measure load	Fuel pressure	Computer /no. of pins	Lambda
D-Jet	1967 VW 311	manifold pressure	pressure 2.0 bar (29 psi)	analog/25	no
L 1-Jet	1974 Porsche 914	air flow	2.5–3.0 bar (36–43.5 psi)	analog/35	some
LH 1-Jet	1982 Volvo	air mass	2.5–3.0 bar (36–43.5 psi)	digital/35	all
Motr.ML.1	1982 BMW 528e	air flow	2.5–3.0 bar (36–43.5 psi)	digital/35	all
LU-Jet	1982 BMW 318i	air flow	2.5–3.0 bar (36–43.5 psi)	analog/25	all
LH 2-Jet	1984 Volvo	air mass	2.5–3.0 bar (36–43.5 psi)	digital/25	all
Motr.ML.3	1984 BMW/Porsche	air flow or air mass	2.5–3.0 bar (36–43.5 psi)	digital/35	all
LH-Motr.	1988 BMW	air mass	2.5–3.0 bar (36–43.5 psi)	digital/35	all

1.2 Applications

Bosch L-Jetronic systems are installed in many makes of cars sold in the U.S., beginning in 1974. Considering licensed versions, L-Jetronic has the largest world-wide installed base of any port-type injection system. The Bosch pulsed systems are developments of L-Jetronic, except D-Jetronic. That leads to an interesting name confusion. To contrast with the original D-Jetronic, a pressure-controlled system, L-Jetronic was described as AFC for Air Flow Control. As it turns out, all Jetronic systems that followed D-Jetronic are air-flow or air-mass controlled, including the continuous systems.

The term "Motronic" originally (in 1979) meant a single control unit controlling L-Jetronic fuel injection and ignition timing. Nowadays, there is more than one type of Bosch Motronic system controlling fuel injection and ignition timing (LH-Motronic, KE-Motronic), so following Bosch nomenclature, I'll use Motronic to mean the original definition: L-Jetronic fuel injection and ignition-timing control. Why do I tell you all this? To help steer you through the various names that auto manufacturers and service people have attached to Bosch systems for 20+ years.

2. L-JETRONIC

In the broad terms I've discussed, L-Jetronic is a fuel-injection system, not an engine-management system. While most L-Jetronic cars have electronic ignition, they do not have electronic management of ignition timing. In most cases, the control unit will affect only the injection pulse-time. Ignition timing will be controlled traditionally—advanced or retarded by fly weights and vacuum diaphragms at the distributor.

Identifying Features

A quick way to recognize L-Jetronic systems is by the curved housing of the air-flow sensor as shown in Fig. 2-1. Note that's also found on most Motronic systems.

Fig. 2-1. Curved housing of air-flow sensor identifies L-Jetronic, also most Motronic.

Depending on its mounting in the engine compartment, you may also see the waffle grid of the curved housing. That design saves weight while maintaining a rigid flat surface inside for vane contact.

L-Jetronic systems seem to require more parts than the familiar carbureted fuel systems, yet several of these are also required by modern feedback-carburetor systems. See Fig. 2-3 for a diagram of the system components.

Pulsed systems are completely electronically controlled. Sensors supply information to the central control unit which then operates the port injectors. In the simplest terms, the air-flow system measures the air intake; the fuel system delivers clean, pressurized fuel to the injectors; the control system adjusts the amount of fuel needed for the various operating conditions and electrically opens the fuel injectors to deliver the fuel corresponding to air intake. In addition, somewhat separate from the control system, is the electrical system. Beginning about 1980 (earlier in California), Lambda control was added to the system. To see how they work together, I'll look at how these parts and systems meet the engine needs discussed in chapters 1 and 2.

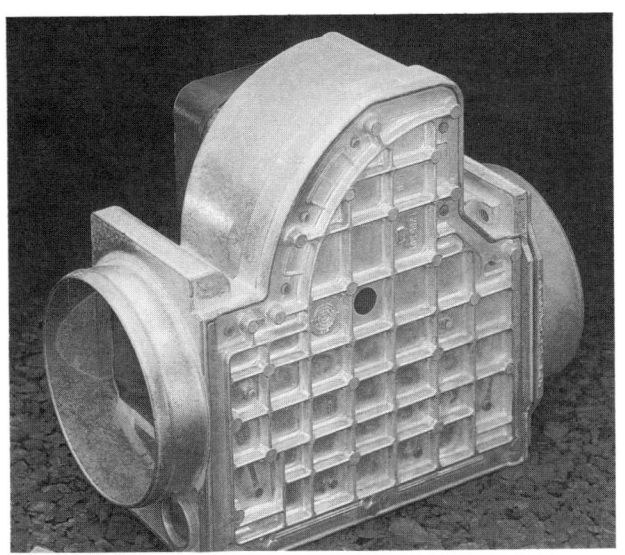

Fig. 2-2. Waffle grid of air-flow sensor maintains a rigid inside surface.

Fuel tank

Control unit

Fuel rail

Fuel pump

Fuel filter

Fuel pressure regulator

Fuel injector

Cold-start injector

Throttle switch

Idle screw

Throttle valve

Air-flow sensor

Air temperature sensor

Mixture screw

Relay combination

Thermo-time switch

Lambda (oxygen) sensor

Auxiliary air regulator

BOSCH

Engine temperature sensor

RPM signal

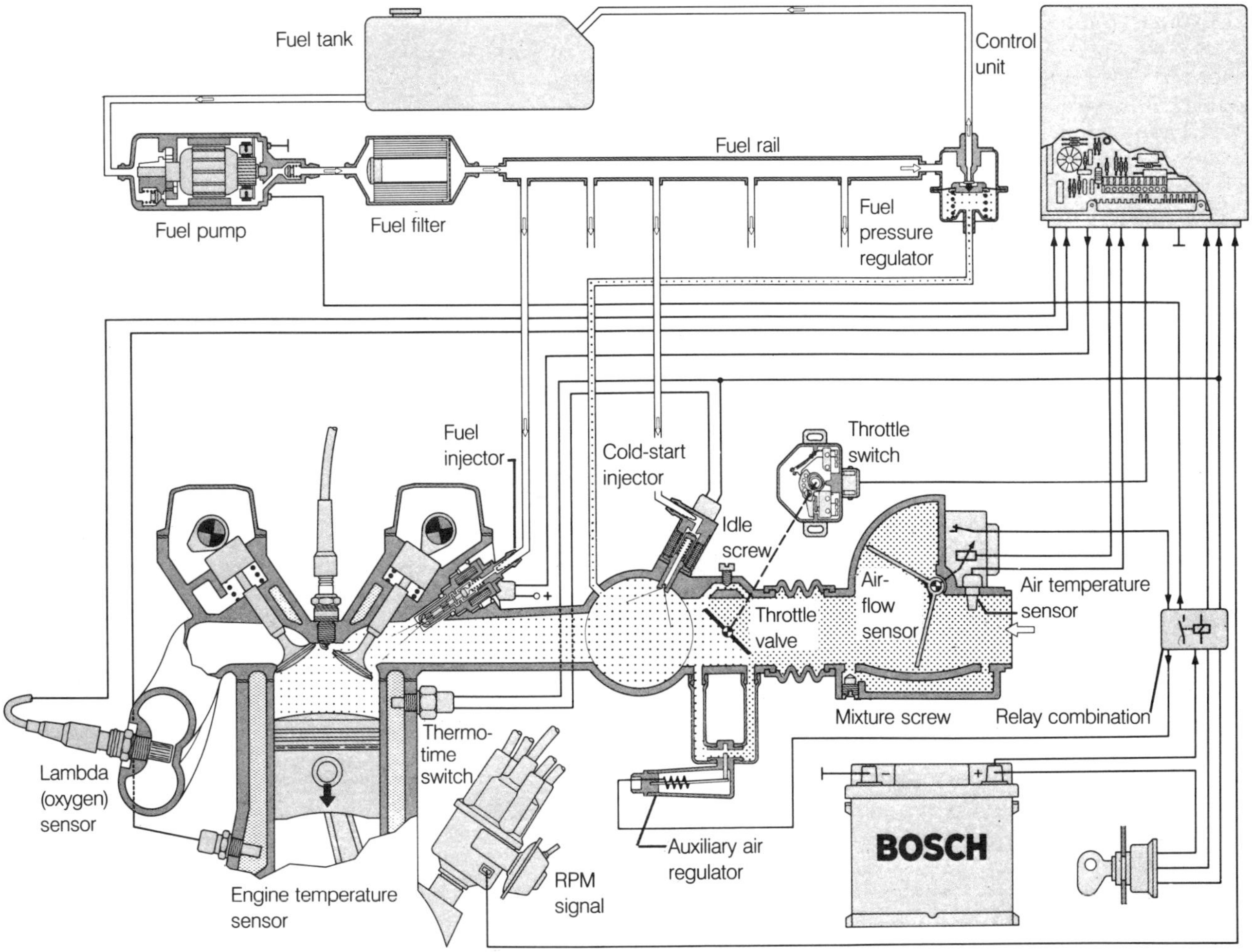

Fig. 2-3. L-Jetronic system diagram. LH-Jetronic is similar, except it has air-mass sensor instead of air-flow sensor.

L-Jetronic

Basic Flow Rate

The design of pulsed fuel injection is based on achieving the ideal air-fuel ratio for normal cruise at part throttle, with a warm engine. The pulse time of the injectors that satisfies the fuel requirements for this condition is known as the Basic Flow Rate. Two engine conditions determine the basic flow rate: load and speed. In measuring the volume of the air passing through the intake manifold into the engine, the air-flow sensor provides a measure of engine load. Engine speed is measured as in a tachometer, by counting the ignition pulses in the primary circuit. From the basic flow rate, compensation times can be added or subtracted for other operating conditions.

2.1 Air-Flow System

The air-flow system measures the amount of air being drawn into the engine (engine load), and sends a voltage signal to the control unit based on that measurement. The air-flow system also regulates idle speed and idle mixture.

Air-Flow Sensor

The L-Jetronic air-flow sensor is a vane-type, so called because its internal vane deflects, or moves, as air is drawn into the engine. The sensor measures the air flow controlled by the regular throttle valve; the sensor does not regulate the air flow. The air vane (also called an air flap) is lightly spring-loaded, and pivots by the force of the air flow as the throttle opens to admit more air. See Fig. 2-4.

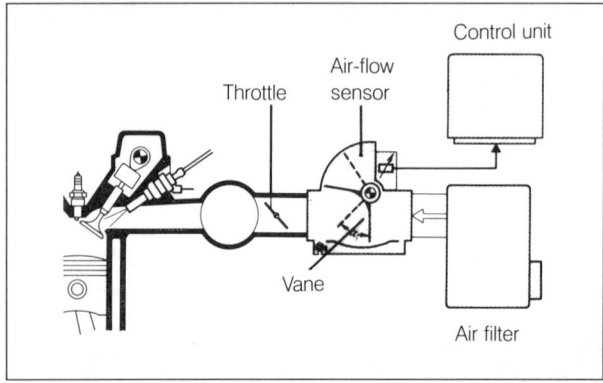

Fig. 2-4. Air-flow sensor air flap measures intake air; throttle controls intake air.

Air-Flow Sensor Design and Operation

If you could remove the waffle-grid cover of the air-flow sensor, you would see the air vane and the damper flap. See Fig. 2-5. The air vane is pushed by the incoming air. In the curved portion of the housing, and mounted on the same shaft as the air vane, the damper flap operates to dampen, or cushion the movement of the air vane by pressing against the air in the chamber. It reduces flutter caused by manifold pressure variations from the opening and closing of the intake valves.

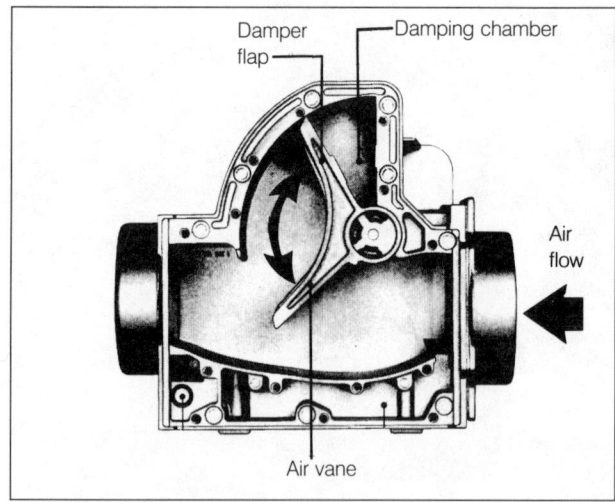

Fig. 2-5. Backfires and pulsations cancel out as opposite forces on the air vane.

Notice the damper flap is the same area as the air-flow vane, and is at right angles to it. This is a clever design: any pulse of increased pressure that rotates the air vane also rotates the damper flap the other way, cancelling the effect of the pressure change. During sudden throttle openings, the air-vane assembly will rotate clockwise rapidly, but its final movement will be cushioned by the damper flap as it squeezes the air in the damping chamber. Bosch tests show the air pressure drop at the sensor is only 0.12 kPa (0.017 psi). That's a fraction of a percent of the atmospheric pressure that forces air into the air system.

The shape of the housing surface opposite the end of the air vane is calculated so the relation between the air passing through and the angle of the flap is logarithmic. That is, a doubling of the air vane angle indicates that air flow has increased 10 times. That means that the most sensitive measurements are at low air flows; at low speeds, measurements are more critical. Maximum air flow is 30 times the minimum.

A moving electrical contact called a wiper is mounted on the same shaft as the air vane. As the vane rotates, the wiper also rotates, crossing a series of resistors and conductor straps on a ceramic base, increasing resistance. See Fig. 2-6. On L-Jetronic air-flow sensors, the resistors oppose current flow from a fixed-voltage input from the control unit. As a result, the more air passing through, the less the voltage signal to the control unit. The control unit is designed to interpret such inverse signals, so the more air there is, the longer the injection pulses are.

In addition to the measured air signalled by the air vane, additional unmeasured air is admitted through a bypass channel to change the idle air-fuel ratio (mixture), as shown in Fig. 2-7. When you turn the bypass screw clockwise, that decreases the amount of unmeasured air. Because the amount of fuel injected is unchanged, decreased air passing into the manifold (for a given vane position) enriches the mixture. Turn the screw counterclockwise to lean the mixture.

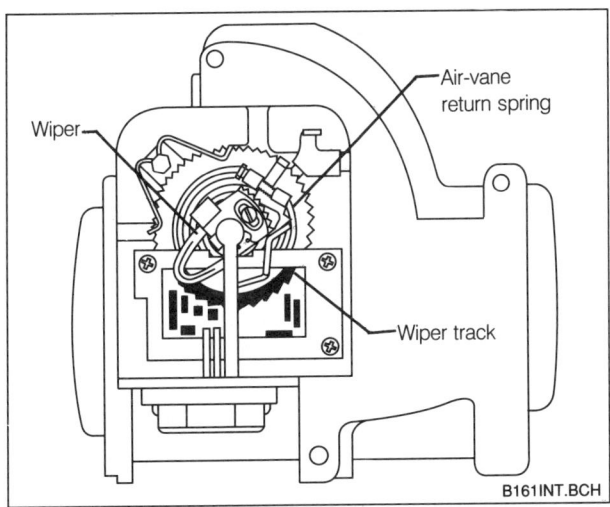

Fig. 2-6. Under the air-flow sensor cover, you can see how the wiper is rotated on the wiper track.

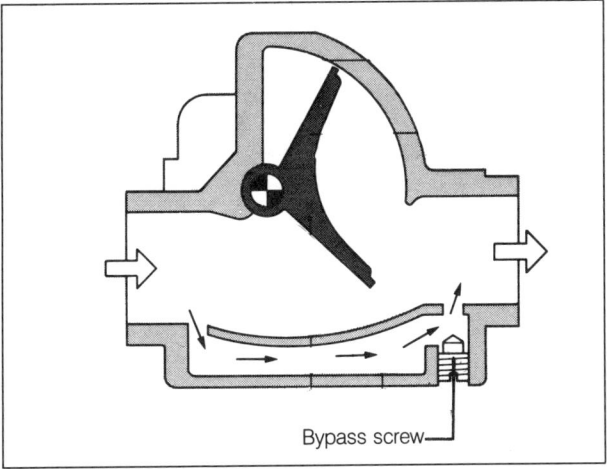

Fig. 2-7. Idle-mixture (bypass) adjusting screw changes bypass of unmeasured air.

Advantages of Sensor Design

What are the advantages of such an air-flow sensor? It provides a direct measurement of the air intake by the engine. By measuring air flow, it tends to compensate for changes to the engine during its service life, such as changes caused by wear, combustion chamber deposits, or valve settings. It permits Exhaust Gas Recirculation (EGR) for emission control without disturbing the measurement of fresh air and the related fuel to be burned with it. In addition, aging of components and varying temperatures have no effect on accuracy because of the resistors in the measurement circuit.

Limitations of Sensor Design

What are its limitations? First, vacuum leaks can be caused by any loose clamp or gasket, or by any slit in the flexible intake boot or vacuum hoses. The air-flow sensor is usually mounted some distance from the intake valves rather than directly on the engine manifold. Any air that enters the intake system between the sensor and the valves is unmeasured air, so-called "false air". Therefore the engine gets no fuel to match that air. The result can be lean mixtures that can cause hard starting, rough idle, low CO, and stumbling; it adds up to poor driveability. Later, in chapter 4, you'll see how to check Jetronics for false air.

Second, the sensor vane measures volume of air intake, but the engine burns weight, or mass of air. Colder air is heavier and requires more fuel than the same volume of warm air. The air temperature sensor corrects for this problem to some degree, as you'll see in **2.3 Control Unit**.

Third, in most cars (those without an anti-backfire valve in the air vane) the vane can be damaged by backfires. The best advice to help prevent this damage when starting the engine is to keep your foot off the accelerator.

Fourth, the air-flow sensor costs more than the pressure sensor, also widely used, particularly in U.S. fuel injection as a measure of engine load. Bosch discarded the pressure sensor in 1974. For more information see **5. D-Jetronic**.

2.2 Fuel System

The fuel system, shown in Fig. 2-8, delivers clean, pressurized fuel to the injectors – usually about one gallon every 2 minutes. As in most newer cars, it is a re-circulating system; the electric pump delivers more fuel than is needed even at maximum consumption, so most of the fuel is returned to the tank. The fuel tank itself is usually pressurized at 7–14 kPa (1–2 psi), controlled by a relief valve in the filler cap. Vapor lock is virtually eliminated because:

• the fuel is cooled by constant recirculation

• the fuel is pressurized, usually at about 2.5 bar (36 psi)

Fuel Pump

The fuel pump, shown in Fig. 2-9, is electrically-driven. Fuel enters at tank pressure through the inlet and is pressurized by the roller cells as shown in Fig. 2-10. The pressure limiter opens and directs fuel back to the tank if fuel pressure in the lines goes over a limit. High-pressure fuel is delivered through a check valve, which closes when the fuel pump stops to hold pressure in the lines. Constantly-pressured fuel lines ensure quicker restarts, and also help prevent vapor lock.

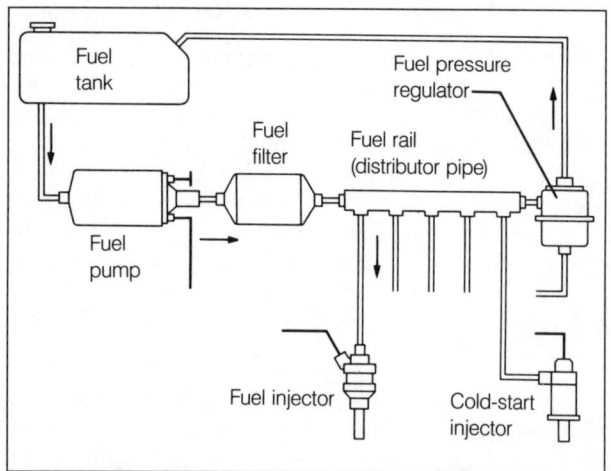

Fig. 2-8. Fuel circulates as pressure regulator at end of fuel rail returns unused fuel to tank.

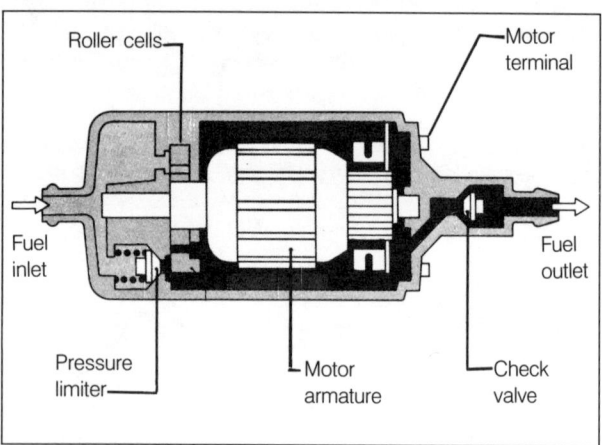

Fig. 2-9. Fuel pump and its electric motor operate surrounded by fuel that cools motor.

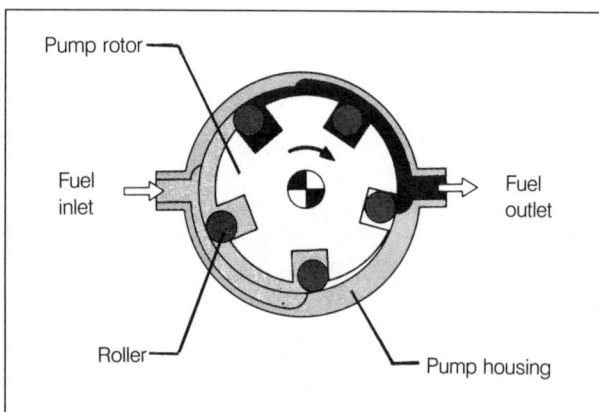

Fig. 2-10. Operation of roller-cell pump. Rotor disc is mounted eccentrically in pump housing. When rotor turns, the rollers are pressed outwards by centrifugal force, and act as a circulating seal. Fuel pressure builds as rollers and disc move fuel through narrowing passages to outlet.

The electric motor actually operates in the fuel in the pump housing. The thought of an electric motor running in gasoline may seem dangerous, but it is safe because the housing never contains an ignitable mixture. There are those who, fearing a burnable mixture, say "never run out of gas with a fuel-injected car," but thirty years of electric pump experience shows they just don't catch fire that way. On the contrary, the gasoline is important—it cools the pump. So if you do run out of gas, just don't crank a long time or you may ruin the pump.

Bosch systems locate the fuel pump next to the fuel tank. This pressurizes the maximum length of the fuel lines to reduce vapor lock. Some cars place the pump inside the tank, or provide a low-pressure electric supply pump in the tank to deliver pressurized fuel to the main electric pump. Electric fuel pumps require relays and safety circuits to ensure that the pumps stop when the engine stops. For more information see **2.4 Electrical Circuits**.

Fuel Filter

In most fuel-injected cars, the fuel filter is located next to the fuel pump. It is much larger than the usual carburetor fuel filter because clean fuel is so important to fuel-injection systems. It is also finer. The paper filter, shown in Fig. 2-11, has a medium pore size of 10 micrometers. It is backed by a strainer that catches any loose particles, and is supported by a plate. The filter is replaced as a complete unit, not as an insert. Vehicle specifications usually call for 30,000–80,000 km (20,000–50,000 mi). When you replace it, observe the direction-of-flow arrows on the filter housing. Some late-model cars use fuel filters that need no replacement unless contaminated.

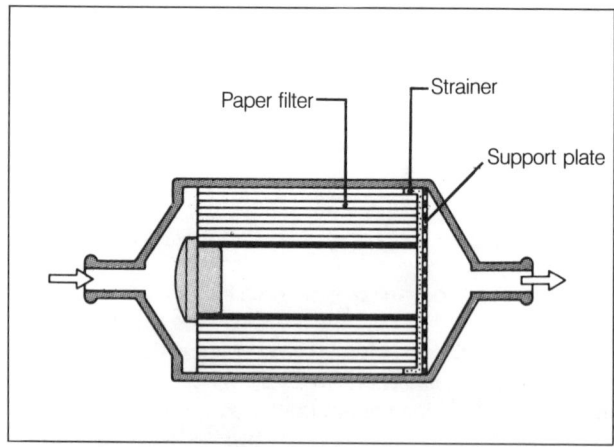

Fig. 2-11. Cut-away of typical fuel filter. Filter is large and fine pore size of 10 micrometers.

Fuel Rail

The fuel rail (also known as the distributor pipe) serves two purposes. The first is to deliver fuel to the injectors. The second is to stabilize fuel pressure at the injectors. You can imagine how the pressures change rapidly in the fuel rail as the injectors pop open and closed. This can affect the amount of fuel injected. But the larger the pipe, the more fuel it stores and the steadier the pressure at the injectors.

On later engines, the fuel rail is larger, a square box, as shown in Fig. 2-12. In the large rail, fuel pressure is more stable at the injectors. On earlier engines, the distributor pipe looks like a typical fuel line, a round pipe leading to all the injectors. In the smaller pipes, with small volume, pressure tends to fluctuate each time the injectors open.

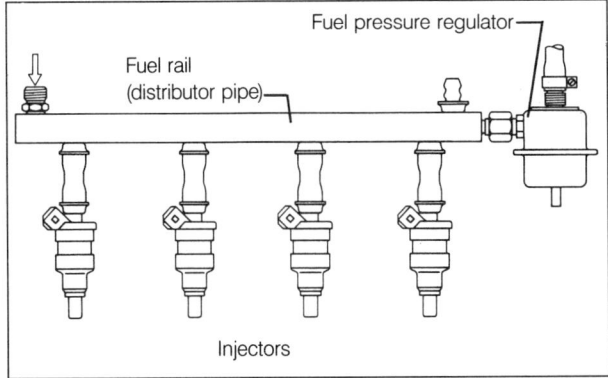

Fig. 2-12. Large fuel rail reduces pulsations as injectors open and close.

Pressure Regulator

Relative fuel pressure in the fuel system is held constant by the pressure regulator, usually mounted on the fuel rail. As shown in Fig. 2-13, spring pressure normally keeps the regulator valve closed. When the fuel pump turns on, fuel pressure presses on the diaphragm to compress the spring and open the valve, returning excess fuel to the tank. The higher the pressure, the more the diaphragm moves away from the return pipe, increasing the volume of the chamber, maintaining the desired pressure. Most systems operate on 2.5 bar (36 psi) gauge pressure, but some higher-powered engines use 3 bar (44 psi) for greater fuel delivery per millisecond.

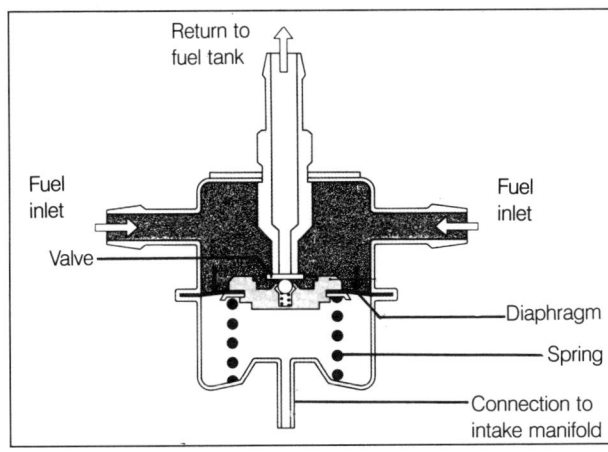

Fig. 2-13. Pressure regulator controls system pressure, usually 2.5 bar (36 psi) above manifold pressure.

Relative Fuel Pressure and Fuel Delivery

For each millisecond of injector pulse time, the amount of fuel delivered through the injector tip depends on the size of the injector opening; that's a fixed factor. But fuel delivery also depends on the relative pressure — the difference between fuel pressure pushing the fuel out into the manifold, and manifold absolute pressure pushing back, as shown in Fig. 2-14.

As you know, manifold pressure changes with throttle opening. If the fuel pressure were constant for all manifold pressures, then at low engine loads, with the throttle partly closed, reduced manifold absolute pressure would increase fuel delivery. To keep that relative pressure constant as the throttle is opened and closed, the fuel-pressure regulator is connected to the intake manifold by a vacuum hose. Manifold pressure acts on the diaphragm to hold the relative pressure constant.

> When thinking about fuel pressure, remember that the ordinary fuel-pressure gauge measures relative pressure — the amount that fuel pressure is greater than barometric pressure (14.5 psi). But barometric pressure is measured from an absolute zero pressure, so 2.5 bar (36 psi) gauge pressure actually means 2.5 bar greater than 1 bar of barometric pressure, or 3.5 bar (51 psi) absolute, measured from an absolute zero. When the gauge is not connected, it reads zero: zero here means the difference between the open connection and barometric pressure.

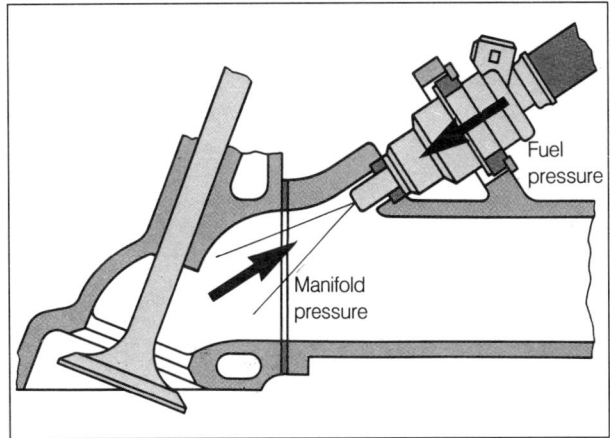

Fig. 2-14. Precise delivery depends on constant relative pressure, that is, difference between fuel pressure and manifold pressure.

For example, with the engine off, manifold pressure is the same as barometric pressure. If you measure fuel pressure with the pump running and the engine off, the regulator controls fuel pressure balancing the forces: fuel pressure on one side of the diaphragm balanced against spring force plus the force of barometric pressure in the manifold. The gauge then reads 2.5 bar, 2.5 bar (36 psi) more than atmospheric. At wide-open throttle (WOT), manifold pressure is close to barometric, so again, the fuel-pressure gauge reads about 2.5 bar.

At idle, absolute pressure in the manifold is about 0.3 bar (0.7 bar less than barometric) — old-timers would say "20 inches vacuum". Now the manifold absolute-pressure pushing the pressure-regulator diaphragm is only 0.3 bar instead of 1 bar. The reduced manifold pressure on the diaphragm allows it to move away from the opening, returning more fuel to the tank, and dropping the gauge fuel pressure in the distributor pipe to about 1.8 bar (2.8 absolute). The relative pressure at the injector tip is still 2.5 bar (2.8 minus 0.3 absolute). That's why fuel delivery per injection is not affected by changes in the manifold absolute pressure. When you understand this principle of fuel delivery, you'll understand the checking of fuel pressures. You'll also understand how greater delivery for off-road and performance operation may be increased by increasing fuel pressures.

Fuel Injectors

The port injectors (located at the intake ports next to the intake valve) are solenoid valves. Each injector is opened by an electrical signal from the control unit, and closed by spring force when that signal stops. See Fig. 2-15. When current is sent through the connection to the winding, electromagnetic force lifts the solenoid, and fuel is delivered. The lift is about 0.15 mm (0.006 in), and takes about 1 millisecond. When the needle valve is closed by the spring, no fuel flows. The pintle on the tip of the needle valve helps to atomize and distribute the fuel.

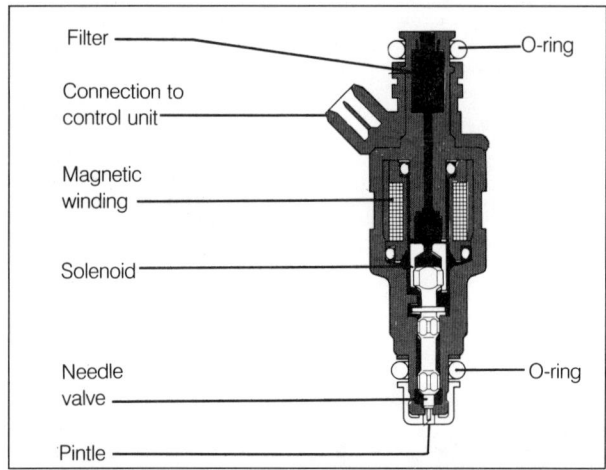

Fig. 2-15. Injectors seal to fuel rail and to manifold by O-rings.

Interestingly, the injectors are electrically-hot all the time the key is ON and the relay combination is closed. For fuel delivery, they are grounded in the final stage of the control unit. In early L-Jetronics (until about 1978) each injector circuit includes a resistor to limit the current flow. In these circuits, the initial current-flow must be great enough to rapidly open the solenoid/needle valve, and then flatten out when only a holding current is needed. Since about 1979, L-Jetronics regulate the current in the control-unit final stage, and so do not need separate series resistors. The resistance is built into the injector itself, using a solenoid winding of brass (higher resistance than the previously-used copper winding). This also requires less power than the series-resistor type.

The injectors are held in the intake manifold with O-rings that tend to insulate the injectors from engine heat and vibration. If the O-rings crack, false air enters, leans the mixture and increases idle rpm. In most engines, the injectors are connected directly to the fuel rail with safety clips. In other engines, they are connected to the rail with fuel lines.

Fuel Metering and Pulse Time

In pulsed injection, the metering of fuel takes place at the tip of the injector as the fuel flows past the pintle. The lift of the needle valve from its seat is always the same distance, so assuming that fuel pressure is regulated as described above, the amount of fuel injected into the engine depends solely on the amount of time the control unit applies voltage to open the injector.

This injector opening time is known as the Pulse Time, and it can vary anywhere from 2 milliseconds to as much as 15 milliseconds or more. The injector takes about 1 millisecond to open, which is counted in the injection time. The closing time is not counted, but it averages out; pulse time is the effective injector open/delivery time.

If, when the control unit grounds the circuit, you measured voltage at the injectors on an oscilloscope, it would look something like Fig. 2-16. Remember that the injectors are always hot, and are grounded in the control unit. Note that when reading pulse time on a scope, it can be referred to as pulse width — the actual width of the pulse-time voltage on the screen.

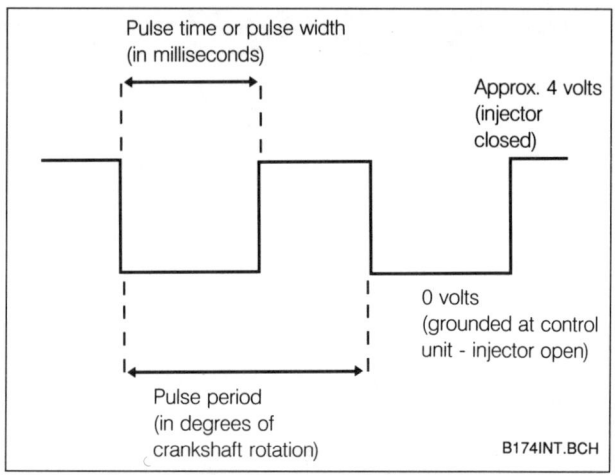

Fig. 2-16. Fuel-injector voltage over time, showing difference between pulse time and pulse period. Voltage supply at injectors is always approximately 4 volts. Oscilloscope shows change in circuit as it is grounded in control unit.

Various operating factors determine pulse time as you'll see below, but there is one other basic factor — called Pulse Period. This is the total time from one opening of an injector to the next opening of the same injector. Normal pulse period for the pulsed systems covered by this book is twice per 4-stroke cycle, or once per crankshaft revolution. Half of the total amount of fuel needed for one firing of a cylinder is injected in each injection. If the pulse period were once per 4-stroke cycle, then pulse time would increase so that all of the fuel was injected at once.

To see the relationship between pulse period and pulse time, let's take a typical engine that idles at 600 rpm and has a maximum rpm limit of 6000 rpm. Dividing each of those revs/ minute by 60 seconds, we can say the engine idles at 10 revs/second, and runs max at 100 revs/second. Fig. 2-17 shows pulse period and pulse time for this engine. Each circle represents 100 milliseconds (ms) of time.

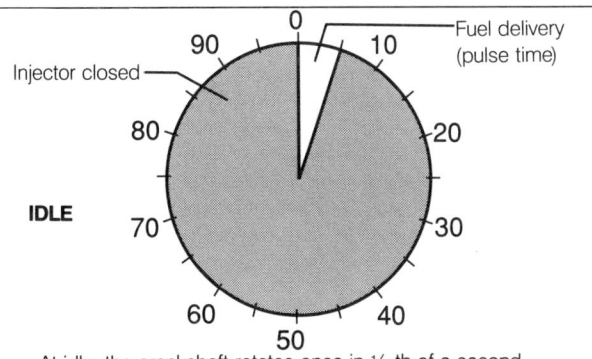

At idle, the crankshaft rotates once in 1/10th of a second, or 100 ms. That's the pulse period. The injectors open once per revolution, for a pulse time of 5 ms.

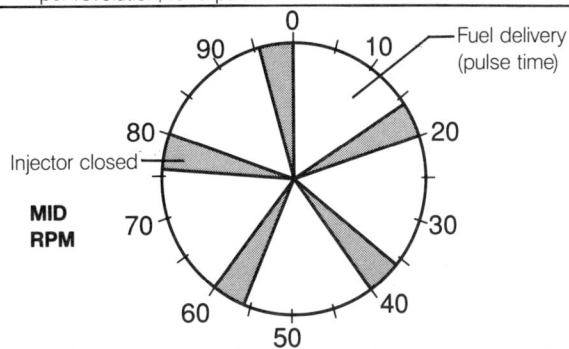

At mid-rpm, say 3000, the crankshaft rotates in a 20 ms pulse period. Full-load delivery for maximum torque at wide-open throttle requires longer injection times than at maximum rpm since at mid-rpm the engine takes in more air per stroke. But a pulse time of 15 ms easily fits into the 20 ms pulse period.

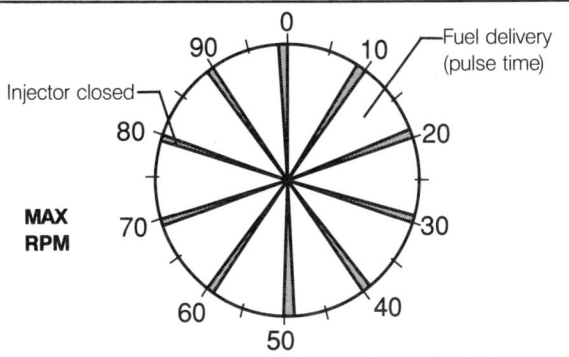

At maximum rpm, the crankshaft rotates once in 1/100th of a second. Pulse period is now 10 ms; in 100 ms the crankshaft rotates 10 times, and the injectors open ten times. If the pulse time is 8 ms, the injectors barely have time to close before they open again.

Fig. 2-17. Relation between pulse period and pulse time. Each circle represents 100 milliseconds (1/10 sec.).

It's important to remember the inputs that control the basic injection pulse time: air-flow volume and rpm. Fig. 2-18 shows that full-throttle injection time varies from about 6 milliseconds to 9 milliseconds, and that injection time follows the torque curve.

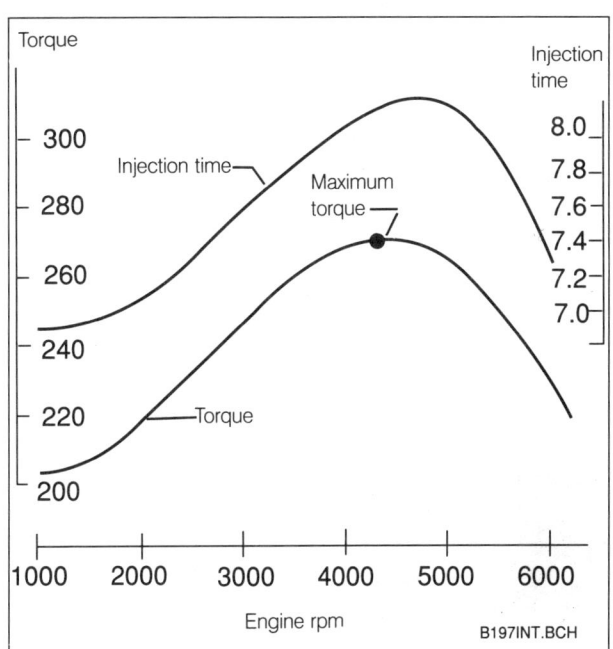

Fig. 2-18. Typical pulse time in milliseconds for full-load matches torque curve.

2.3 Control Unit

The control unit is sometimes called the "brain" of the control system. It receives input of signals from various sensors, processes those signals, computes the pulse-time for the injectors, and completes the circuits in a series of pulses to deliver the required fuel.

The control unit is a complicated piece of electronic equipment. If you open it – and that is not recommended – you'll find it full of transistors, integrated circuits, and printed circuit boards, as shown in Fig. 2-19. The heart of the control unit is a microprocessor chip, about the size of a thumbnail as shown in Fig. 2-20.

In most cases, the control unit is sensitive to heat and vibration, so it is usually located in the passenger compartment away from the engine. A common location is the kick panel to the right of the passenger's feet. You may also find it behind the glove box, under the passenger seat, or in the fresh air compartment under the hood.

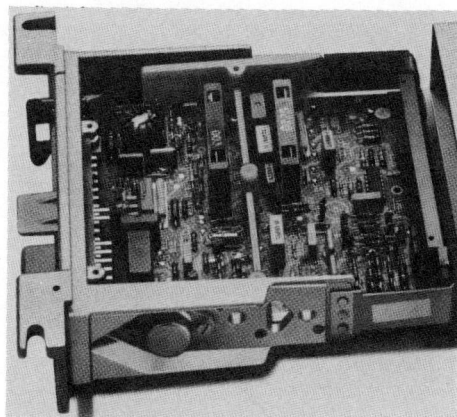

Fig. 2-19. Control unit does the data processing for fuel-injection management.

Fig. 2-20. The chip is the heart of the control unit. This integrated circuit is about the size of a thumbnail.

Modern control units are protected from EMI (Electro-Magnetic Interference) — stray radio frequency signals that can scramble the control-unit brain. I've heard a story about early pulsed systems, before proper shielding was provided: A driver in a Volkswagen Beetle, running on the freeway beside an 18-wheeler, called the truck on his CB radio. When the 18-wheeler answered, using CB power boosted way over the legal limits, the CB signal shut down the VW engine. Story, remember.

Fig. 2-21. Jetronic car is tested under transmitter antenna to check EMI shielding.

Control Unit Inputs and Outputs

Fig. 2-22 is a block diagram showing the basic processing functions of the control unit for the inputs it receives. The rpm signal from the ignition circuit enters the control unit through a shaper and a divider. As shown in Fig. 2-23, from each ignition pulse, the pulse shaper generates a rectangular pulse (voltage on/voltage off), 4 per each 4-stroke cycle. This is divided by two to provide one injection pulse per crankshaft revolution; for a 6-cylinder engine, it is divided by 3.

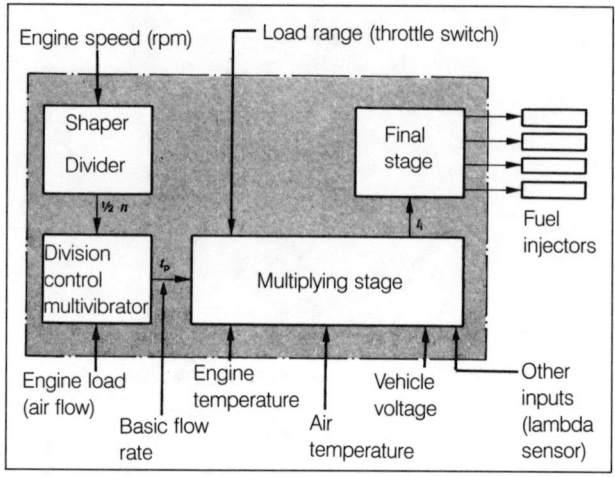

Fig. 2-22. In control unit, basic flow rate is compensated for variables including engine temperature, air temperature, vehicle voltage.

In the division-control multivibrator, the output of the shaper (rpm divided by two) is combined with the load-signal input from the air-flow sensor, resulting in a basic-pulse time. This corresponds to the basic-flow rate, the quantity of fuel to be injected for each injector opening.

The multiplying stage processes the compensation inputs, such as those from the two temperature sensors, engine temperature and air temperature; from the throttle-switch, closed or full-open signals. In later cars, generally those since 1980, it also processes inputs from the lambda sensor. The control unit calculates compensation factors, applying them to the basic pulse-time. In most cases, this will enrich the mixture: when the engine is cold, the injection pulse-time may be two to three times the basic-pulse time.

Compensation for battery voltage is also necessary. This is because battery voltage may vary from 14 volts while the alternator is charging, to 10 volts or less during cold cranking. The lower the battery voltage, the less power to the injector solenoids, and the less fuel injected, perhaps during a cold start when the engine needs it most.

In a 4-cylinder system, the final stage grounds 4 injectors at the same time. Control units for 6-cylinder engines may have two final stages, but they operate together, firing all injectors simultaneously. With two injections per four-stroke cycle, each delivery is half of the total required.

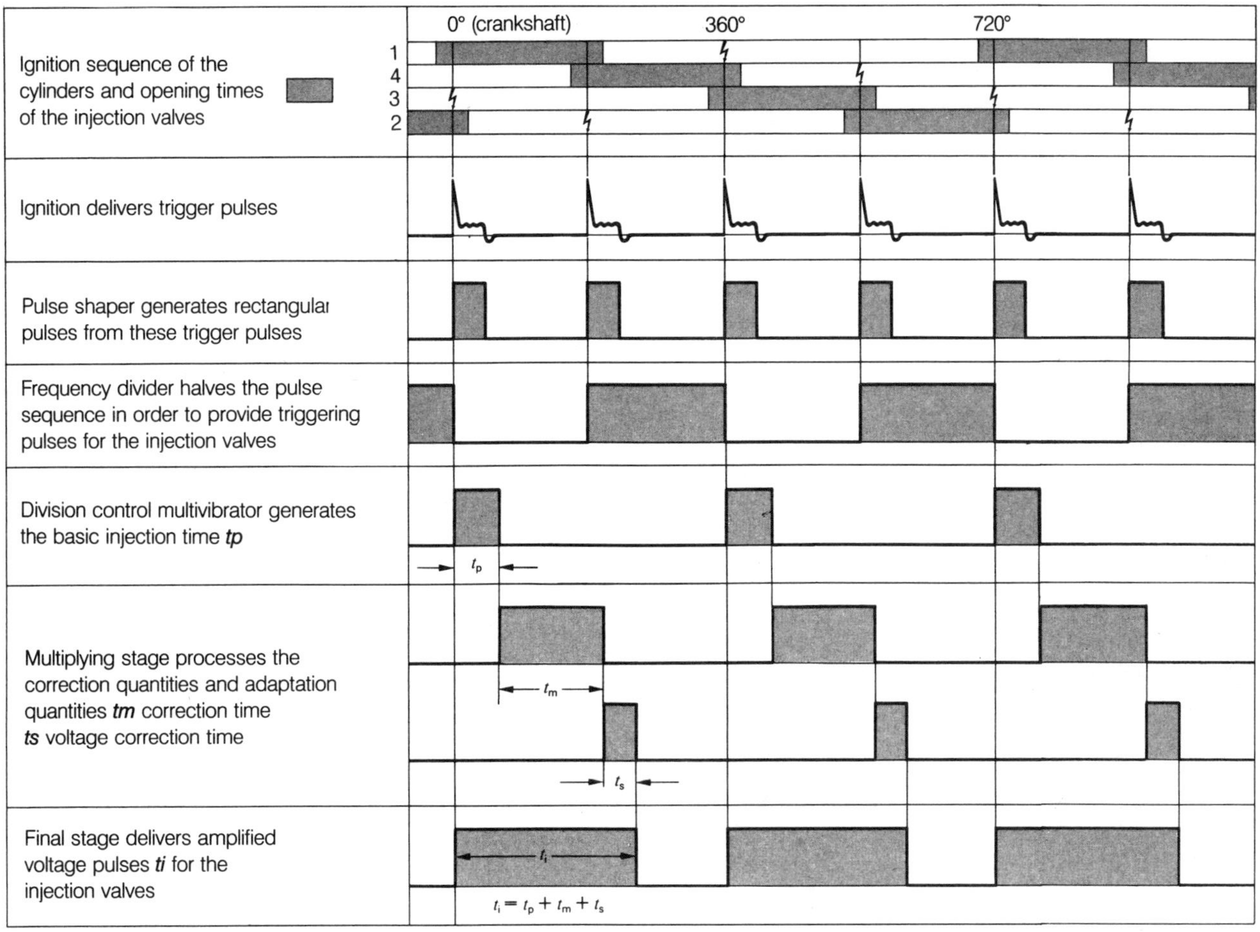

		0° (crankshaft)	360°	720°
Ignition sequence of the cylinders and opening times of the injection valves	1 4 3 2			
Ignition delivers trigger pulses				
Pulse shaper generates rectangular pulses from these trigger pulses				
Frequency divider halves the pulse sequence in order to provide triggering pulses for the injection valves				
Division control multivibrator generates the basic injection time t_p				
Multiplying stage processes the correction quantities and adaptation quantities t_m correction time t_s voltage correction time				
Final stage delivers amplified voltage pulses t_i for the injection valves		$t_i = t_p + t_m + t_s$		

Fig. 2-23. Each basic injection pulse-time is compensated and corrected.

Although it may seem as if the control unit passively receives inputs from the sensors, in some cases its role is an active one, as shown in Fig. 2-24. For example: voltage output from the control unit (terminal 5) is fed to the engine temperature sensor. The sensor reduces the flow of current back to the control unit (terminal 13) which the control unit measures. The colder the engine, the more resistance, and the less current back to the control unit. In the same manner, a voltage signal is sent to the air-flow wiper track, and the return signal is measured. Note that the cold-start injector, thermo-time switch, and auxiliary-air device operate independently of the control unit.

RPM Signal. Engine rpm, to help determine the basic flow rate, is one of the most important inputs to the control unit. Bosch pulsed systems are synchronized, that is, injector pulses are synchronized with crankshaft rotation. All L-Jetronic injectors fire twice per 4-stroke cycle, or once per crankshaft revolution. As you saw earlier, the total amount of fuel delivered depends on how many injections there are per minute, and how long the injectors are open for each injection.

In the control unit, air measurements alone could determine the amount of fuel required for the ideal air-fuel ratio. But, as you'll see, injector opening-times are computed for many conditions and air-fuel ratios, some of which depend on engine rpm. So engine rpm is one factor in determining the basic flow rate.

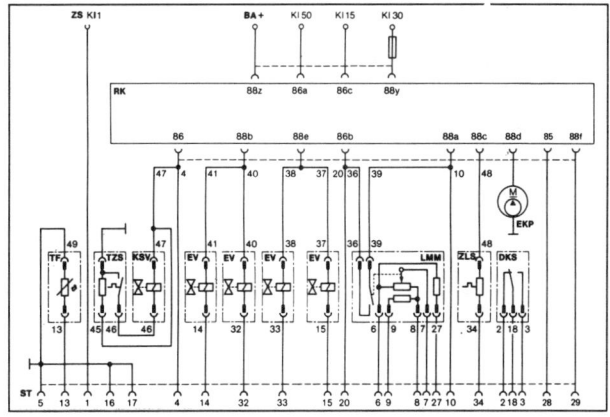

Fig. 2-24. Example of control-unit connections for L-Jetronic system with regulated final stage and no resistors. For component identification, see **Table b** below.

The L-Jetronic rpm signal comes from the primary ignition pulse, the current that turns the coil on and off. For most cars, that is a signal from the negative terminal of the ignition coil.

The ignition circuit sends four signals for each 720° of crankshaft revolution (each 4-stroke cycle). In a 4-cylinder engine, this means that the ignition circuit closes every 180° of crankshaft revolution (once for each distributor revolution). If you look at Fig. 2-23, above, you'll see the ignition pulses for the firing order at the top of the diagram, and the opening time of the intake valves.

Analog/Digital

Early control units were analog-type; later are digital. You are bound to run into these terms when troubleshooting, so it's important to know the difference. Analog measures continuously; an analog display changes in direct proportion to the input, such as an ordinary odometer. If you've gone two-thirds of a mile since it measured an even mile, the tenths digit will read about two-thirds. On the other hand, digital measures in steps. The digital clock shows exact steps, nothing in between as shown in fig. 2-25. Each step can be extremely accurate, and does not change with wear or aging factors.

Fig. 2-25. Analog measures continuously; digital measures in steps, and each step is exact.

The important point is: steps versus continuous change. See Fig. 2-26. Analog accuracy is limited, as indicated by the curved line measuring the signal of the straight line. But analog measurement can be more accurate than digital measurement that uses large steps. By increasing the number of steps, digital accuracy is usually greater than analog.

Some inputs are analog; for example, consider the changing current flow from the resistor of the engine-temperature sensor. That analog signal must be changed in the control unit by an analog-to-digital (A/D) converter. Some inputs are digital, for example the rpm pulses from the ignition primary.

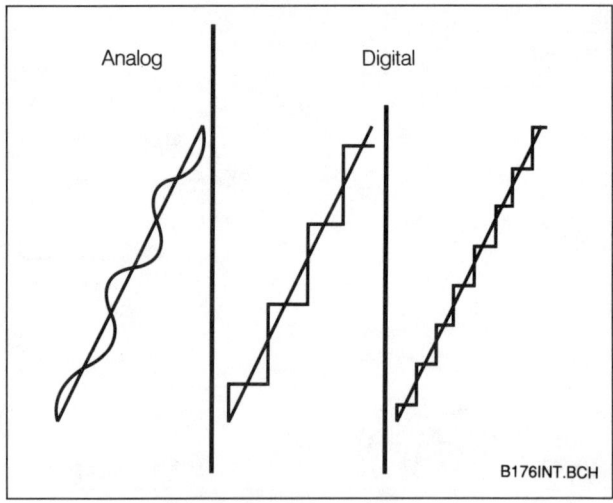

Fig. 2-26. Analog accuracy is limited by many variables; digital limited by number of steps.

Some outputs are digital, as you'll see later. These are a series of pulses sometimes described as percent dwell. Analog outputs from digital control units are handled by digital-to-analog (D/A) converters.

Control-Unit Connectors

Inputs and outputs of the control unit are generally handled by a multiple-connector plug, as shown in Fig. 2-27. Some control units provide 35 pins for the connectors, while others, with simplified circuitry, operate with 25-pin connectors. When reading circuit diagrams or making tests at the control-unit connector you'll need to understand the German abbreviations that often identify the connector terminals. **Table b** lists the common abbreviations and their English equivalents.

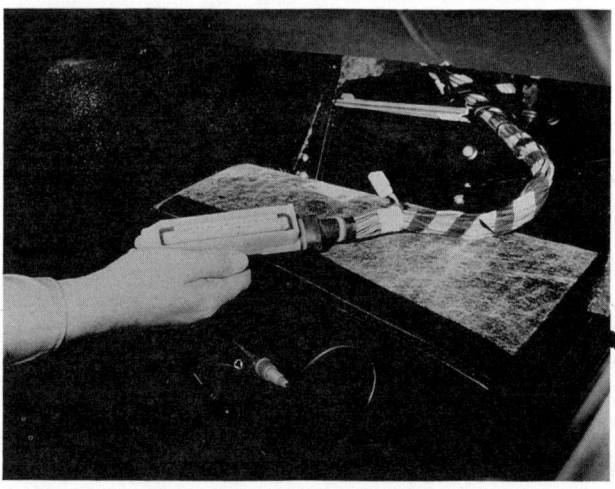

Fig. 2-27. Multiple-pin connector handles all inputs and outputs to control unit pins.

Table b. Fuel-Injection Wiring Harness Terminal Identification

Abbr.	German Word	English Equivalent
TF	TEMPERATURFUHLER	Engine-temperature sensor, NTC II
TZS	THERMOZEITSCHALTER	Thermotime switch
KSV	KALTSTARTVENTIL	Start valve
EV	EINSPRITZVENTIL	Injector
LMM	LUFTMENGENMESSER	Air-flow sensor
ZLS	ZUZATZLUFTSCHREIBER	Auxiliary-air device
DKS	DROSSELKLAPPENSCHALTER	Throttle-valve switch
EKP	ELECTRISCHE KRAFTSTOFFPUMPE	Electric fuel pump
RK	RELAISKOMBINATION	Relay combination
ZS	ZUNDUNG, SPULEN	Ignition coil
BA	BATTERIE	Battery
ST	MEHRFACH-STECKER	Multiple-plug to control unit

2.4 Electrical Circuits

In addition to the electronics of the control system, L-Jetronic requires simple electrical circuits to power the control unit, fuel pump, and other components.

Control Relay

When the ignition is turned on, battery power flows through the ignition switch to a relay, or a combination-relay. This relay then closes to provide current to the control unit; it also feeds the cold-start injector, and the auxiliary-air valve. Thus, do not replace a control unit to "fix" these parts. See Fig. 2-28.

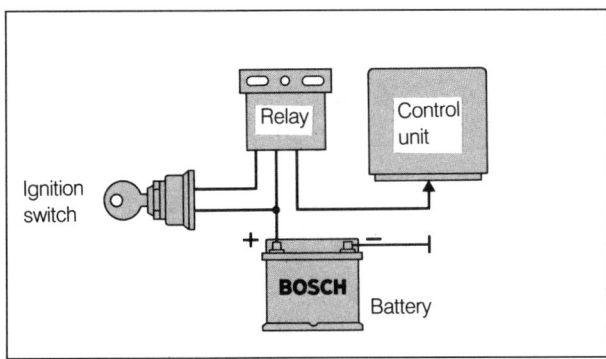

Fig. 2-28. System power flows through relay powered by ignition switch.

Fuel-Pump Safety Circuit

The relay also controls the fuel-pump safety circuit. The START position of the key starts the fuel pump during cranking. But the pump will stop running in 1–2 seconds unless it receives a signal that the engine is turning. This helps prevent flooding, or a fire in the event of an accident. See Fig. 2-29. In most L-Jetronics, the relay receives the rpm signal from the ignition circuit and shuts down the pump if the engine is not turning fast enough: 4-cylinder engines, faster than 225 rpm; 6-cylinder engines, faster than 150 rpm. Early L-Jetronics (before 1978) do not use the rpm signal; instead, a safety switch in the air-flow sensor closes when the air vane begins to open.

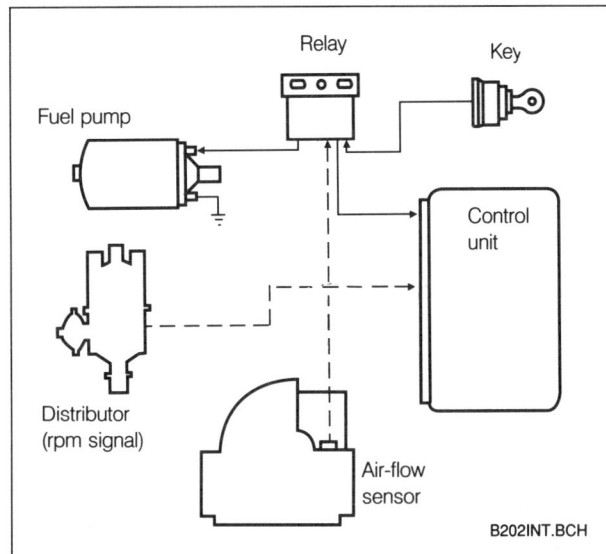

Fig. 2-29. Pump safety circuit is usually controlled by rpm signal from primary ignition circuit; in early L-Jetronics, control is by air-flow sensor switch.

Service technicians can bypass the safety circuit to run the fuel pump without running the engine: on early cars by reaching in to move the air vane with the ignition on; on later cars by jumping the rpm safety circuit either at the relay set or at the air-flow sensor. For more information see chapter 5.

2.5 Operating Conditions

For all of the conditions under which the engine operates, the control system computes the following:

- Main variables: engine load and speed for the basic flow rate. Each basic flow rate can be modified, or compensated based on inputs from sensors that signal changes in operating conditions.

- Compensation variables: such as starting, cold operation, and special load conditions, and

- Precision compensation: tweaking the other compensations for overrun or coasting, and mixture (lambda) control for emission-reduction.

All compensations are added to, or subtracted from the basic flow rate. So let's look at each of the conditions and see what sensors and signals affect L-Jetronic through the control unit. We'll also see some auxiliaries that operate independently of the control unit.

L-Jetronic

Starting

As we've seen, depending on the temperatures, the engine may need additional fuel, and also additional bypass air around the closed throttle. How does L-Jetronic compensate for all the different starting conditions? How does the system operate without any setup by the driver; how can you simply turn the key to start it?

These systems have no choke or fast idle cam to set. As the car sits, the engine changes temperature due to changing weather, or cooling after running. Its sensors signal the control unit about engine temperature and other conditions that could affect starting. Other components inject additional fuel or allow additional air.

Engine Temperature Sensor. The most important sensor is the engine temperature sensor, shown in Fig. 2-30. It is a semi-conductor resistor, also known as a thermistor. You may find service instructions referring to it as NTC II. NTC stands for Negative Temperature Coefficient, meaning the sensor's resistance goes down as the temperature goes up. This is the opposite of most resistors; hence the term "negative." See Fig. 2-31.

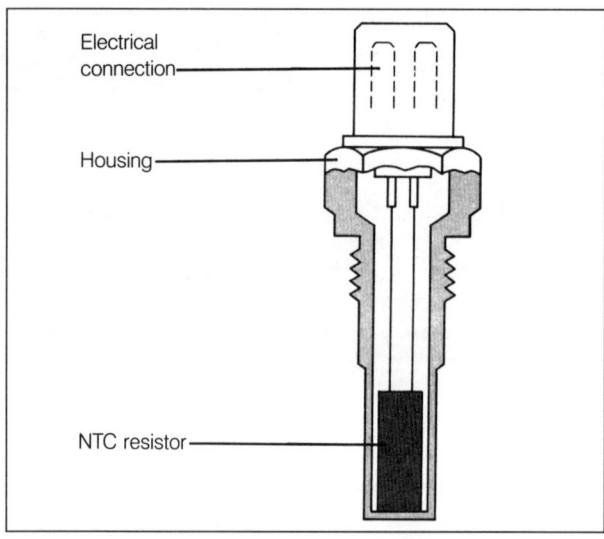

Fig. 2-30. Engine temperature sensor is a solid-state resistor, sometimes called a thermistor. NTC means Negative Temperature Coefficient.

Suppose the engine is in the shop and has not run recently, say for several hours. In round numbers, the resistance at ordinary indoor temperatures is about 3Kohms (3,000 ohms). If the engine has run recently, long enough to fully warm up, resistance is only about 400 ohms. If the control unit applies a fixed voltage to the NTC resistor, it will receive a smaller signal back as input from a cold resistor with higher resistance than from a warm one. From a cold input signal, the control unit will compensate by adding to the basic pulse time as shown in Fig. 2-32.

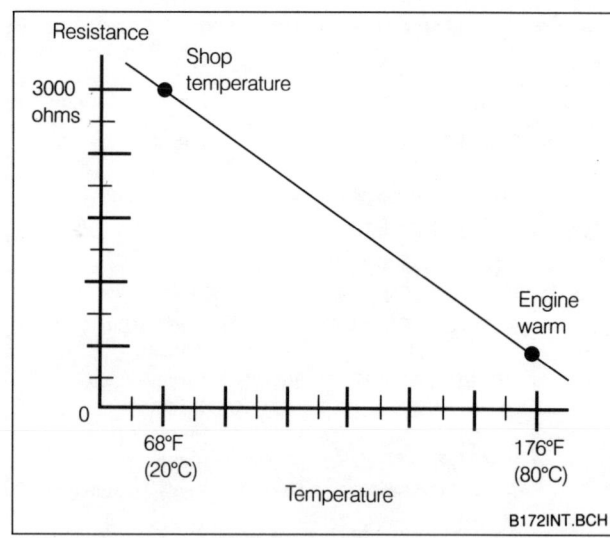

Fig. 2-31. NTC resistance is lower when engine is warm.

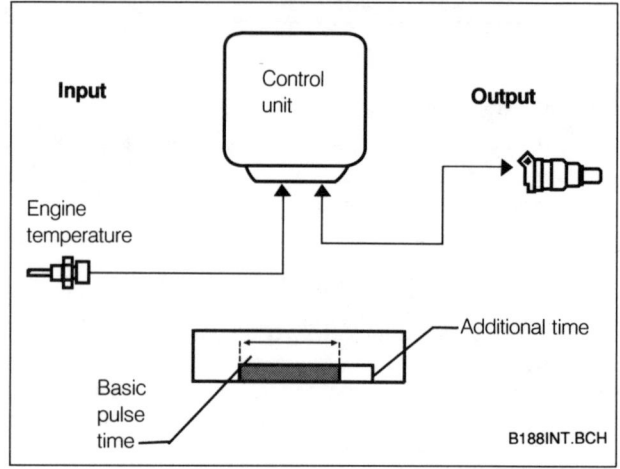

Fig. 2-32. Cold engine temperature sensor causes increased pulse time.

On water-cooled engines, look for the engine temperature sensor, sometimes called NTC II, to be mounted in the cooling system, usually in a fitting near the block, or in the thermostat housing. See Fig. 2-33. Be careful not to confuse the engine temperature sensor with the thermo-time switch or the coolant temperature sender for the instrument-panel temperature gauge.

The NTC sensor is quite simple: an electrical connector on a housing, enclosing the thermistor. The first L-Jetronic came on an air-cooled engine, so on those, you'll find NTC II in a cylinder barrel or a head.

Air Temperature Sensor. With the air cleaner removed, you can see the Intake Air Temperature Sensor in the air-flow sensor as shown in Fig. 2-34. Known as NTC I, this sensor also has a negative-temperature coefficient. Remember, the control unit sends a signal which is changed by the air temperature sensor. Cold air, weighing more, needs more fuel, so a cold-air signal also adds fuel to the basic-pulse time. The air-fuel ratio relates weight of air to weight of fuel. The combination input signals of volume (from the air-flow sensor), corrected for temperature, are the same as inputs of weight or mass of air passing into the engine.

Fig. 2-33. Engine temperature sensor (also called Temp II or NTC II) is often mounted in thermostat housing. In this BMW, temperature sensor (**2**) is mounted with thermo-time switch (**1**) and engine temperature sender (**3**).

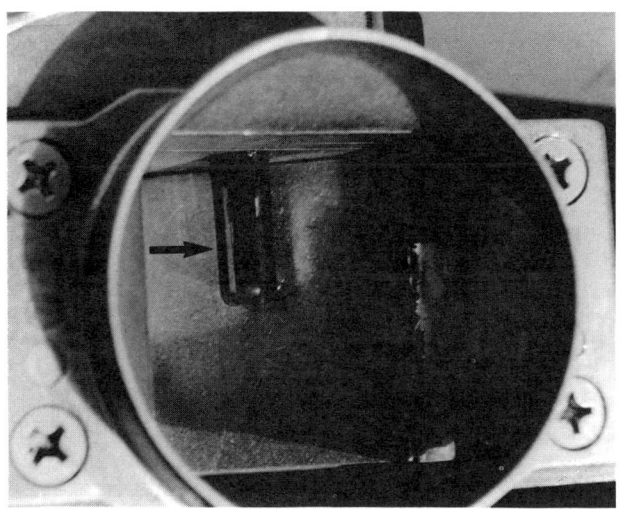

Fig. 2-34. This air-temperature sensor (**arrow**) (also called Temp I or NTC I) signals intake air temperature.

Cold Starting Enrichment. There are two ways that additional fuel can get to the engine during cold starting. The first is an auxiliary or cold-start injector which is controlled by the thermo-time switch. The second method is the Start Control System in the control unit which adjusts the pulse time of the fuel injectors as needed. Most L-Jetronics use the cold-start injector. Some combine that with start control; and a few use only start control.

One system missing from most fuel-injection engines is the Intake Air Heater, or Heated Air Cleaner. For emission control and driveability, carbureted and throttle-body-injected engines heat the intake air to about 100°F (40°C). Warmed air tends to prevent condensation of the air-fuel mixture on the intake walls, which causes stumbling and increased emissions from the lean mixture. Because L-Jetronic delivers fuel at the intake ports, it can compensate fuel flow as necessary, enriching for cooler air without wall wetting. Without the Intake Air Heater, L-Jetronic delivers cool air, with greater density, producing more power in cold conditions.

The cold-start injector (also called the cold-start valve) is energized during cold starting. When the key is turned to start, a relay delivers power to the start injector. A thermo-time switch grounds that circuit, causing fuel delivery based on switch temperature. This auxiliary injector does not operate through the control unit, and is not usually affected by the engine temperature sensor.

If installed the cold-start injector is easy to find, mounted on the intake manifold downstream of the throttle, as shown in Fig. 2-35. It has an electrical connector and a fuel line. Whenever the system is pressurized, the start injector is pressurized; that means it is under pressure whenever the engine is running, and for some time afterward. Why is that important? If this injector leaks, the engine runs rich.

Fig. 2-35. Cold-start injector (**1**) delivers fuel into intake manifold to enrich mixture for all cylinders. Electrical connection is at **2**, fuel connection at **3**.

The cold-start injector is solenoid-operated, something like the port injectors, but it holds open for seconds rather than milliseconds. At system pressure, fuel enters through the fuel fitting; when current flows through the electrical connections, the solenoid lifts the armature from the valve seat. Fuel flows to the nozzle where it is swirled. That swirl action provides a fine vaporized mist that fills the manifold between the closed throttle and the intake valves.

The injector receives voltage whenever the key is in START position, but it only operates when it is grounded by the thermo-time switch. Think of this as a switch, not a resistor. It limits the open-time of the start injector to prevent flooding during cranking. See Fig. 2-36. Inside the housing is a bimetal switch arm surrounded by a heater winding. When the bimetal is cold, it closes the switch contact. The bimetal is a strip of two different metals bonded together; as it gets hot, it bends to open the contact.

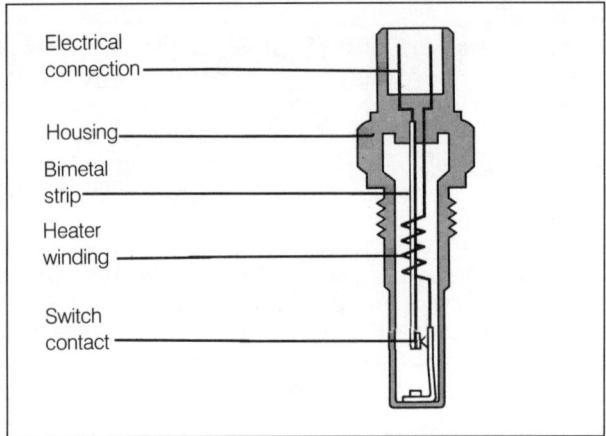

Fig. 2-36. Thermo-time switch bimetal opens start-injector circuit when coolant is warm, preventing flooding.

When you turn the key to START, that powers the start injector, and also powers the heater. If the engine is cold-cold, the closed switch grounds the circuit and the start injector delivers fuel. But as the heater winding warms the bimetal, it bends away and opens the grounding circuit, ending the start injection. Even at the coldest temperatures, the delivery time will be less than 10 seconds. Of course, delivery stops when you release the key from START. If the engine does not start for some reason, the already-heated bimetal will stay open, preventing excess fuel delivery. If the engine is warm-cold when you crank it, the bimetal is already bent to open the circuit as a result of coolant temperature. The result, no fuel delivery from the start injector.

Consider this auxiliary fuel injector in two terms: it must operate when the engine is cold, or the engine won't start. It must remain closed without spray or drip when the engine is warm, or the mixture will be rich. A leaking cold-start valve will cause hot-starting problems as fuel leaks into the manifold.

In the start control system, the control unit lengthens the injector pulses to the port injectors when the key is turned to START, based on the input from the engine temperature sensor. The colder the engine, the longer the pulses. To prevent flooding from repeated attempts to start, the control unit

includes a recovery-time factor. This determines how many seconds must elapse before ⅔ of initial start-fuel quantity can be delivered.

In some cars, the start-control impulse is also influenced by the rpm during cranking, as shown in Fig. 2-37. The start-control impulse (t2) is superimposed on the basic pulse-time (t1). When the engine is cold-cold, cranking rpm is likely to be low. If so, the reduced-rpm signal causes the system to double the pulses, two per crankshaft revolution, and to add large compensation to the basic pulse-time. If the engine is warm, but cranking rpm is low, perhaps because of low battery voltage, less compensation is added, still 2 pulses per revolution. When the engine is warm, usually above 140°F (60°C), there is only 1 pulse per revolution.

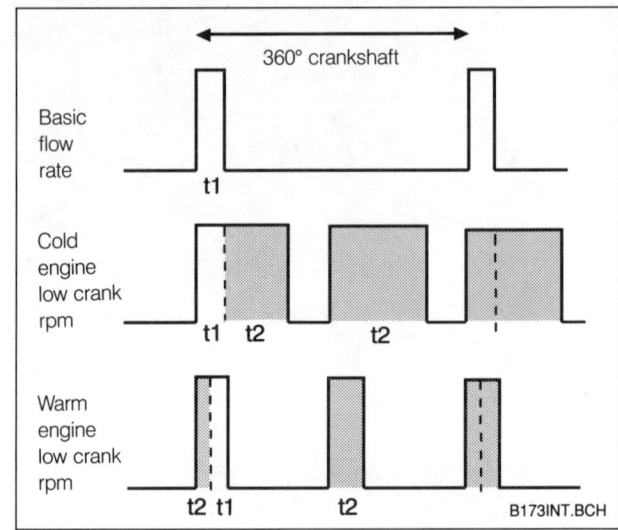

Fig. 2-37. On some systems, pulse time increases if engine is cold and turning slowly; pulse time decreases if engine is warm and turning slowly.

Auxiliary Air Valve. The auxiliary air valve helps the engine overcome the mechanical drag of a cold engine by adding additional air flow as shown in Fig. 2-38. It does this without racing the engine, as is common on engines with a fast-idle cam. Usually located on top of the cylinder head or valve cover, the auxiliary air valve can also be on the block because it is influenced by engine temperature. Look for it to bypass the throttle body with an auxiliary passage of air at barometric (atmospheric) pressure to the intake manifold at reduced manifold pressure. It is this pressure difference that causes the air flow when the valve is open.

The air bypassing the throttle through the auxiliary air valve is measured air; it has been measured by the air-flow sensor, so the required fuel is being delivered for injection. With the cold engine receiving extra air-fuel mixture, the extra friction resistance is overcome. Cold-idle rpm is increased, but will vary according to the engine manufacturer's calibration, as well as oil viscosity, engine condition and other variables. It is not rpm-controlled, as we'll see in later systems with idle-speed stabilization.

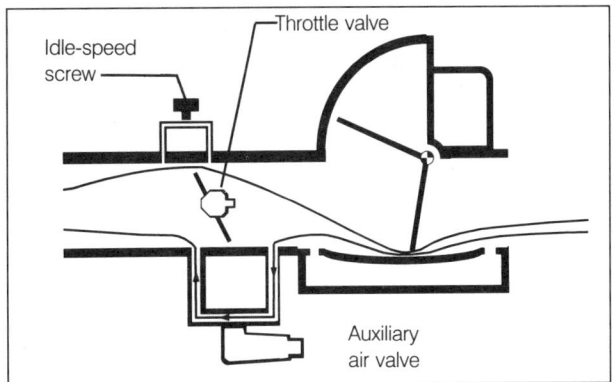

Fig. 2-38. Auxiliary air valve bypasses closed throttle when cold.

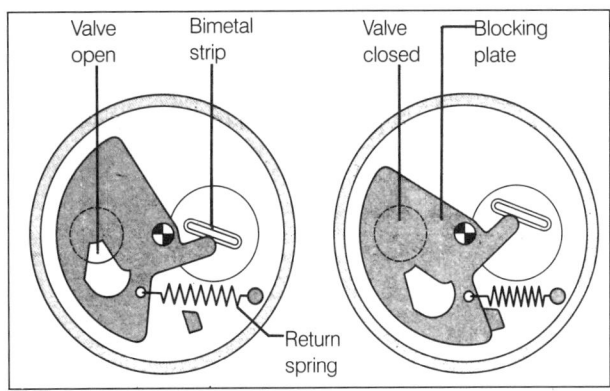

Fig. 2-40. Bypass opening closes when bimetal strip bends and spring pulls rotary valve.

Two types of auxiliary air valves are used. Most are electrically heated, while a few early types are coolant heated. When the valve is cold, the blocking plate pivots to allow bypass air through the opening. When it is warm, the plate closes the opening.

Operation of the electrical type depends on the position of a bimetal arm, wrapped with an electric heating element, as shown in Fig. 2-39. When cold, the bimetal arm moves the blocking plate for maximum bypass air. When warm, the bimetal closes the opening. See Fig. 2-40. By shaping the hole in the plate, the designer can control the amount of bypass air for each engine temperature so idle rpms are close to the desired speeds.

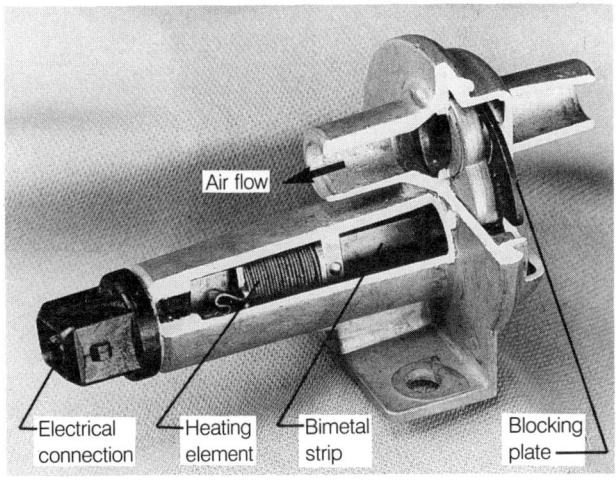

Fig. 2-39. Electric heating element closes auxiliary-air bypass in a maximum of 8 minutes from freezing.

The position of the bimetal depends on engine temperature and the length of time current flows through the heating element. When the engine is warm, the bimetal is influenced primarily by the block temperature. When the engine is cold and the plate is open, the heating element is the main influence on the bimetal. Where the delivery of auxiliary fuel for starting is generally timed in seconds, the delivery of auxiliary air is timed in minutes. From the coldest, the heater will close the bypass in about 8 minutes, faster than block warm-up would close it.

Proper operation of the auxiliary air valve is important to driveability:

- for cold starts, and post-start, this valve must be open

- when the engine is warm, it must be closed to prevent fast-idle

In chapter 4, you will see two ways to check if the auxiliary air valve is functioning properly.

Post-Start and Warm-Up

You've already seen how the cold-engine idle rpm is affected by the auxiliary-air bypass. For post-start and warm-up fuel enrichment, pulse time of the port injectors is increased, but much less than during start control.

Bosch considers post-start, sometimes called after-start, as the first several seconds after the cold engine fires. The engine is running, but it needs a hefty enrichment factor. When the key is released from START to RUN, that ends the enrichment from the injectors, as much as 250%. For a start from ordinary temperatures, 70°F (20°C), the engine might need an enrichment factor of 150%, decreasing rapidly with time. This compensates for poor mixture formation and fuel coating of the manifold walls. The richer mixture increases torque, and therefore driveability.

After 20–30 seconds, the engine is considered into the warm-up phase, and the enrichment factor tapers off. Remember, driveability must be balanced with emission control; some engines can fail the EPA cold run test in the first 30 seconds. Warm-up driveability improves with more enrichment (or slower reduction of enrichment), but at the cost of high emissions.

Acceleration

How do L-Jetronics supply the acceleration enrichment ordinarily delivered by the accelerator pump of a carburetor?

L-Jetronic

Part-throttle. For part-throttle acceleration, the mixture is enriched by over-swing of the air vane. When you open the throttle quickly, the vane swings beyond the required position. This over-swing adds compensation pulse-time for a second or so until the vane stabilizes for the new air flow. During warm-up, the engine temperature sensor operates through the control unit to add additional enrichment compensation to the over-swing.

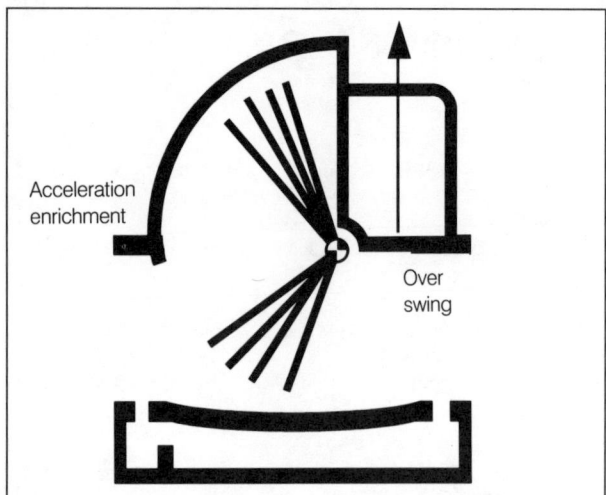

Fig. 2-41. Acceleration enrichment momentarily increases pulse time by over-swing with sudden throttle opening.

In some engines, over-swing enrichment is enhanced with electronic acceleration enrichment. The control unit evaluates how fast the air-flow sensor signal is changing. In effect, it looks at the velocity of vane movement. Above a certain voltage jump, acceleration enrichment begins, adding to the pulse-time. Electronic acceleration enrichment is modified or cancelled:

1. when the engine is warm

2. during starting

3. during manual shifting

4. at higher air flows

Full-throttle. Full-throttle acceleration enrichment is handled differently. When you floor the throttle, over-swing enrichment is delivered. But more importantly, a full-throttle signal is sent by the switch mounted on the throttle shaft. See Fig. 2-42. Sometimes you'll find microswitches out in the open; usually they are enclosed in a single unit as shown in Fig. 2-43. The full-load contact is closed by the cam on the throttle shaft, signalling the control unit to add enrichment. Full-throttle acceleration requires a rich mixture, so the control unit switches the system from closed-loop operation to open-loop, with the mixture determined by factory-programmed air-fuel ratios.

Fig. 2-42. Throttle switch is usually mounted directly on throttle shaft.

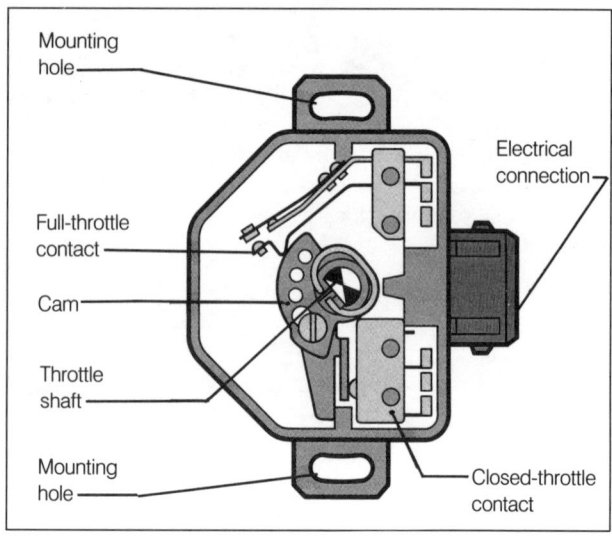

Fig. 2-43. Throttle switch signals full-throttle and fully closed throttle.

Idle and Drive-Off

Idle enrichment is usually signalled by the closed-throttle switch, shown above in Fig. 2-43. This switch was originally called the idle-switch, but in recent cars it signals closed throttle during coasting, and that's not idle, hence the name change. The switch is also closed by the cam on the throttle shaft, when the throttle is fully closed. When the computer receives an input that the throttle is closed, it adds compensation pulse-time to the basic flow rate, enriching the idle mixture. In chapter 4, you will see how simple it is to check the throttle switches for continuity.

Increased fuel for drive-off from idle can improve driveability, particularly when the engine is cold, something like an accelerator pump improves drive-off of cold carbureted engines. In some cars, to improve driveability, when the closed-throttle contact circuit opens as the throttle is opened from idle, a special circuit adds compensation fuel, usually temperature-compensated.

Coasting Cut-Off

In later systems, fuel is cut off during coasting, or deceleration, when closed-throttle is signalled. This improves emission control and also fuel economy. In the control unit, the closed-throttle input is combined with inputs from engine temperature and rpm to determine the beginning of cut off and the end of it. Remember, as rpm drops toward idle, the closed-throttle switch may first signal for zero fuel, then at idle will signal for enrichment.

Anti-Search Circuitry

In some systems, transition from one mode to another is smoothed by anti-search circuitry. This part of the control unit looks for sudden changes in rpm, either deceleration or acceleration. It keeps the engine from searching for the proper flow during the transition, improving driveability.

RPM Limitation

When some engines approach red line, the computer compares engine rpm to a programmed limit-rpm. As soon as engine rpm rises above the limit, the injectors receive pulse-time signals once per every two crankshaft revolutions, cutting fuel flow in half. As soon as rpm drops, every-revolution signals resume. You'll feel as if the the car is first decelerating, but if you hold the throttle open, you'll feel the engine surge as rpm cycles around the limit. See Fig. 2-44. In some cars, drivers report rpm limitation differs slightly from the tachometer red line. During a quick run-up through the gears, good drivers shift before the rpm-limitation cut-off.

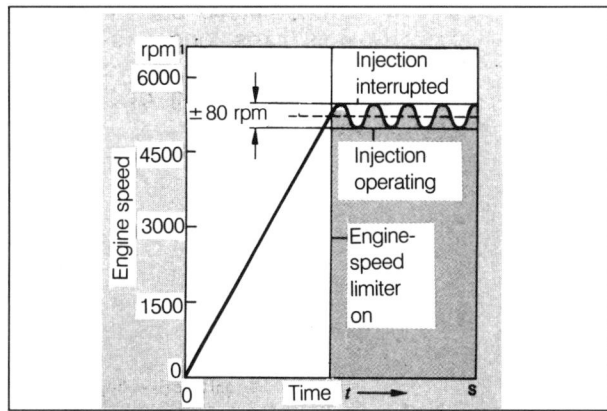

Fig. 2-44. Engine-speed limiter holds rpm at maximum allowed ±80 rpm, only one injection per 4-stroke cycle.

Pulse-Period Limitation

Recall the earlier discussion of pulse period, the time available for injector-open pulse time. Pulse-period limitation circuitry operates under two conditions to limit pulse time:

1. Lower limit: at high revs, in overrun or coasting, minimum pulse time must be limited to prevent excess Hydro Carbons from too-lean mixtures;

2. Upper limit: at low revs, during full acceleration, maximum pulse time must be limited to prevent too-rich mixtures, causing stumbling, depending on temperature.

Lambda Control

You'll remember from chapter 2 the discussion of the ideal air-fuel ratio (lambda) and its relation to emissions. On most systems, the air-fuel ratio for best emission control is achieved by sensing the oxygen content of the exhaust gas with a lambda sensor, shown in Fig. 2-45. The lambda sensor's signal is monitored by the control unit, which then adjusts pulse time to maintain the ideal air-fuel ratio. The system operates closed-loop.

Fig. 2-45. Lambda sensor looks something like a spark plug installed in exhaust pipe or exhaust manifold to sense oxygen content of exhaust gas. The closer to the exhaust valves, the faster it heats up.

Lambda Sensor Design and Operation. The lambda sensor is essentially a small battery that generates a voltage signal based on the differential between the oxygen content of the exhaust gas, and the oxygen content of the ambient air.

A cutaway view of the lambda sensor is shown in Fig. 2-46. The tip of the sensor that protrudes into the exhaust gas is hollow, so that the interior of the tip can be exposed to the ambient air. Both sides of the ceramic tip of the sensor are covered with metal electrodes that react to create a voltage only if the ambient air has a higher oxygen content than the exhaust and the ceramic material is hotter than 575°F (300°C)

L-Jetronic

When these conditions are met, voltage is generated between the two sides of the tip. The voltage is usually about 1 volt. But if the engine is running lean, the exhaust gas has about the same amount of oxygen as the ambient air, so the lambda sensor will generate little or no voltage; if the engine is running rich, the oxygen content of the exhaust will be much lower than the ambient air and the sensor voltage will be larger. See Fig. 2-47.

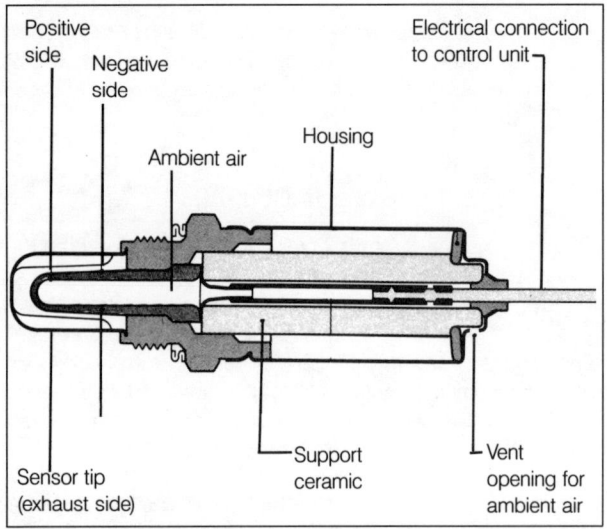

Fig. 2-46. Cutaway view of lambda sensor.

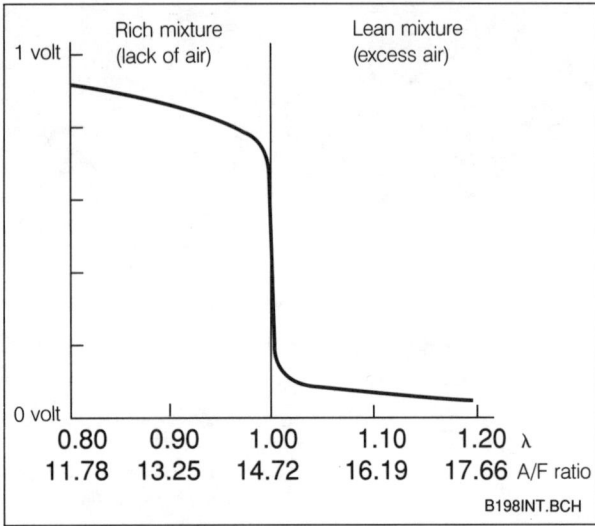

Fig. 2-47. Rich mixture and low content of oxygen in exhaust causes voltage output from oxygen sensor. Remember, rich = voltage.

Some cars have a lambda sensor that has a heating element built-in to speed warming of the sensor to improve the driveability and reduce the emissions of a cold engine. On a cold engine, it may take 90 to 120 seconds for an unheated lambda sensor to get warm enough to start generating voltage, while a heated sensor may be warm enough after 10 to 15 seconds.

Lambda Closed-Loop Control. Recall the discussion of open-loop/closed-loop systems in chapter 2. The lambda sensor and the control unit form a closed-loop system that continually adjusts the air-fuel ratio by means of the fuel-injector pulse time. For example, the sensor generates a high voltage because the mixture is rich, so the control unit reduces pulse time to lean the mixture. Sensor voltage falls, so the control unit increases pulse time to enrich the mixture. Sensor voltage increases, etc....

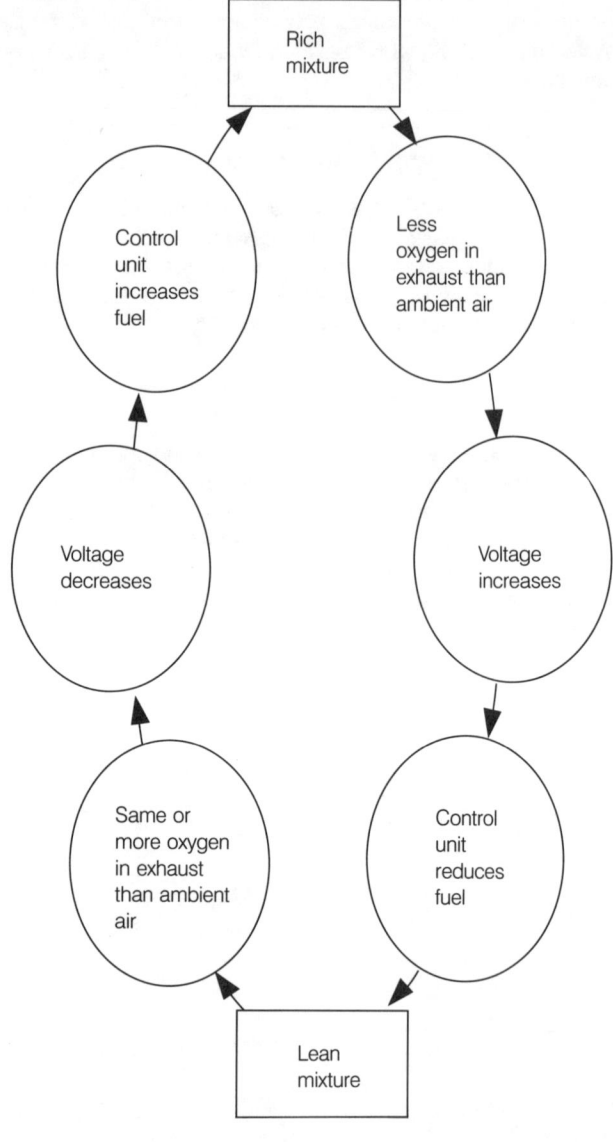

Fig. 2-48. Since about 1980, fuel-injected engines operate closed-loop most of the time.

The lambda sensor voltage is always fluctuating as shown in Fig. 2-49, so it is hard to maintain the exact point at which the air-fuel ratio is ideal. Instead, the ratio tends to oscillate to either side of the ideal ratio, but the oscillation is so fine (about 0.1% change in the air-fuel ratio) that it is not noticeable in engine performance. The rate of the air-fuel ratio oscillation is related to the quantity of exhaust passing the sensor. At idle, the cycle from lean to rich and back again may take 1 to 2 seconds. At cruising speed, the cycle may happen several times a second.

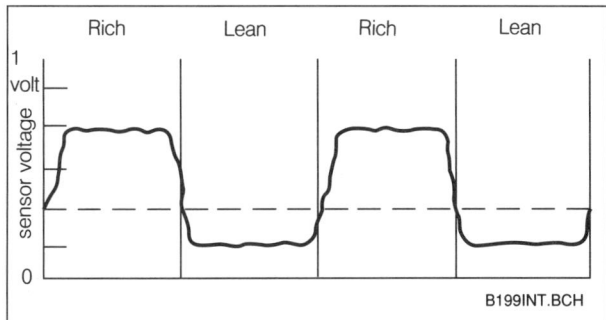

Fig. 2-49. Closed-loop lambda sensor voltage cycles back and forth from slightly rich to slightly lean. In chapter 4, you'll measure the effects of this cycling.

This closed-loop system can compensate to some degree for changes in the engine over time. For example, if a valve is leaking slightly, or if there is an intake air leak, the lambda sensor senses the change in combustion and brings the system back within its design limits. This has been described as having a skilled technician under the hood, continuously tuning the mixture for best operation under all conditions. Changes beyond the system's range, though, can still lead to driveability problems.

When the oxygen sensor is cold and not generating a voltage signal, the control unit is programmed to operate open-loop at a programmed injection rate. The same thing happens if you disconnect or cut the lambda sensor wire, or if the sensor is fouled by leaded gasoline. This becomes important when you are trying to make closed-loop adjustments at idle, but the sensor cools off because not enough exhaust is passing it. Many service procedures depend on closed-loop operation, so remember that the sensor has to be warm enough.

As you'll see in the service chapters, the lambda control system is properly adjusted when engine-out CO is the same whether the system is closed-loop or open-loop. This adjustment point allows the system its full range of compensation for operating conditions.

L-Jetronics have been used since 1974. Engines adapt easily to L-Jetronic. In 1980, when tighter U.S. emission limits were mandated, virtually every European car imported to the U.S. which had used carburetors in 1979 switched to L-Jetronic. Those that did not, switched to K-Jetronic.

2.6 LU-Jetronic

LU-Jetronic is a refinement of L-Jetronic, used in few cars in the U.S. In fact, the U stands for U.S.; it is a lambda-control version of a refined Jetronic used in Europe: LE-Jetronic. Its main features are optimized circuitry in the control unit, and a start control system. The control unit is analog. You'll find it in:

- a few BMW 3 series, used before switching to Motronic

- 1983–84 California AMC Alliances when the Bendix central fuel-injection system system could not meet California emission limits

- a few late-model Renault Fuego and 18i

3. LH-JETRONIC

LH-Jetronic is a further refinement of L-Jetronic. The main difference between the two systems is the method of measuring air flow on LH-Jetronic. Instead of an air-flow sensor, LH systems use something called an air-mass sensor, which you'll learn more about later. All LH-Jetronics use lambda control, and the control unit is digital, not analog.

Volvo was the first user of LH-Jetronic systems on the 1982 Volvo GL with the 2.3 liter 4-cylinder engine. By 1987 all Volvo engines, including the 2.6 liter V6, were using LH-Jetronic. The first Volvo system is identified as LH 1. In addition to the differences mentioned above, it uses a vacuum switch to signal inlet pressure (instead of a throttle switch), and uses an auxiliary air valve, the same as on L-Jetronic.

Since 1984, LH systems are identified as LH 2. Instead of the auxiliary air valve, LH 2 has an idle-speed stabilizer for automatic idle speed control. The LH 1 vacuum switch is replaced by the throttle switch. Since 1988, LH-Motronic combines the air-mass sensor with control of ignition timing as well as adaptive control and diagnostics. See **4. Motronic Systems**.

3.1 Air-Mass Sensor

The air-mass sensor is completely electronic. It depends on the measurement of current flowing through heated wires to measure air flow. It is also known as the hot-wire sensor because of its heated-wire design, hence the "H" in LH. It has several advantages over the vane-type air-flow sensors of L-Jetronic.

1. It measures air mass, or weight, so it requires no correction for changes in density due to temperature or altitude. The air-fuel mixture ratio depends on mass: so much weight of fuel mixed with so much weight of air. Measuring mass eliminates the need for compensation sensors: air temperature, and altitude. It also reduces correcting computations in the control unit.

2. It has no moving parts. That means mechanical simplification. It responds even faster than the moving vane of the air-flow sensor. Measurements follow changes in air-mass in 1–3 milliseconds.

3. It offers insignificant resistance to the passage of air. Even at maximum air flow, drag force on the wire is measured in milligrams.

Air-mass measurement by hot-wire improves driveability, stability, and reliability. It is used in racing. In my opinion, air-mass sensing will probably supplant the measurement of air flow by vane-type sensors.

Air-Mass Sensor Design and Operation

Underhood, between the air cleaner and the manifold, you'll see a simple black plastic cylinder with an electronic box, as shown in Fig. 3-1. If you remove it and look inside the protective screens, you may be able to see the small platinum resistance, or hot wires, that are suspended inside the cylinder so that the intake air can flow over them. See Fig. 3-2. How fine are these wires? The diameter is 70 micrometers—that's less than 1/10 millimeter, finer than a human hair. By careful design of the sensor and its mounting, the fine wires survive automobile vibration. In the unlikely event a wire should break, the warm engine runs, though without fuel compensation, in a limp-home mode. You can simulate limp-home mode by driving the car after pulling the air-mass sensor connector of a warm engine.

Fig. 3-1. Look for air-mass sensor between air cleaner and intake manifold.

The hot-wire system depends on measurement of the cooling effect of the intake air moving across the heated wires. Suppose you had a fan blowing across an electric heater. With a small movement of air past the heated wires, the cooling effect is small. With more air moving past the heated wires, the cooling effect is greater.

LH control circuits use this effect to measure how much air passes the LH hot wire. The hot wire is heated to a specific temperature differential 180°F (100°C) above the incoming air

when the ignition is turned on. As soon as air flows over the wire, the wire is cooled. The control circuits then apply more voltage to keep the wire at the original temperature differential. This creates a voltage signal which the control unit monitors; the greater the air flow (and wire cooling) the greater the signal.

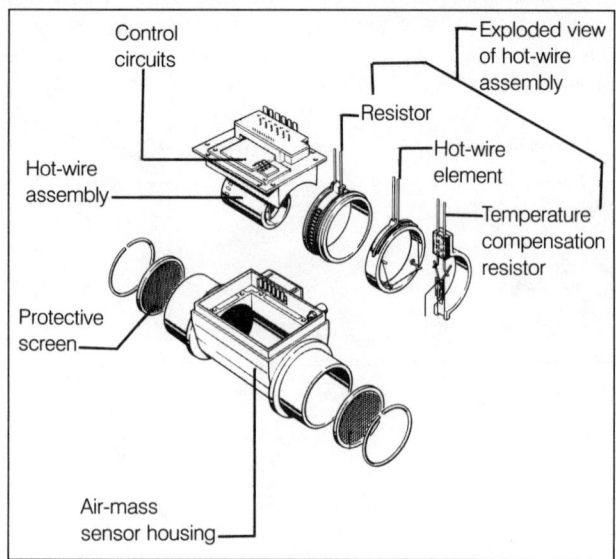

Fig. 3-2. Air-mass sensor includes hot-wire assembly and control circuits.

Control Circuits. The LH control circuits use a Wheatstone bridge, as shown in Fig. 3-3. The hot wire is one leg of a bridge circuit whose output voltage is held to zero by regulating the heating current. The hot wire, known as Rh, changes resistance with temperature. Incoming air passes over the hot-wire, Rh, and also over another resistance wire, Rk. The same voltage is applied to both wires. In series with Rk are two fixed resistances, R1 and R2; in series with our hot wire, Rh, is fixed resistance R3.

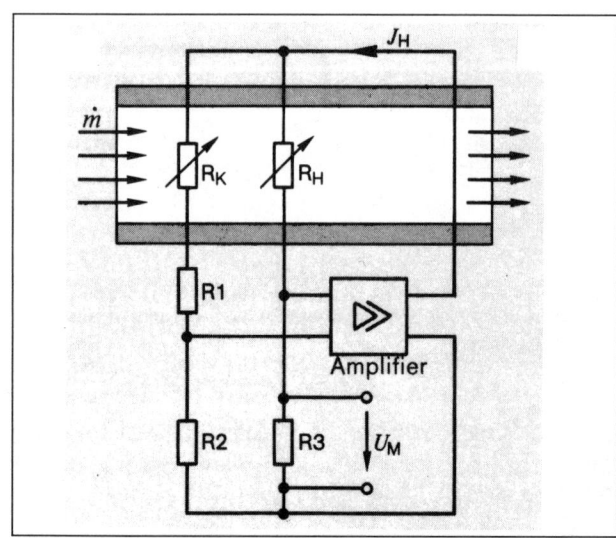

Fig. 3-3. Wheatstone bridge principle explains maintenance of hot-wire temperature.

The Rh wire is controlled to be 180°F (100°C) hotter than the intake air flowing through. For example, if the air is at freezing, 32°F (0°C), the wire will be heated to 212°F (100°C). On a hot day, if the air is at 86°F (30°C), the control circuits heat the wire to the same 180°F (100°C) difference, to 266°F (130°C).

The air mass changes when the driver changes the throttle opening:

1. more air passes over both resistance wires in the air-mass meter;

2. both Rh and Rk are cooled by the increased air mass;

3. Rh decreases its resistance, because of its Positive Temperature Coefficient;

4. current flowing through Rh increases more than the current through Rk;

5. that unbalances the bridge circuit;

6. comparator increases its output;

7. amplifier increases current (Jh) to bring Rh back to its original resistance, and thus its original temperature back to 180°F (100°C) above ambient temperature of the intake air;

8. the heating current is measured as a voltage drop (Um) across the fixed resistance R3;

9. this voltage drop is a measure the air mass and is used as the output signal to the control unit.

All this happens is 1–3 milliseconds.

The air mass can also change because air temperature or altitude changed its density.

Control Unit Inputs and Outputs. LH-Jetronic control of injection pulse time is the same as in L-Jetronic. Low engine speeds mean a smaller air flow. As the air mass increases, the air-mass sensor control circuits provide an output voltage signal to the control unit, and pulse time is adjusted accordingly. Note that in contrast to L-Jetronic, the LH-Jetronic air-mass signal increases with air flow.

For limp-home operation, pulse time is fixed. Each pulse is delivered to the injectors with no compensation. For any rpm above idle, the control unit is programmed to deliver fixed pulses, typically 7.5 milliseconds. Of course, the faster the engine turns, the more pulses are delivered and the more fuel is injected. When the closed-throttle switch completes its circuit, each fixed pulse is a little shorter, typically about 5 milliseconds to keep the idling engine from running too rich.

Shut-Off, Hot-Wire Cleaning. With the air-mass sensor connected for normal operation, the hot wire does not glow. It does not look hot, any more than the heated wires in a rear-window defroster do.

But shortly after you shut off the engine, the control system will heat the wire red hot for about one second, hot enough to burn off any dirt. See Fig. 3-4. Because of the hot-wire cleaning, LH is ready to start every time with a clean wire.

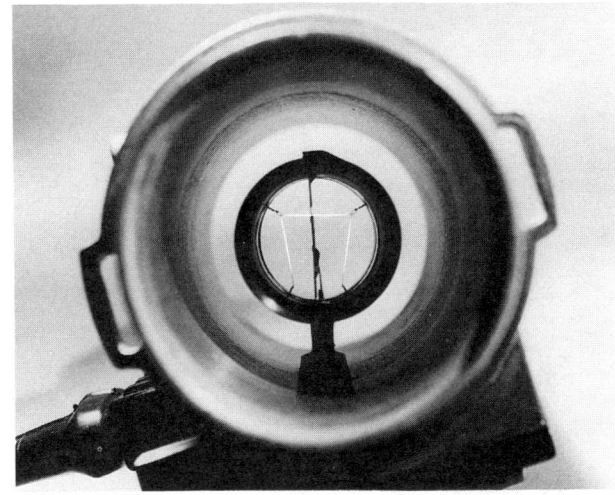

Fig. 3-4. LH hot wire glows at 1800°F (1000°C) during burn-off at engine shut-down.

How's this for attention to detail: There's no burn-off unless the engine has run above 3000 rpm; that usually happens normally each time the car is driven. That way, restart won't normally happen with an 1800°F (1000°C) hot wire. Further, there's no burn-off if the engine stops running at less than 200 rpm; the engine would not be running at such a low rpm, so it has stalled. This prevents the hot wire from glowing after an accident.

In normal operation, don't suspect the wire is inoperative if it fails to glow. The wire is so fine you may not even see it unless the light is right, so don't suspect it is broken if you can't see it right away. You can measure the output signal, as described in chapter 4.

Mixture Adjustment. You'll find the idle-mixture adjustment screw on the side of the air-mass sensor, under a seal. Electronic circuits inside this sensor are not serviceable. When you adjust idle mixture, you adjust the voltage signal from the air-mass sensor by turning a potentiometer, a variable resistor. Because it is an electrical adjustment, it feels different; it turns more turns than the mechanical adjustment of L-Jetronic air-flow sensor.

3.2 Idle-Speed Stabilizer

On most LH-Jetronics (LH 2 and later) the idle-speed stabilizer replaces the auxiliary-air valve. See Fig. 3-5. Idle-speed stabilizers can have different shapes, but they all work the same. They change the amount of air that bypasses the closed throttle. You'll remember this is measured air so it changes the amount of air and fuel to the engine, which changes idle speed, not mixture.

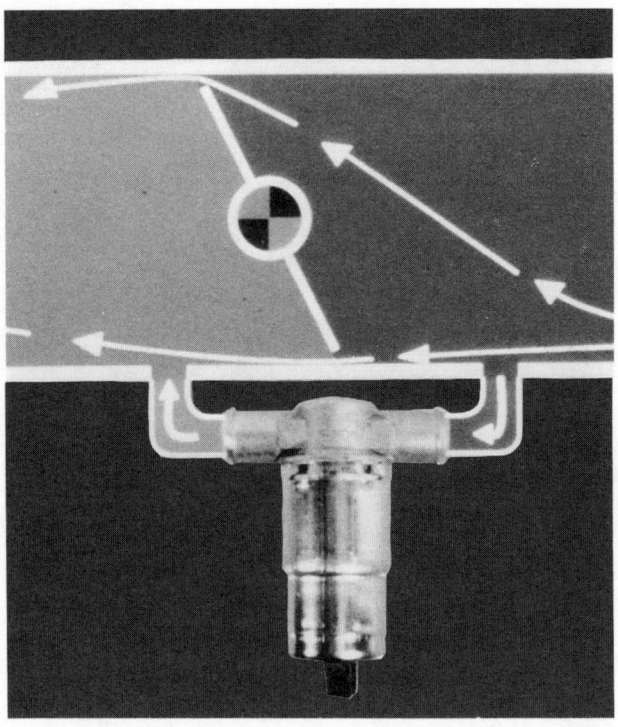

Fig. 3-5. Idle-speed stabilizer allows measured air to bypass closed throttle.

In a cutaway in Fig. 3-6, you can see the rotary valve that turns slightly to change the opening for the bypass air. The stabilizer may look like a motor, but it's not. Signals from the control unit drive the valve toward open.

Bosch also calls the idle-speed stabilizer an actuator. On other cars, a similar device could be called an idle-air control, idle-speed compensator, or idle-speed control. The distinction is whether it actually controls rpm by measuring rpm—a closed-loop system—or on some preprogrammed basis—an open-loop system.

Experts can be fooled by other terminology. Some call idle-speed stabilization CIS—Constant Idle Speed. For our purposes, CIS means Continuous Injection System.

Dwell Signals

The opening of the idle-speed stabilizer depends on the percent dwell—the on-off ratio—of the digital control signals from the control unit. Note that digital signals are on-off, nothing in between. Measuring the on-off ratio tells you the percentage of time that current is on, something like a dwell meter measures ignition-points closing time. The greater the on-off ratio, the further open the stabilizer and the more air will bypass the throttle. For a normal warm engine at idle, the stabilizer is open part-way with a typical dwell of 32%. You can use a dwell meter to measure the on-off ratio, as discussed in chapter 4.

Fig. 3-6. Idle-speed stabilizer valve controls rpm. Smaller opening (top) reduces idle rpm. Medium opening (middle), normal rpm. Larger opening increases idle rpm.

RPM Stabilization

The idle-speed stabilizer is a closed-loop governor that holds a steady rpm. If load is added to the engine—such as when the lights are turned on, increasing alternator drag—the control unit increases the on-off ratio; that keeps the engine speed steady at normal idle. It can keep engine rpm steady for other loads too, such as power steering, or shifting the automatic transmission from Neutral to Drive.

On some cars with automatic transmission (A/T), the idle-speed stabilizer operates to reduce creep at stoplights. While the transmission is in Neutral, engine idle rpm might be governed at 1000 rpm at a temperature of −4°F (−20°C). When the engine is warm, Neutral idle rpm might be 750. But while the A/T is in Drive, the stabilizer may cut back the idle rpm to 670. This luxury touch pleases some drivers, and is possible if the engine does not exhibit roughness at the lower revs.

On the other hand, the stabilizer can work to increase rpm. Turning on the air conditioner (A/C) usually requires higher idle speeds to insure adequate A/C cooling. The input signal to the control unit may be one or more switches:

- driver's actuation of the A/C switch

- closing of the circuit for the A/C magnetic clutch

- pressure increase in the high-pressure side of the compressor

Idle-speed stabilization can also avoid an rpm drop when switching on the A/C. The engine control unit talks to the A/C control, and delays energizing the A/C compressor for about ⅓ second after the A/C is switched on at idle. This gives the stabilizer time to hold or increase idle rpm as necessary for the load of the A/C compressor. On the other hand, while in Drive, A/C operation will not raise the rpm, which could cause creep of a stopped car idling in Drive.

When you service the engine, it can be important to know the difference between:

- closed-loop idle-speed stabilizer

- open-loop systems, such as:
  - Bosch auxiliary-air valve
  - fast-idle cam
  - fixed-position throttle positioner

While open-loop systems increase the air flow into the engine, with most of those, rpm will vary according to engine temperature, engine condition, weight of oil in the crankcase, whether the transmission is in Drive or Neutral, and many other variables. Unless rpm is sensed, fast-idle rpm must usually be set higher to handle unforeseen variations, higher rpm than would be necessary on an engine with an idle-speed stabilizer.

Because the idle-speed stabilizer usually operates closed-loop on the basis of engine rpm, it needs to increase rpm less for cold conditions. This saves stop-light fuel and reduces emissions. In closed-loop, it will maintain idle rpm with small-load changes. As rpm drops, the stabilizer receives control unit dwell signals to bypass extra air to keep the engine from stalling.

Cranking the engine will cause an open-loop signal to increase the air flow as needed. Cold engine temperatures will increase idle rpm, usually only a few hundred rpm from the normal warm idle speed.

When the car coasts, the stabilizer will receive a brief open-loop increase-bypass signal, based on how rapidly the air-flow signal drops toward zero. Functioning in a manner similar to a throttle dashpot, this provides a gradual drop in manifold pressure, helping to reduce the emission of Hydro Carbons.

Because the idle-speed stabilizer automatically controls engine rpm, you must be alert when making an rpm drop test for the injectors or ignition. The fact is, with the idle-speed stabilizer working, you probably won't get any rpm drop; that can be confusing if you don't realize what the stabilizer does. For example, suppose you disconnect one injector. When the stabilizer is working, the engine does not slow down the way you'd expect it to. As soon as the rpm starts to drop off, the stabilizer opens, to increase rpm back to normal.

Adaptive Control

Beginning with some 1984 systems, idle-speed stabilization can be adaptive. That means that, during the lifetime of the engine, it will adapt the idle rpm setting to compensate for changes in internal friction. Adaptation takes place as the engine idles for a few minutes. This dwell value is stored in the control unit non-volatile memory which then becomes the center point for increase/decrease rpm signals.

Idle Speed Adjustments

Although idle speed is automatically controlled, in some cars you may still need to set the basic idle speed (lower than curb idle) with a knob, changing the amount of bypass air. You have to ground the idle-speed stabilizer as mentioned above, so it is fixed. When you remove the grounding, the engine will seem to take a gulp of air as the stabilizer rapidly opens; then rpm steadies at normal idle.

On the newest systems, idle-speed stabilization has become so well controlled by adaptive circuits in the control unit that no adjustment is provided for idle mixture or idle rpm. For more on adaptive circuits, see **4. Motronic Systems**.

4. MOTRONIC SYSTEMS

In simple terms, Motronic is an engine-management system with a single control unit for control of ignition timing as well as fuel-injection. Many of the sensors important for fuel injection are also needed for ignition-system control, as shown in Fig. 4-1, so the integration of the two systems can accomplish many things that L-Jetronic and LH-Jetronic can't:

;bl;Integrated control of fuel injection and ignition can manage the engine better than control of either one alone. That is, timing is sometimes dependent on the air-fuel ratio, and vice versa; also, emissions can be reduced by coordinated control

- Engine control can be based on actual needs of each engine model based on large amounts of engine-test data during different operating conditions stored in the Motronic Read Only Memory (ROM)

- Many additional operating functions can be provided. Important from the standpoint of service are the adaptive functions, and the self-diagnostics for troubleshooting

- Motronic advantages are better driveability and fuel efficiency, and reduced emissions. More specific benefits:

1. fuel savings achieved from best combination of mixture and timing

2. dependable starting, cold or hot

3. stable idling at reduced rpm

4. relative freedom from maintenance

5. good torque characteristics, allowing longer gear ratios

Applications—Identifying Features

Motronic systems use the same pulsed fuel injection principle as the L-Jetronic systems described earlier, so many of the functions and components are the same. This section covers those Motronic functions and components which are additional to basic L-Jetronic fuel injection.

Underhood, how do you tell a pulsed Motronic system? Most "L" Motronic systems use the vane-type air-flow sensor as in L-Jetronic. Some Motronic systems also operate with the air-mass sensor of the LH-Jetronic, as in BMW 750i V12 (known as LH-Motronic). So checking these components won't immediately identify a pulsed Motronic system.

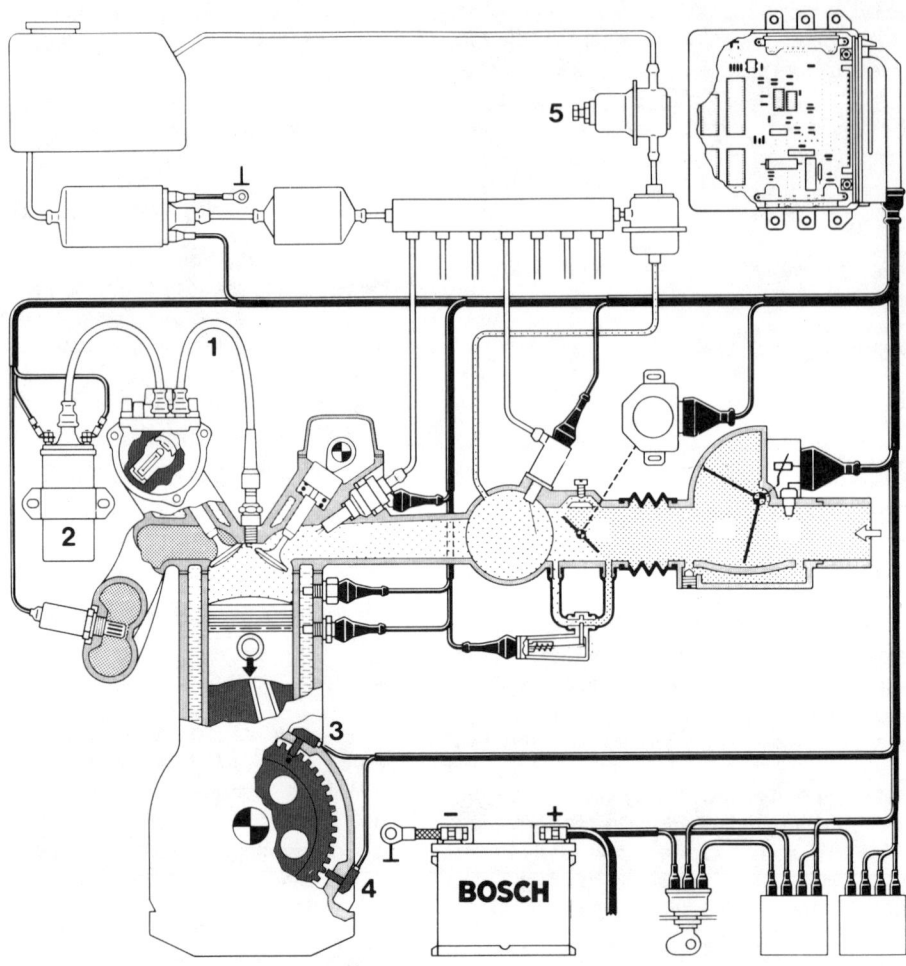

Fig. 4-1. Motronic schematic shows ignition system (**1, 2**), crankshaft signalling (**3, 4**), and pulsation damper (**5**) added to L-Jetronic.

Try looking at the distributor as shown in Fig. 4-2; a Motronic distributor will not have vacuum hoses — or centrifugal weights inside. But even this can be confusing. Some 6-cylinder Porsche 911 engines use centrifugal weights to advance the rotor during high rpm for better contact between the rotor and the cap contacts, but not to affect ignition timing. Several cars, such as Porsche 928 and Peugeot 505STX have separate electronic ignition controls, no vacuum hoses or weights, and are not considered Motronic systems. Don't be confused by systems that control distribution of ignition-firing pulses, so-called distributorless ignition.

Fig. 4-2. Most Motronic distributors are simple; no vacuum hoses, no centrifugal weights. With distributorless ignition, the control unit times and distributes power to multiple coils.

The best way to identify Motronic is to check your vehicle specs. With modern, complex engines, you cannot guess about engine management systems; you must have the vehicle shop manual.

4.1 Ignition Timing Control

I'm talking here about more than electronic ignition — the systems that have replaced the points and condensers of yesterday. I'm talking about microcomputer control of ignition-advance angle for every plug firing, or at most, every two plug firings. The millisecond response time of electronic ignition-advance control is far faster than the traditional mechanical flyweight/vacuum advance systems.

In a mechanical system (and that includes electronic ignition), as rpm increases, centrifugal weights advance timing. Changing load (manifold pressure) can further change timing advance with a vacuum diaphragm. As rpm increases, there is less dwell time for the coil to charge when perhaps the engine needs more spark energy.

Traditional curves for distributor timing show the limited control of timing advance possible. A timing point that is proper for one combination of rpm and load is probably wrong for other combinations of centrifugal weight position and vacuum

diaphragm action. Once the vacuum control or the centrifugal weights reach their limits, advance control is fixed. Mechanical timing has been patched with more vacuum hoses, temperature and delay valves, and other servicing headaches; it's still a compromise. Even at its best, this control falls far short of the precise and rapid timing needed by today's cars.

In Motronic systems, however, the control unit processes a number of inputs as shown in Fig. 4-3, and then adjusts timing for all conditions based on its internal data map.

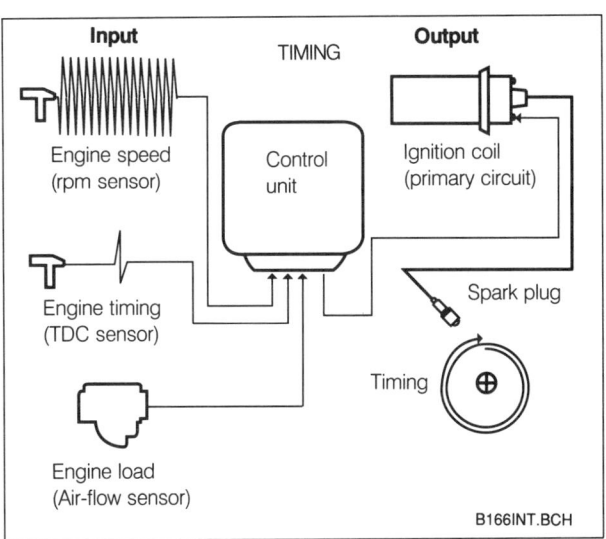

Fig. 4-3. Timing is controlled by the control unit by inputs of rpm, TDC, and load. Control unit controls charging of coil.

Timing Data "Maps"

To determine precise timing-advance requirements, each family of engines is tested to learn the best timing for each condition of speed, load, and other variables in heat and cold, on the dynamometer and in the mountains. The goal is to find the timing for best power, for best economy, all the while meeting emission limits.

The result of these tests is a series of data "maps", as shown in Fig. 4-4. Literally thousands of data points from these tests are stored in the computer memory of the Motronic control unit ROM (Read Only Memory) for readout during engine operation. As you may know, a ROM cannot be changed. For any combination of engine load and rpm, the control unit can supply the best ignition timing. For example, for an rpm-input signal of 2000 rpm, at a load signalled by the air-flow sensor, the computer would look up the timing advance angle, let's say it should be 22 degrees BTDC.

But control is even more precise. Suppose the rpm were 2050, and the memory contained only data points for 2000 and 2100; then the computer would look up both 2000 = 22 degrees BTDC, and 2100 = 24 degrees BTDC, and interpolate. It would calculate an advance for the 50 rpm difference between

2000 and 2050, and would output timing of 23 degrees BTDC. In the control unit, timing is computed so fast that Motronic can adjust timing for every firing of each spark plug!

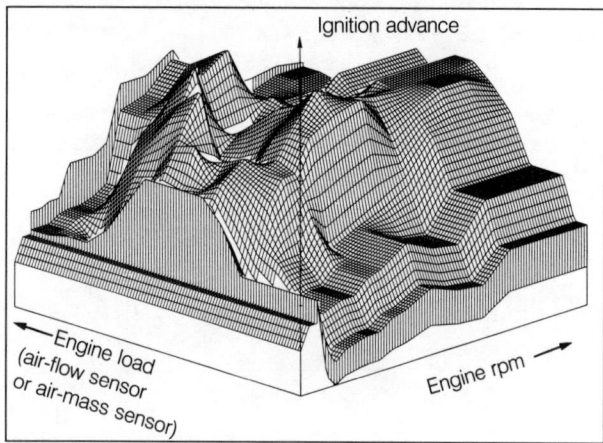

Fig. 4-4. Ignition advance map shows electronic timing control according to engine load and speed. Maps are symbolic of thousands of data points stored in control unit memory. Don't confuse these maps with term MAP, commonly applied to Manifold Absolute Pressure.

Distributor

Because it does not control timing or signal rpm, the Motronic distributor's only job is to distribute the secondary, that is to send the control-unit-timed spark to the proper cylinder.

While timing and dwell were formerly dependent on each other, Motronic memory provides separate dwell-angle data, based on battery voltage and rpm. Later, we'll see how timing and dwell control improve starting and other variations, but here are the principles.

Dwell-Angle Control

As engine rpm increases on mechanical advance systems, there is less dwell time for the coil to charge between firings, resulting in a fall-off of coil voltage. In Motronic systems, dwell angle is electronically controlled so the coil receives the proper current at the time of plug firing. The more battery voltage, the less dwell angle needed. On the other hand, as rpm increases, more dwell angle is needed for the time to charge the coil. By controlling dwell angle, the coil is charged properly for each ignition firing, no more, no less. The objective is to provide the required secondary power at the plug at the moment of firing with minimum losses in the ignition output transistor and the coil.

To reach the nominal value of primary current at the moment of firing, the dwell angle is changed according to the battery voltage as shown in Fig. 4-6. When battery voltage is less, dwell time is increased. In effect, the dwell control answers the question: "When should the primary circuit be closed for the

optimum time for the coil primary current to rise to proper value at moment of opening primary circuit?" In addition, the control-unit output stage limits current so that if, due to rapid engine speed changes as in acceleration, current reaches the nominal value before the ignition point, current is held constant.

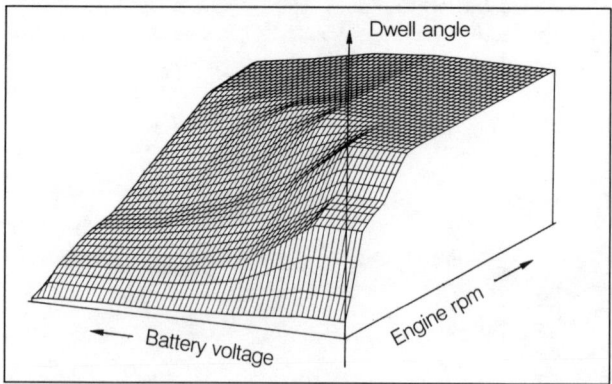

Fig. 4-5. Dwell angle is controlled for battery voltage and engine speed, according to memory map in control unit.

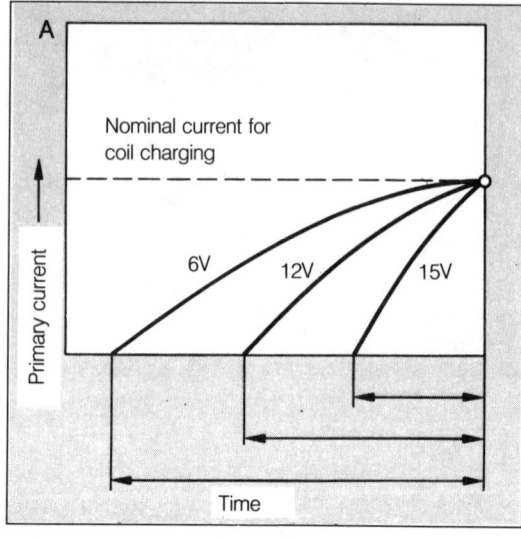

Fig. 4-6. The lower the battery voltage, the longer it takes to charge the primary in the coil.

So engine rpm, and battery voltage are inputs to the control unit. From dwell-angle data in the ROM, control unit output is the dwell, or charge-time control, that conserves energy and prevents overheating the coil. Also, if the rpm signal indicates less than 30 rpm, as when the engine stops or the key is left on with the engine stopped, primary cut-off prevents coil overheating.

Now that you understand the basics of Motronic ignition control, let's look at the rest of the Motronic system and draw similarities and differences from the Jetronic systems described earlier.

4.2 Air and Fuel Systems

The basic flow rate is still determined by the air-flow signal and by the rpm signal. But in Motronic systems, the rpm signal comes from the flywheel rpm sensor instead of from the distributor. Also, the pulsed Motronic air-flow sensor signal increases with increasing air flow, a direct relationship. Remember, in the L-Jetronic air-flow sensor, the signal decreases with increasing air flow.

Pulsation Damper

The Motronic fuel system is quite similar; see the L-Jetronic section for details of the pump, distributor pipe, pressure regulator, and injectors. Some Motronic cars add a pulsation damper on the fuel rail, to reduce pressure fluctuations and noise in the system. It looks similar to the pressure regulator as shown in Fig. 4-7, so learn to tell them apart. Although each has a hose connection on its base, the pulsation damper hose does not have a pressure function; it provides for a possible rupture in the diaphragm, dumping the fuel into the manifold instead of spilling it on the engine. How else do you tell the difference? The pressure regulator is on the high pressure side; the damper is on the low pressure, or return side.

Fig. 4-7. Fuel pressure regulator (right) and fuel pulsation damper (left) look similar, so be careful when servicing them.

4.3 Control Unit

Motronic systems make extensive use of other data maps stored in ROM for many of the control unit functions, including:

1. fuel injection

2. lambda closed-loop control

3. warm-up

4. acceleration

5. ignition timing

6. dwell angle

7. EGR

8. idle-speed control

Data "Maps"

The main map in the control unit is a set of lambda memory-points used to determine the desired air-fuel ratio for the fuel injection pulses. The lambda memory map, shown in Fig. 4-8, is derived from lab and road tests, and modified for the requirements of the vehicle and the country where it will operate:

● at part load, maximum economy and minimum emissions

● at wide-open throttle, maximum torque while avoiding knocking

● at idle, maximum smoothness

● during throttle opening, maximum driveability

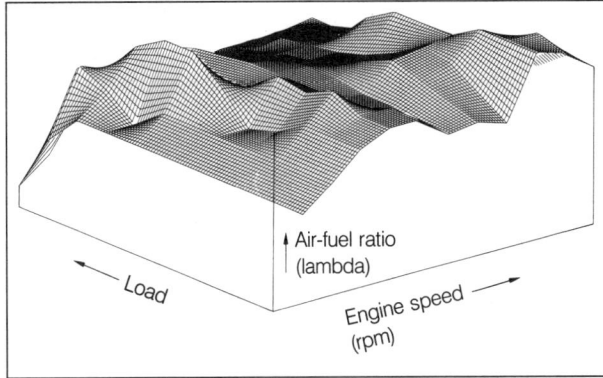

Fig. 4-8. Lambda "map" establishes lambda, or air-fuel ratio for each load/speed point.

Air quantity and the requirements for control of injected fuel can vary due to many factors such as manifold-pressure variations resulting from individual piston pumping, and intake-valve opening/closing. As shown in Fig. 4-9, idling manifold absolute pressure can vary by 5 kPa from an average 42, over more than 10% plus or minus. For this and other variations, the lambda map insures the best possible adjustment of the air-fuel ratio as corrected by inputs that control fuel injection, and without affecting other operating points.

Air flow and rpm set up basic pulse time. Other inputs can change that basic pulse time, and compensate for other variations. Later, you'll see how each of these works for each driving condition.

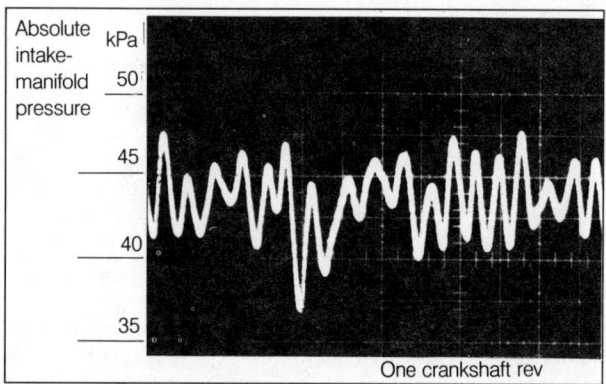

Fig. 4-9. Manifold pressure fluctuates rapidly (as much as 10%) as a result of intake valves opening and closing. This can cause changes in fuel-injection requirements.

Timing Signals: RPM, TDC

For the most accurate measure of engine timing and speed, Motronic systems read the position of the crankshaft directly, instead of from the ignition system as in L-Jetronic. Special sensors, shown in Fig. 4-10, pick up signals from the flywheel teeth. Taking RPM and TDC timing signals from the crankshaft avoids inaccuracies from gear-lash or belt-drive such as when rpm and timing are determined in a camshaft-driven distributor, causing "spark scatter".

Fig. 4-10. RPM sensor (**1**) sends engine-speed signals from flywheel teeth; TDC sensor (**2**) sends one pulse per passing of set screw (arrow) each crankshaft revolution.

The rpm sensor (also called the engine-speed sensor) is an inductive-pulse sender that picks up pulses from a toothed wheel, usually the flywheel. The rpm signal can be displayed on a scope just as it is sent to the control unit, one blip or spike for each tooth as shown in Fig. 4-11. It is so accurate it can sense an rpm change while the crankshaft turns only a few degrees.

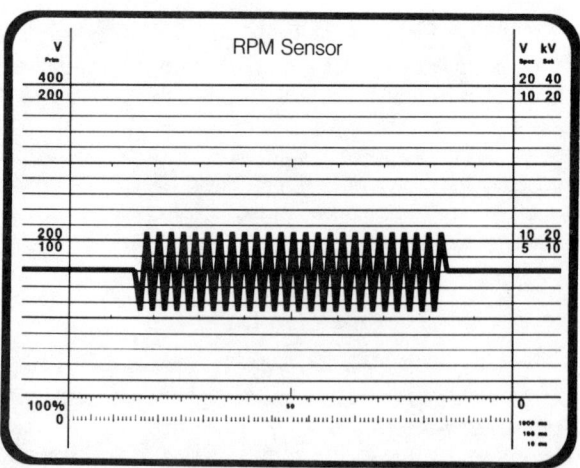

Fig. 4-11. RPM sensor scope pattern shows one pulse per flywheel tooth.

The TDC, or reference-mark sensor (reference from cylinder #1 TDC), is triggered by a set screw on the flywheel. Each time the screw passes the TDC sensor, the sensor signals one blip for each crankshaft revolution as shown in Fig. 4-12. Both sensors are magnetic, with a soft iron core that stores the magnetic field. When a tooth in the flywheel or the reference pin moves through the magnetic field, the change induces an electrical voltage in the winding. This voltage is the input signal to the control unit. The sensor is known as a passive diffusion-field sensor because it does not require a current supply.

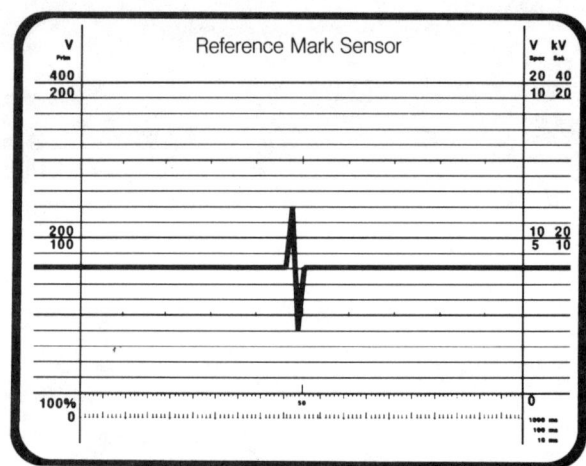

Fig. 4-12. TDC sensor scope pattern shows one pulse per crankshaft revolution.

One of these flywheel sensors provides input of rpm to the control unit. The other sensor provides input of TDC reference. The air-flow sensor provides input of engine load. From the control unit ROM, an output signal to the coil primary sets timing advance and dwell for the next spark firing.

Some engines operate with only one sensor that combines both functions, using a toothed timing-wheel instead of the flywheel. Two missing teeth signal TDC, as shown in Fig. 4-13.

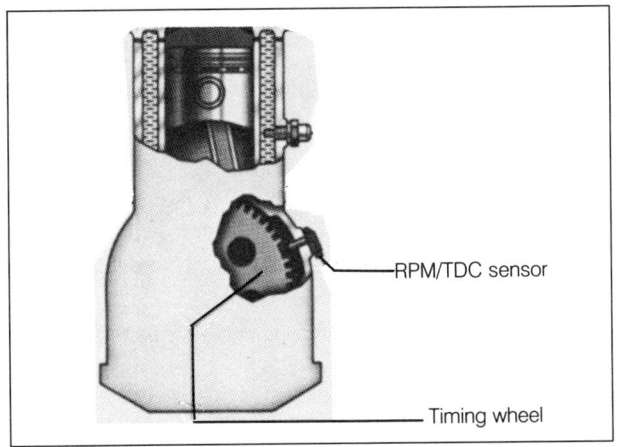

Fig. 4-13. In some Motronics, special timing wheel replaces starter gear ring or is mounted on front of crankshaft. Single sensor picks up rpm from teeth, and picks up TDC from gap in teeth.

Injector Pulse Patterns

On the scope, Motronic injector patterns look different from other Jetronic injector patterns because the digital control signals are different. See Fig. 4-14. The small blip on the left is the TDC reference signal. The injector is opened by the first negative, or grounding pulse, and held open by a series of smaller pulses. When the injection signal ends, a sharp counter voltage shows as a plus spike.

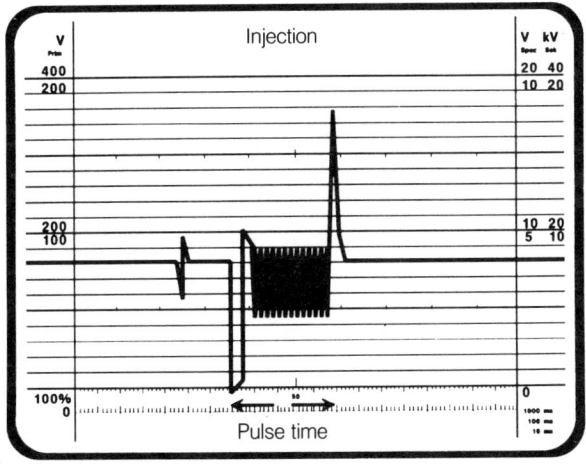

Fig. 4-14. Injection pulse on scope shows large pull-in signal and pulse time. Notice signal drops to 0 (ground) to open injector.

Adaptive Circuitry

Some Motronic control units (those identified as ML.3) include adaptive circuitry so your engine is controlled based on its individual operating conditions, and on how you drive it. It adapts or learns what is best for you and this engine, which

may be different from the engine tested to develop the basic data for the memory maps.

If the concept seems hard to handle, think of it in this way: Remember from earlier discussion that closed-loop control depends on the oxygen sensor signal that fine tunes the air-fuel ratio, and that the air-fuel mixture is properly adjusted when the output of CO is the same whether the system is open-loop or closed-loop. In effect, the adaptive circuits do the same thing. The control unit looks at the base pulse-time signals during closed-loop, then adapts the open-loop base pulse-time signals to match. The result is less worry about intake air leaks, as the system adapts to such variations. It also adapts to the thinner air at higher altitudes, reducing the need for an altitude sensor.

The adapted values are stored in the control unit's volatile memory — that means that if the battery cable is disconnected for any reason, even something as remote as installing the anti-theft alarm, that adapted value will be lost. If the value is lost, you may find the car drives differently when it's first restarted. Driving it for about 10 minutes should give it time to re-adapt. Adaptive time is carefully considered: anything shorter than 10 minutes would cause the control unit to be adapting to every little change. Anything longer than 10 minutes would cause concern to the driver: "What did they do to my car when they put in that alarm system?" It pays to know when someone disconnects your battery.

> Drivers who drive their adaptive-engine GNX or other hot cars to the dragstrip often find they've lost a few tenths of a second in strip times as the engine adapted to the street. Their times picked up when they ran the strip a few times to re-adapt.

Canister Purge Control. You may know that the evaporative fuel control system stores fuel vapors in a charcoal canister to be drawn into the running engine upstream of the throttle, as shown in Fig. 4-15. Motronic ML.3 considers the adaptive function in the canister-purge control in the control unit. The air-fuel mixture is controlled to lambda 1 by the lambda sensor system as the control unit slowly operates the canister-purge valve. During any purge of the evaporative canister, the adaptive value is forced to neutral by a so-called "forget" function, and the adaptive circuits are turned off. Adaptive control is allowed only when no canister flow is feeding fuel vapors into the manifold. All this adds up to proper emptying of the evaporative fuel vapors without interfering with driveability.

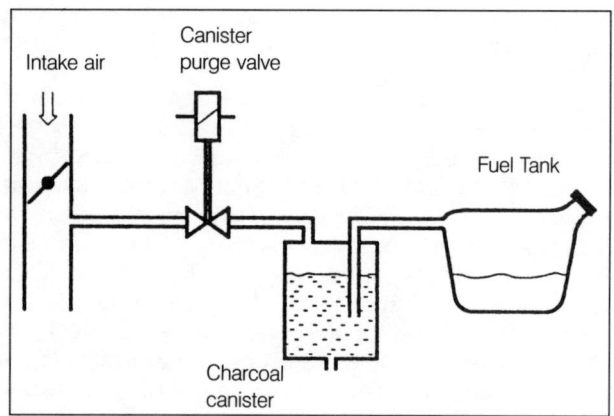

Fig. 4-15. Motronic ML.3 controls canister-purge valve to admit fuel vapors to the intake manifold without affecting adaptive system.

Sequential Injection

Sequential injection means delivering fuel separately from each injector in sequence – in firing order. With an increase in computing power, and an increased demand for idle smoothness and reduced emissions, Bosch Motronic ML.3 can provide sequential injection. Each injector is timed to cut off just before the intake valve opens. During acceleration, the ECU can deliver additional enriching pulses, cylinder by cylinder for the rapidly changing conditions.

In spite of its increased cost, you'll see more pulsed Motronic sequential injection because:

- each individual injector pulse can be much longer than simultaneous injection, as long as 720° crankshaft rotation less intake valve opening, permitting lower idle speeds

- less variation between cylinders in air-fuel ratio for smoother idle and reduced emissions

- injector fuel-rail pressure is more constant because only one injector opens at a time, meaning smoother operation

- improved acceleration enrichment because it can be applied to the "next cylinder"

- soft rpm limitation by reducing fuel flow cylinder by cylinder

Some people think "sequential" means injecting the fuel just as the intake valve opens, but it is not that simple. Depending on the engine, the injection may be timed before the valve opens. It is all based on the desired stratification in the manifold, swirl effects in the cylinder, emissions, and other considerations.

4.4 Idle Speed Control

Most Motronics control idle rpm by a combination of the idle-speed stabilizer and ignition timing. The stabilizer is described in the section on LH-Jetronic. Inputs include rpm, closed-throttle signal, and engine temperature. The control unit sends on-off, or digital signals to the idle-speed stabilizer. Early Motronics use the auxiliary air valve to increase air flow during warm-up, also described in the L-Jetronic section. In these, cold-engine idle rpm is increased according to temperature; it is an open-loop system.

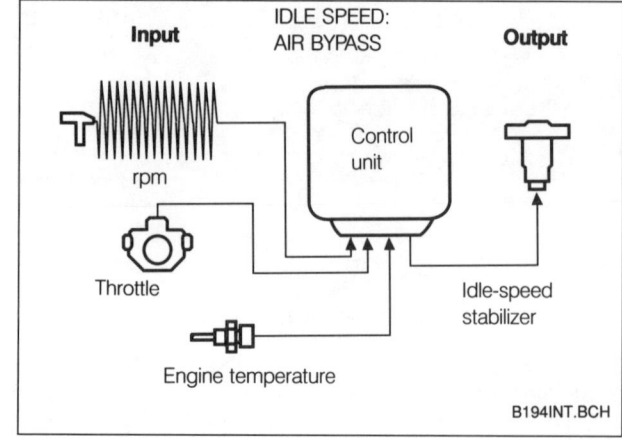

Fig. 4-16. Idle-speed control by idle air bypass. Idle-speed stabilizer handles coarse rpm corrections.

Ignition Timing

Even before the idle-stabilizer acts on air flow, Motronic idle rpm is first stabilized by changing ignition timing. If idle rpm falls, the control unit advances ignition timing to increase rpm. On the other hand, if rpm rises, it retards timing to cut back rpm. Ignition timing handles small rpm changes, handles them in milliseconds. The idle-speed stabilizer handles larger changes, and takes a bit longer.

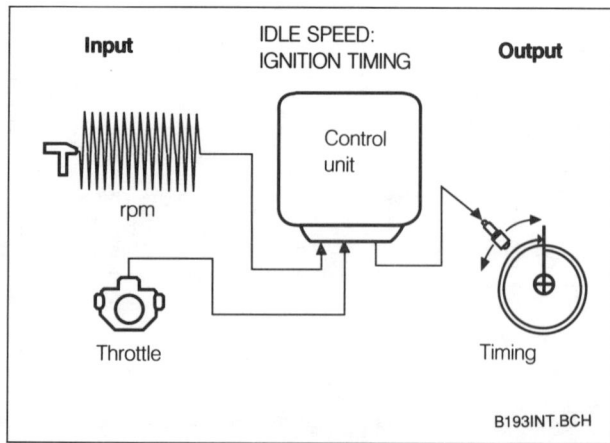

Fig. 4-17. Idle-speed control by ignition timing responds in milliseconds.

4.5 Operating Conditions

In this section, let's look at operating conditions to see how Motronic controls injection and ignition as well as idle-speed bypass air for good engine management.

Starting

Start Control, Fuel. In most Motronics, start enrichment is provided by the port injectors rather than the separate start injector/thermo-time switch. For start control, the important inputs are rpm and engine temperature. The control unit monitors cranking rpm, and also the number of revolutions since the time cranking began. The injection pulses may be longer than normal, once per crankshaft revolution, as you've seen in L-Jetronic.

But at low engine temperatures, the control unit may deliver instead several shorter injection pulses per revolution to improve starting. Remember, these are 10 millisecond pulses in a pulse period, or crankshaft-revolution time of 200ms. To prevent flooding, fuel quantity will cut back after a measured number of revolutions, or after the engine has reached a cranking rpm that is temperature related, for example 200–300 rpm.

Start Control, Ignition Timing. With a cold engine at low cranking speeds, timing will be controlled near TDC. For normal cranking rpm, a large advance in timing might fire too early, damaging the starter. Further, starting would be difficult if not impossible. With a cold engine at higher cranking rpm, however, timing will be advanced for better starting.

For hot starts, or high intake air temperatures, timing will be retarded because a hot cylinder can fire easier while the piston is rising at cranking rpm. Retarded timing prevents knocking that can occur in high-compression engines, and may be masked by the starter sound.

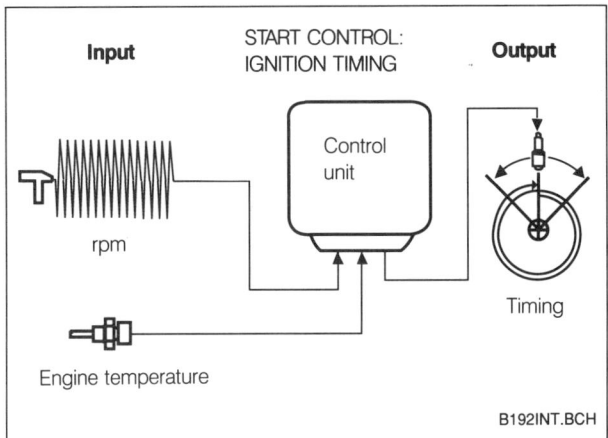

Fig. 4-18. Start control sets ignition timing based on rpm and engine temperature.

Start Control, Air. As in L-Jetronic, the idle-speed stabilizer is normally driven open more by increased dwell signals to provide extra bypass air for cold starting.

Post-Start

Post-start—the 5–30 seconds when you want the engine to keep running after you release the key from CRANK—is affected by engine temperature as well as a timer in the control unit. Colder engines get more fuel injected. Also, colder engines get more timing advance. Based on the start-up temperature, and the time since start, post-start enrichment is gradually reduced. The idle-speed stabilizer maintains rpm.

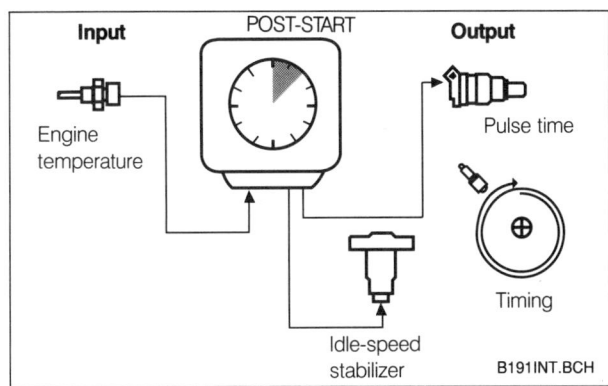

Fig. 4-19. Post-start temperature and time determine injection pulse time, ignition timing, and idle-speed stabilizer bypass opening.

Warm-Up

During warm up, the most important input, besides engine load and rpm for basic pulse time, is engine temperature. The control unit includes a ROM map of warm-up characteristics to apply a correction factor based on load and engine speed, greater at low loads and rpm than at high loads and rpms. See Fig. 4-20. Control unit outputs include injection pulses for the proper mixture, ignition timing for driveability (advance under part-load acceleration, retard on deceleration to reduce HC emissions), and idle stabilization to help keep the engine running at idle.

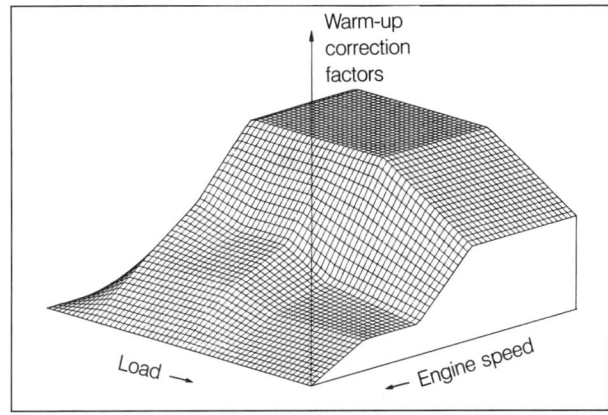

Fig. 4-20. Warm-up correction is greatest at low speeds and low loads, represented by the back of this "map." At the front, correction is least for high loads and high rpm.

Remember, during warm-up, the temperature of the lambda sensor is important. Unheated lambda sensors can be heated faster by changing ignition timing. Motronic retards timing to cause a hotter exhaust. That heats the oxygen sensor much quicker and also heats the catalytic converter faster for more efficiency.

Closed-Loop Operation

Based on the inputs of a warm engine, and oxygen-sensor voltage when hot, the system normally operates closed-loop. The oxygen sensor fine-tunes the fuel-injection delivery for proper air-fuel ratios.

For every combination of inputs, the control unit looks in its memory for the best timing, the best dwell angle, the best lambda, or air-fuel ratio. On the basis of engine temperature, Motronic controls ignition timing with independent calibrations for starting, idling, deceleration, and acceleration, both part-load and full-load or WOT (Wide Open Throttle). For example, control of timing advance at idle reduces the need for idle fuel enrichment.

Part-Throttle Acceleration

Just as in L-Jetronic, the overswing signal of the air-flow sensor increases fuel delivery, compensated by temperature. For air-mass sensors, the rate of increase of the voltage signals acceleration. As shown in Fig. 4-21, during part-throttle acceleration, the normal pulse for steady cruise increases for just about one second while you open the throttle. Then the pulse cuts back even as the engine picks up speed; this saves fuel and reduces emissions.

In addition, ignition timing can be retarded to avoid the brief acceleration knock that could occur for the first few cycles of engine acceleration. Ignition timing control also reduces the formation of NOx, which is normal with acceleration.

Rate of ignition timing change is also controlled;

● to avoid knocking, the control unit allows fast change

● to reduce jerk during transition, the control unit normally changes ignition advance gradually

Full-Load Acceleration

Full-load enrichment is signalled by the throttle switch. As long as the throttle is full-open, longer injection pulses will be delivered for enrichment. Rich mixtures reduce the tendency to knock. This enrichment is based solely on rpm, and programmed from engine tests. The air-flow signal is ignored.

At the same time as rpm and throttle position control full-load injection enrichment, air temperature and engine temperature supply inputs to control the advance of ignition timing for

best acceleration without approaching knock. For engines without knock sensors, timing advance is based on engine test data stored in the timing map. The engine can operate close to the ignition-advance limit, developing maximum torque over the entire rpm range with least chance of knocking based on three factors:

1. the timing curve can be programmed specifically for each engine operating point

2. the system operates with narrow tolerances and freedom from mechanical wear

3. timing is compensated according to engine temperature and intake-air temperature

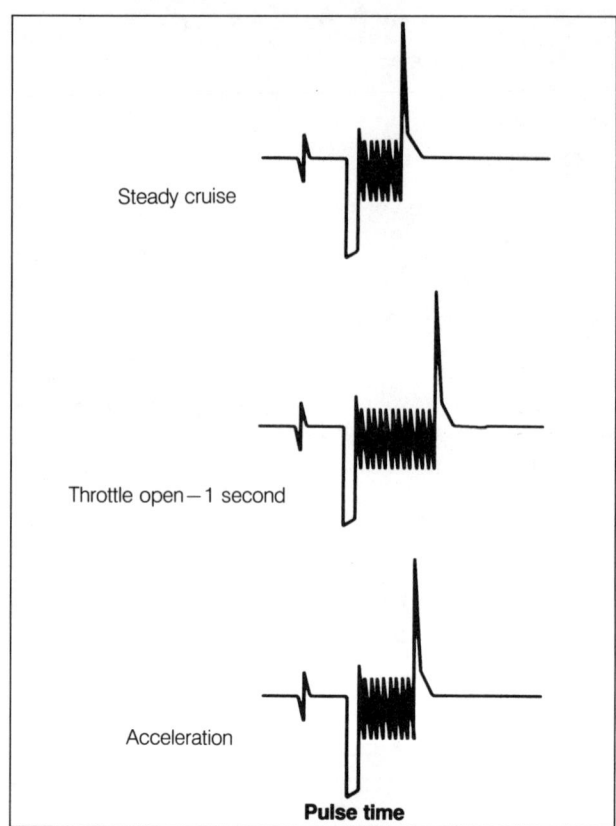

Fig. 4-21. Part-throttle acceleration enrichment is limited to the time the throttle is opening, usually about one second.

In some turbocharged cars, a second air-temperature sensor is mounted in the manifold downstream of the turbo or the intercooler to sense the temperature of the air entering the engine. This is a special fast-response sensor overriding the first air-temperature sensor to handle the quick changes in temperature that can occur during full-throttle acceleration.

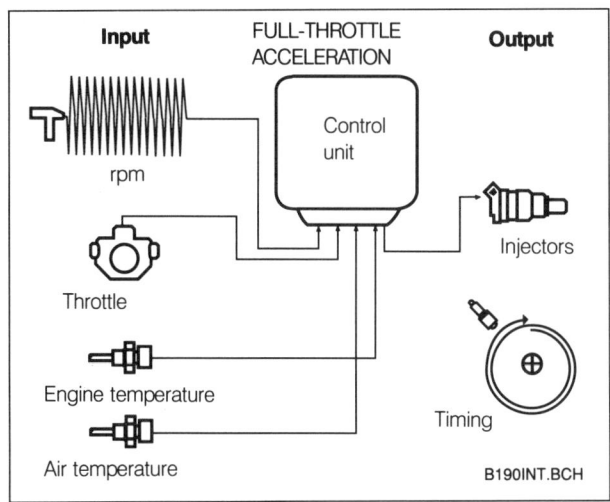

Fig. 4-22. Full-throttle acceleration controls injection and ignition timing, based on inputs of rpm, throttle switch, air temperature, and engine temperature.

Knock Control

Programmed timing accuracy, even with compensations, can be improved by a knock-sensor system that causes timing to advance to just before the point where knock could damage the engine. With a knock sensor controlling timing advance in a separate closed-loop system, power output can be maximized without endangering the engine.

Some engines, particularly turbocharged engines, include knock sensors to pick up vibrations from the engine at the first signs of knock. See Fig. 4-23. Some systems use two control units, mounted inside the passenger compartment, one for Motronic, and a related knock control regulator, called KLR. Knock-sensing inputs help both control units to work together. The first and fastest output changes the timing. If the engine is turbocharged, the second output controls the boost. See Fig. 4-24. In the Motronic known as ML.3, knock sensing and engine management are combined in one control unit.

When the engine knocks, it vibrates with characteristic frequencies of 5–10 kilohertz together with corresponding harmonics. These vibrations must be sorted from other engine vibrations in a recognition circuit in the control unit. The Motronic control unit has accurate information on crankshaft position and firing order so it is possible to determine which cylinder is knocking (one cylinder usually knocks before the others). Further, circuits operate with such speed that ignition timing can be retarded only for the knocking cylinder, and advanced for the firing of the next cylinder. Each knock signal can modify timing in milliseconds. When ignition retard eliminates the knock, timing is advanced slowly in steps to the original value or until knocking again occurs.

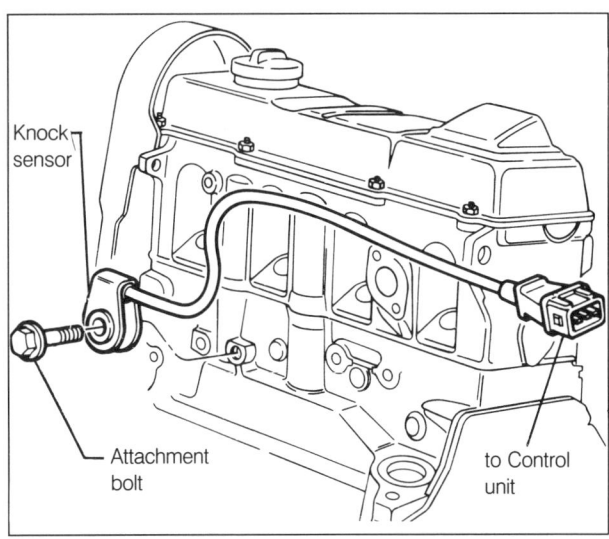

Fig. 4-23. Knock sensor is fastened to selected part of block or head. It is coated with plastic to reduce the effect of changes due to engine temperature.

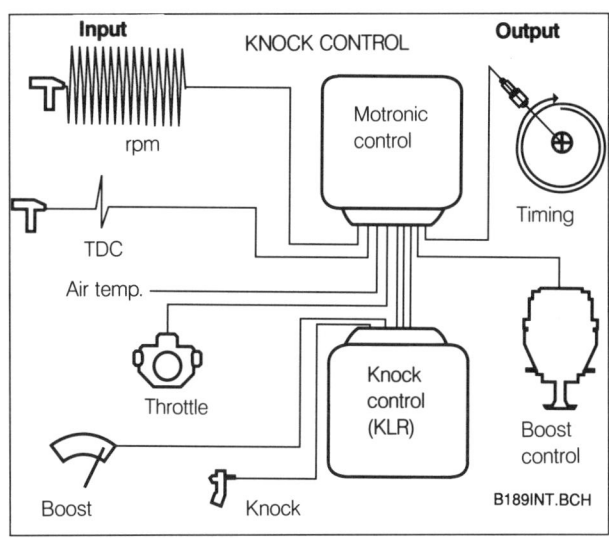

Fig. 4-24. Knock control uses several inputs to control ignition timing and boost, sometimes with a separate knock control computer, known as KLR.

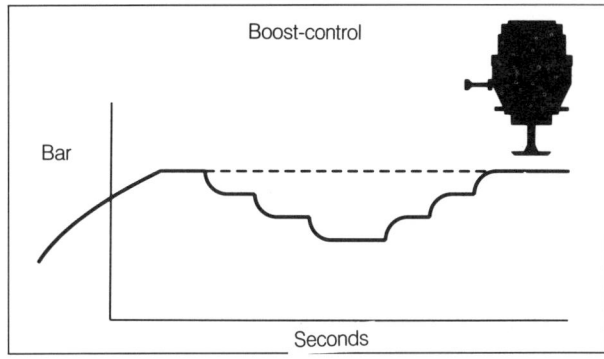

Fig. 4-25. Boost control cuts back pressure in steps until knocking stops, then gradually restores boost.

Motronic Systems

On turbocharged engines, if knock continues for seconds, boost control reduces manifold pressure. Knock control permits more boost, and higher compression ratios for greater power. Automatically, it tends to adjust for the octane-rating of the fuel being burned.

With precise control of ignition timing and turbo boost, engines can be designed with higher compression ratios for greater power output. On all engines, the knocking limits depend on many factors:

- intake air temperature
- engine temperature
- engine deposits
- combustion-chamber form
- mixture composition—A/F ratio, and stratification
- fuel quality
- air density

For years, drivers have known that using higher-octane fuel did not add to engine power unless ignition timing was adjusted at the distributor to take advantage of the improved anti-knock index. Now, with knock sensors and closed-loop ignition-advance control, power output can depend on the anti-knock index of the fuel being burned. It is not unusual to see engine power specifications include the anti-knock index of the fuel to be used.

RPM Limitation

If the rpm signal is greater than max-allowable rpm stored in the computer memory, the control-unit signals a cutback of the fuel-injection quantity. A scope set to read four pulses for simultaneous injection for four cylinders shows the limitation cutback of every other pulse, one pulse every other crankshaft revolution. See Fig. 4-26. If you press the engine into the rpm-limitation range, you will feel a surge as the limitation of fuel injection cuts in and out.

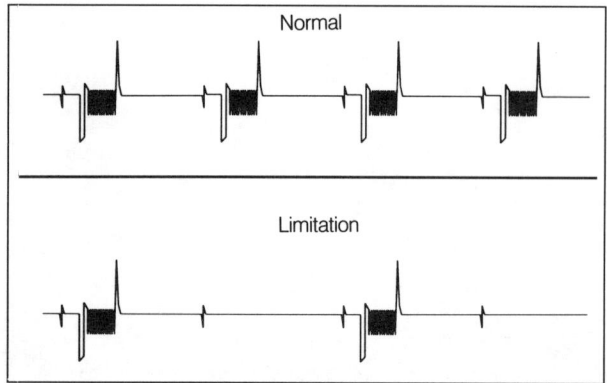

Fig. 4-26. Normal 4 pulses per two crankshaft revolutions, or two 4-stroke cycles. RPM limitation cuts every other injection pulse.

Coasting Cut-Off

Signals of rpm, closed throttle, and engine temperature control coasting cut-off of injection. On the scope during coast-down, the single-injection pattern looks blank except for the TDC pulse. See Fig. 4-27. No fuel is being injected. As rpm approaches idle speed, the normal pulse will show again. If the engine is colder, normal fuel injection returns at some higher rpm, to prevent stalling. See Fig. 4-28. Motronic adds one refinement: as fuel-delivery resumes, perhaps as the driver resumes speed, ignition timing is gradually advanced to smooth the transition from fuel cut-off to cruise or acceleration.

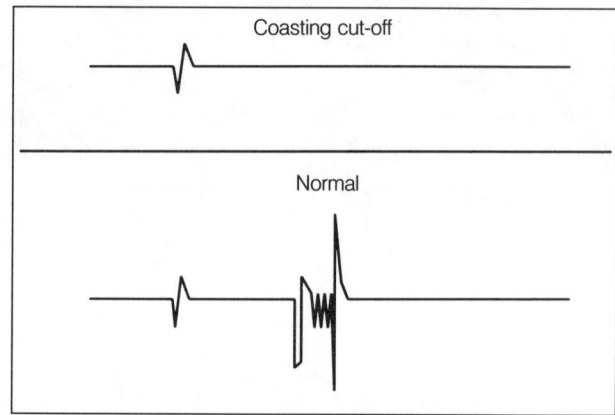

Fig. 4-27. Scope pattern for coasting cut-off shows only TDC pulse until injection resumes to prevent stalling.

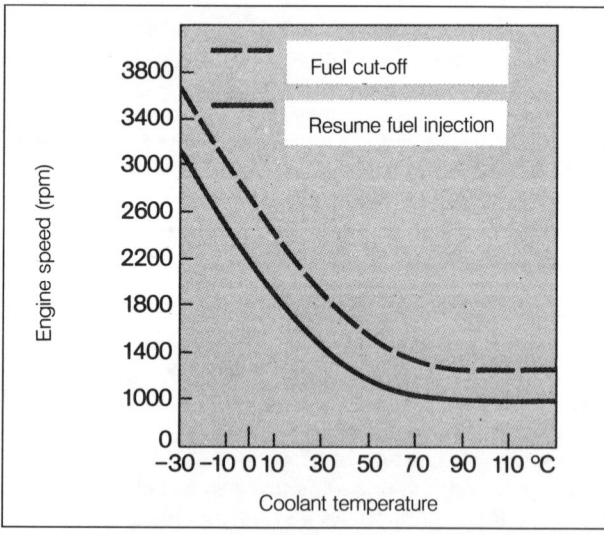

Fig. 4-28. Warm engine cuts off and resumes fuel at lower rpm than cold engine. RPM difference between cut-off and resume fuel prevents hunting.

Electronic Throttle – Drive-By-Wire

Some of the newest Motronic systems have an electronic throttle (also known as a drive-by-wire system). The electronic throttle has no mechanical link between the accelerator pedal and the throttle valve. Instead, as shown in Fig. 4-29, an accelerator sensor picks up your movement or position of the accelerator pedal. It signals the control unit about pedal movement, and the control unit signals the servomotor on the throttle shaft to open. The electronic throttle (Bosch calls it EGAS) may sound like something none of us needs, but it has many benefits.

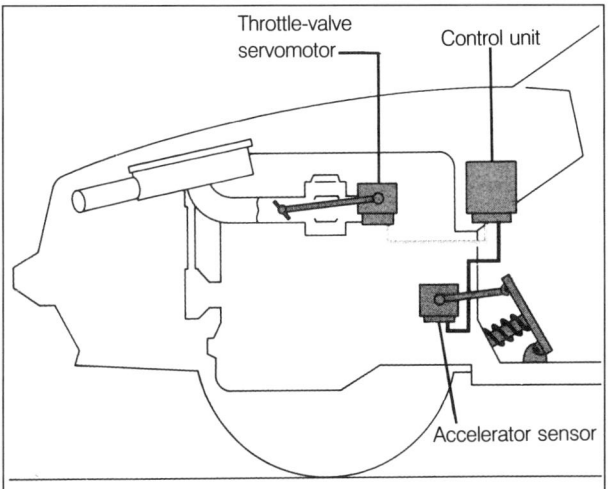

Fig. 4-29. Electronic throttle sends accelerator signal to Motronic control unit. DC motor moves throttle valve, modified by engine temperature, idle rpm, and maximum rpm.

The throttle opening signal may be modified according to engine rpm and engine temperature. It can provide simplified cruise control. It can also control minimum rpm, replacing the idle-speed stabilizer, and control maximum rpm, replacing the alternate-injector cut-out.

But there's more. For traction control, the electronic throttle links with the ABS (Anti-lock Braking System). The same wheel-speed sensors of ABS also feed the ASR (Anti-Slip Regulator). When any driving wheel starts to slip, a slight brake application prevents that wheel from slipping so much that the differential delivers no power to the other driving wheel. If both driving wheels show signs of slipping, the electronic throttle cuts back power for maximum traction. Traction control gives the car the max acceleration the tires can deliver to the road. If laying down tire smoke is your thing, avoid ASR. But for making the fastest takeoff, you cannot beat it.

Can you trust the electronic throttle? Will it provide unintended acceleration? The system checks its safety circuits before start-off, and reports defects to the driver. If a defect is found, a limp-home circuit may disable some of the functions, but it will allow you to get to the house or the shop. It may well be more reliable than some cable actuated throttles.

Diagnostics

Beginning with Motronic ML.3, diagnostic systems in the control unit provide troubleshooting information. Each time you turn the key ON, the computer checks itself. Then it checks the sensors. Any defects are stored in the fault memory for later readout by a diagnostic device. During servicing or troubleshooting, the actuators can be driven by short pulses so the technician can hear or feel the actuation to pinpoint the problem. See Fig. 4-30. You'll see more about diagnostics in chapters 4 and 6.

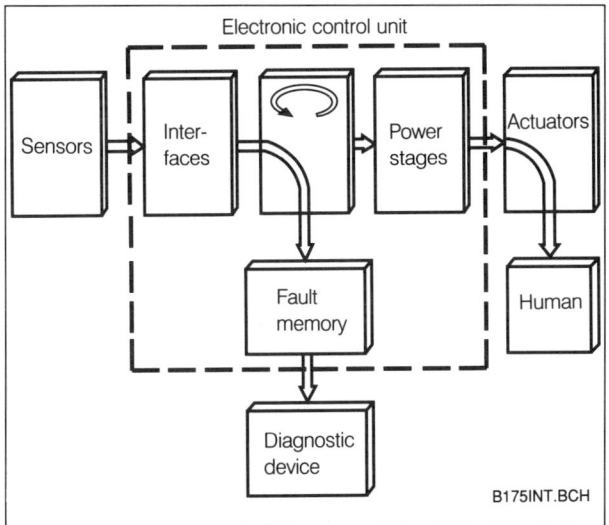

Fig. 4-30. Diagnostics store trouble codes in fault memory for readout by diagnostic device.

The more you study the manuals for different applications, the more you'll realize how well engines can be managed by Motronic injection and ignition control, and idle-speed stabilization.

5. D-Jetronic

D-Jetronic was the first of the Bosch pulsed injection systems. I write about it at the end because the most recent cars with Bosch D-Jetronic were the 1975 Volvo 164E and the 1975 Mercedes 450. In 1976, both changed to K-Jetronic. It's most interesting to see how L-Jetronic evolved from D-Jetronic. At the 1967 introduction, Bosch called it ECGI, Electronically Controlled Gasoline Injection, contrasting this new electronic system with a long background of mechanically-driven injection pumps.

Based on a cross-licensing of the Bendix Electrojector system, which was briefly used in the U.S., D-Jetronic depends on sensing manifold pressure as an indication of engine load.

You can refer to the L-Jetronic section for details concerning most of D-Jetronic. The major differences are:

● manifold-pressure sensor senses engine load

- trigger-contacts in the distributor synchronize injection pulses

- Injectors are operated in two groups
  - 4-cylinder, 2 groups of 2: VW, Porsche, Saab, Volvo
  - 6-cylinder, 2 groups of 3: Mercedes, Volvo
  - 8-cylinder, 2 groups of 4: Mercedes

> Bosch now regards manifold-pressure sensing as a less accurate measure of load than measuring air flow, as in L-Jetronic, and in K-Jetronic, or measuring air mass, as in LH-Jetronic. For certain Bosch racing applications, however, manifold-pressure sensing is sometimes used. Manifold-pressure sensing is widely used by GM, Ford, Chrysler, AMC, and some Toyotas, but the trend in passenger cars is away from manifold-pressure sensing.

Manifold Pressure Sensing

As shown in Fig. 5-1, the pressure sensor is connected to the intake manifold between the throttle and the intake valves. The air-temperature sensor signals for correction for colder, denser air.

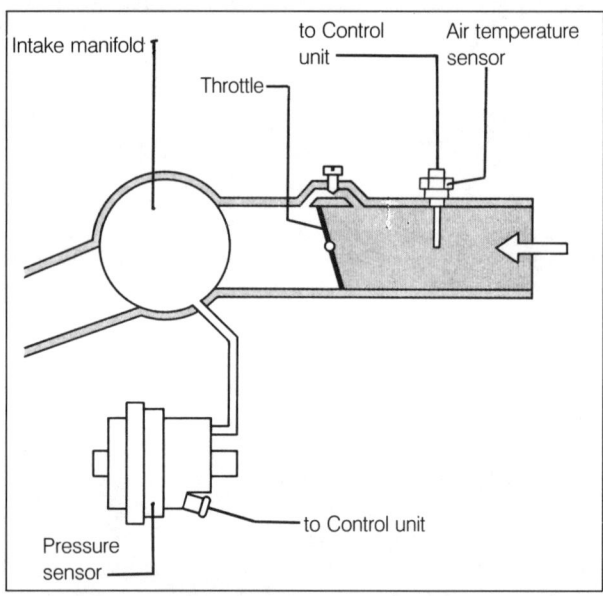

Fig. 5-1. Pressure sensor signals manifold absolute pressure (MAP) downstream of throttle.

As shown in Fig. 5-2, the pressure sensor contains two pressure diaphragm cells that expand and compress according to pressure. The inside of one is vented to the atmosphere; the outside of both is connected to the intake manifold pressure. When manifold pressure increases, the cells compress and pull an iron core armature into a coil, changing the electrical signal to the control unit. Increasing manifold-pressure

signals increasing load so the control unit increases the pulse-time of the injectors. This is a succession of analog actions, requiring no digital computation. D-Jetronic control units are analog. When manifold pressure is low, as at idling, the diaphragm cells expand, pushing the core into the coil; the electrical signal to the control unit reduces the pulse-time, reducing the fuel injected per stroke.

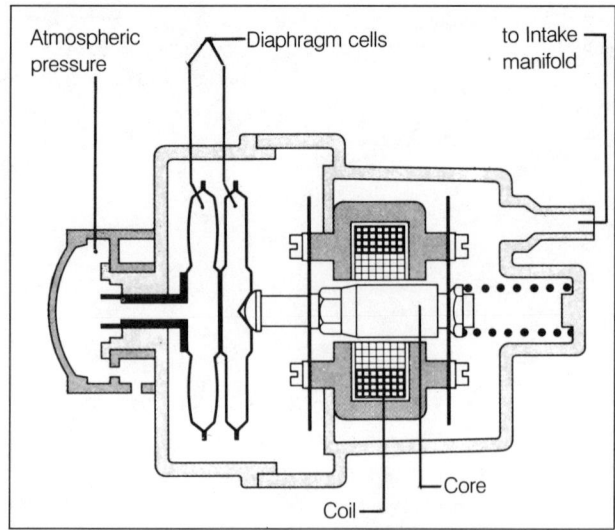

Fig. 5-2. Cutaway view of manifold pressure sensor.

Altitude, or density compensation is provided by the venting of one cell to the atmosphere. In the part-load range, both manifold pressure and atmospheric pressure are reduced with altitude, so the fuel-injection signal is tailored to the thinner air.

Trigger Contacts

Injection timing is determined by a set of trigger contact points in the distributor. See Fig. 5-3. These contacts are mounted below the centrifugal weights of the ignition advance mechanism. They are operated by a special cam on the distributor shaft. Bosch makes the distributor, too. The trigger contacts provide a signal of crankshaft position for timing, and of crankshaft rotation for rpm input. Because they are mounted below the centrifugal weights, the trigger contacts are not affected by distributor ignition timing action.

D-Jetronic injectors are triggered in two groups. Each 4-stroke cycle, D-Jetronic fires each group once instead of firing all twice as in L, LH, and Motronic systems.

Acceleration

Acceleration is compensated at the throttle switch. See Fig. 5-4. This switch contains the full-load contact and the closed-throttle contact as in L-Jetronic. But in D-Jetronic, it also contains a contact path for acceleration enrichment. As the throttle is opened during acceleration, the signal enriches the mixture by increasing the pulse time, and also by increasing the number of injection pulses.

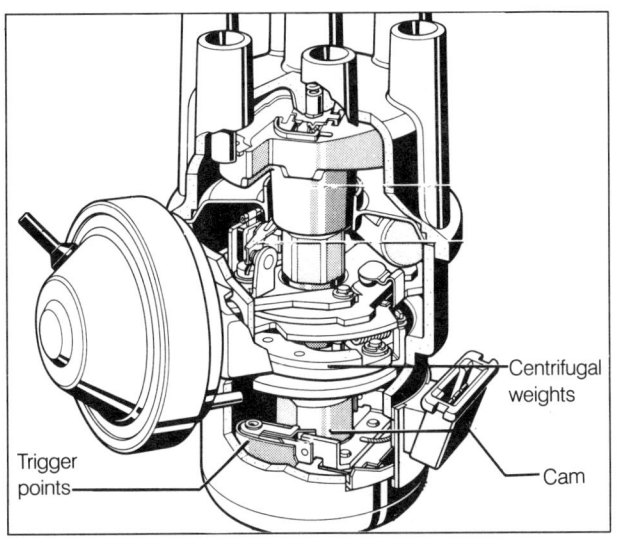

Fig. 5-3. D-Jetronic trigger contacts are built into the distributor.

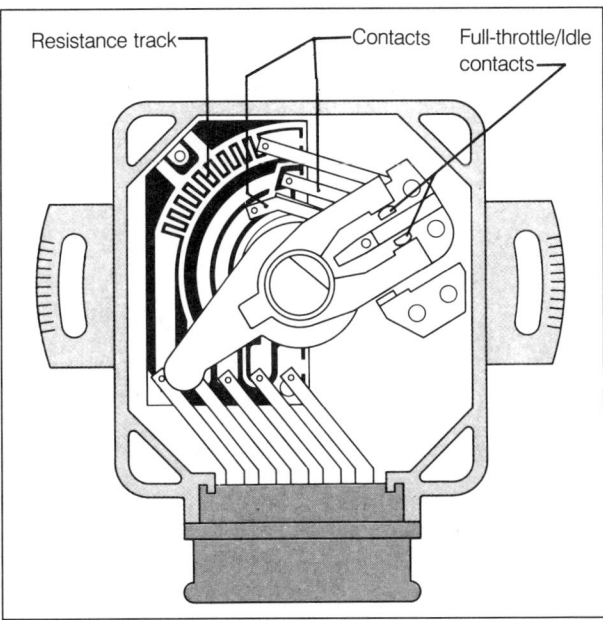

Fig. 5-4. D-Jetronic throttle-position sensor compensates for part-throttle acceleration.

Pressure Regulator

The final difference in D-Jetronic is the fuel pressure regulator shown in Fig. 5-5. It maintains constant pressure at 2 bar, less than the 2.5 or 3 bar of L-Jetronics. Unlike L-Jetronic, fuel pressure can be adjusted by the nut on the bottom of the regulator to control pressure on the spring. Increasing the fuel pressure will enrich the mixture and increase power, but not equally in all rpm ranges. Also unlike L-Jetronic, fuel pressure does not need to be related to manifold pressure because the system is already controlling pulse-time, and therefore mixture, according to manifold pressure.

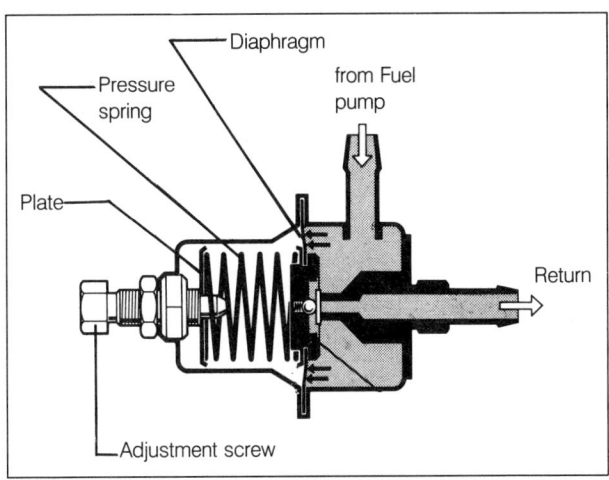

Fig. 5-5. D-Jetronic fuel pressure can be adjusted at regulator.

6. Digifant II

Digifant II is a Volkswagen cousin to Bosch Motronic, controlling pulsed fuel injection, electronic ignition, and idle speed from the same control unit. It includes a knock sensor for ignition-timing control. First use is on 1988 8-valve engines in Golf and Jetta, replacing Bosch CIS.

Similarities to Motronic:

- vane-type air-flow sensor on un-supercharged engines (VW Corrado uses pressure sensor)
  - A/F mixture is adjusted by bypass of unmeasured air
  - intake air temperature sensor

- electrically-opened pulsed injectors

- ignition timing controlled by control unit

- coolant-temperature sensor

- switches on throttle shaft
  - closed throttle (idle)
  - WOT

Similarities to CIS:

- idle-speed stabilizer valve with throttle bypass adjusting screw

- fuel system: pump, relay, filter; no accumulator

Digifant II differences from Motronic or CIS are few:

- system pressure is specified with engine idle: 2.5 bar (36 psi) instead of engine off. Actually, engine-off pressure (comparable to Bosch specs) is 3.0 bar, 0.5 bar higher than many pulsed systems

Digifant II

- service port for fuel-pressure testing

- idle-speed stabilizer circuit delivers fixed (higher) idle rpm when coolant temperature sensor is disconnected
  - measuring current flow to stabilizer at idle shows fluctuation 400–460mA, closed loop, causing 790–850 rpm. Disconnecting sensor shows fixed 430mA, causing 930 rpm, open-loop

- ignition timing is set with coolant sensor disconnected for fixed idle rpm

- control unit increases idle rpm for hot start

Chapter 4

Pulsed Injection – Troubleshooting & Service

Contents

2   PULSED INJECTION – TROUBLESHOOTING & SERVICE

TABLES

1. INTRODUCTION

In this chapter you'll see the troubleshooting and general service procedures that apply to Bosch pulsed (EFI) fuel injection systems. These include L-Jetronic, LU-Jetronic, LH-Jetronic and pulsed Motronic systems. For more information on how these systems work, see chapter 3, as well as the general discussion of fuel injection in chapters 1 and 2. Unless otherwise indicated, all service procedures are intended to apply to all pulsed systems covered by this book.

These systems are not mysterious. When troubleshooting is approached logically and systematically, most problems can be easily fixed. However, fuel injection is not serviced as carburetors once were; it does not respond to simple adjustments with a screwdriver while a practiced ear listens for the smooth-idle point. You must understand the function of each component, how to isolate a problem, and how to correct it. That is nothing more than standard troubleshooting.

There are many tests in this chapter. You'll make some when the engine is warm, and others when the engine is cold. In most systems there are ways to send hot or cold engine-temperature signals to fool the fuel-injection control unit. But if you are testing the system "as is", remember that "engine cold" means shop temperature; the engine has not run for several hours. "Engine warm" means normal operating temperature; the upper radiator hose is too hot to hold. "Engine cold-cold" means near freezing; the car has been outside in winter weather for several hours.

Fig. 1-1. Though engine compartment may look intimidating, you can troubleshoot and repair fuel injection problems if they are approached logically.

1.1 Tools

In addition to a basic selection of good quality tools for working on your car—wrenches, sockets, screwdrivers, pliers, and a timing light—some specialized tools will also be required.

The most important tool you can have is the factory repair manual for your car. It contains all the specifications, component locations and procedures for your specific model. When combined with the detailed testing procedures outlined in this book, you'll understand what you're doing, and why you're doing it.

Many of the electrical tests in this book call for the measurement of resistance (ohms) or voltage of sensitive electronic components. A digital, high input impedance voltmeter will register millivolts and milliamps, and will not overload electronic components.

A test light is also needed to measure current. For component protection it should not have an incandescent bulb, but should be of the high input impedence type, such as an LED (Light Emitting Diode) test light.

A fuel pressure gauge, such as the one shown in Fig. 1-2, is required for fuel pressure tests. And finally, access to an exhaust-gas analyzer (or CO meter) is required for some tests, and for others (though not required), it will make the tests or adjustments more accurate.

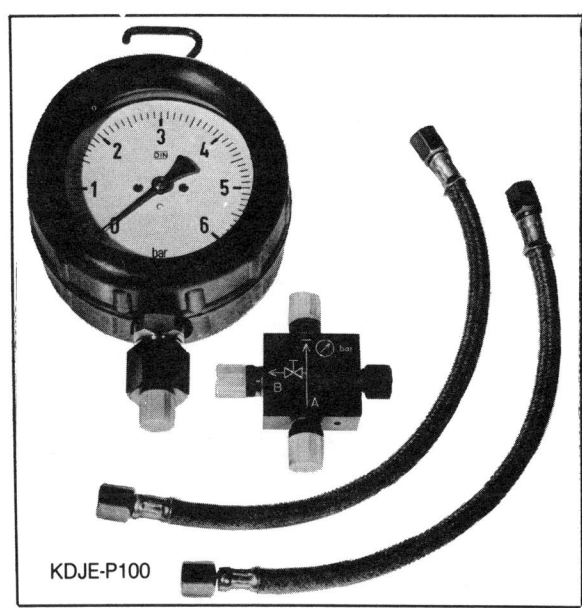

KDJE-P100

Fig. 1-2. Bosch fuel pressure gauge. Bosch gauge reads to 6 bar (87 psi), but for pulsed systems, any gauge that reads to 4 bar (60 psi) is fine.

1.2 Safety

If you haven't already, read the general Warnings at the beginning of this book, and follow basic safety rules, as well as those specific to fuel injection.

WARNING ——

●*Gasoline fuel is one of the most concentrated sources of energy around. Keep any spilled fuel away from hot engine parts, and confine the fuel spray during any injector testing or opening of fuel lines. Do not smoke or create sparks when fuel is present, and always have a fire extinguisher handy. Work in an area that is well ventilated.*

● *Fuel injection systems operate under pressures much higher than other fuel systems. See Fig. 1-3. Confine the fuel spray during any injector testing or opening of fuel lines to minimize the chance of a fire.*

● *Remove jewelry, metal watches, and watch-bands; if one of those shorts a circuit, you may wear the scar the rest of your life.*

● *An engine has the power to crush you. If you run the engine of a car with an automatic transmission while testing, do not trust your life to the PARK position of the lever. Set the parking brake and chock the drive wheels. Avoid working in front of the bumper whenever possible.*

Working Practices

For many tests, such as compression checks or cylinder balance checks, either disable the fuel injection system or keep the duration of the test short. Cranking a fuel-injected engine without starting it can deliver raw fuel into the cylinders, and from there into the exhaust system. On cars with catalytic converters the converter may overheat and melt down when the engine is restarted.

To prevent straining or twisting fuel lines when disconnecting them, use two wrenches as shown in Fig. 1-4. Hold one hex fitting with one wrench while loosening or tightening the other fitting with the other wrench.

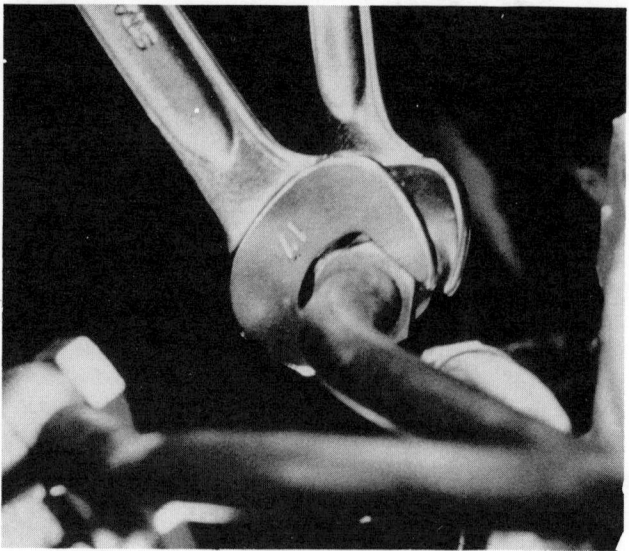

Fig. 1-4. Correct way to open fuel injection lines. To prevent damaging lines, use two wrenches. Clean fuel line union before opening.

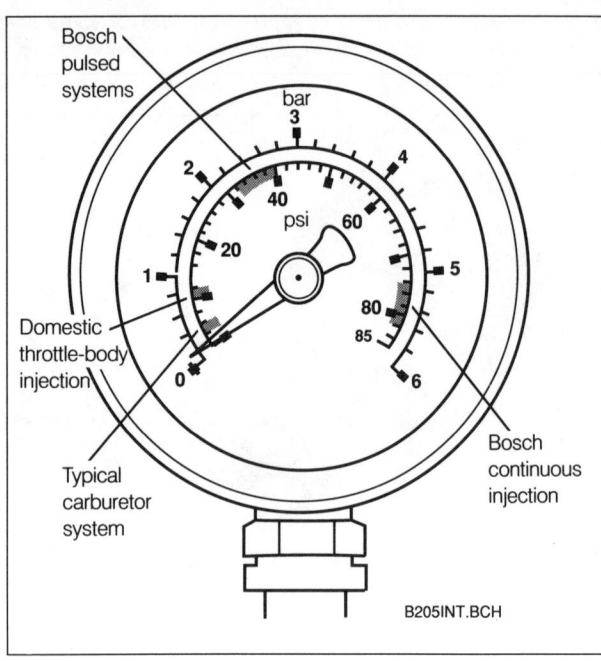

Fig. 1-3. Bosch fuel injection systems operate at high pressures.

Always use new gaskets, O-rings and seals when reconnecting lines or installing components. Many of these seals are designed to crush on tightening. If a crushed seal is reused, it may leak immediately or worse, it may develop a leak later as you drive.

After any work on the fuel-injection system, check idle rpm and the air-fuel mixture as described in **6.Idle Speed** and **7. Mixture (CO) and Lambda Control**.

Avoid excess voltage or voltage spikes to the control unit. Most fuel injection systems since 1980 are protected from surge and overvoltage, but watch for the following conditions which may damage any system:

● Check for disconnected or loose battery connections. Alternator output goes up as sensed battery voltage goes down, and an open battery circuit will cause the alternator to deliver excess voltage that could damage the control unit, as well as the car's wiring harnesses.

1.3 General Precautions

To ensure that you don't damage the engine or fuel injection system inadvertently, follow these general practices whenever working on the car.

Cleanliness

Dirt is the first enemy of fuel injection systems. Even minute particles can clog the small orifices of the components. Before you open a fuel fitting, wipe it clean with a solvent.

If the system is open, avoid using compressed air and don't move the car unless necessary. If you leave the job unfinished, cover removed parts and system openings with plastic, not cloth. When installing new parts, unwrap them just before installation.

● Avoid the use of high-voltage battery boosters or chargers. Anything greater than 16 volts is potentially harmful. Be aware that some service stations use 24-volt boosters to turn over engines in sub-zero weather.

● Do not disconnect a booster or charger with the engine running with a minimal electrical load. This can cause a voltage spike. This may seem strange if you've just managed to start a cold, dead engine, but add load for the moment it takes to disconnect the cables by turning on the lights and blower or rear-window heater to load the boosted-car electrical system.

● Do not disconnect or reconnect the wiring-harness connector to the control unit with ignition ON. This can send a damaging voltage spike through the control unit.

2. TROUBLESHOOTING

This four-part section covers the troubleshooting of pulsed fuel injection systems, based on observed problem symptoms that may be the result of a fault with fuel injection.

The first part of the section is a troubleshooting table that will help you to narrow down your tests to specific components or areas of the system. The three other parts of this section are general troubleshooting procedures which are not component-specific, but which are fairly easy to perform. In most troubleshooting, performing these simple checks first may save you any further testing of the system.

It is not possible to directly test the electronic control unit without highly specialized equipment. If you troubleshoot all other areas of the system and they check out ok, only then should you suspect that there may be a fault in the control unit.

2.1 Troubleshooting Basics

The basic function of the fuel injection system is to supply and meter the correct amount of fuel to the engine in proportion to the amount of air being drawn into the engine to achieve the optimum air-fuel mixture. Any problems with electrical connections, air intake sensing, or fuel supply will cause poor running. Any troubleshooting should begin with simple and easy checks of the tightness of system wiring and the integrity of the air intake system. Proceed from there to more involved troubleshooting.

Generally, fuel injection problems fall into one of four symptom categories: cold start, cold running, warm running, and hot start. Warm running is the most basic condition. Before troubleshooting a condition in any other category, make sure that the system is working well and is properly adjusted for warm running.

To simplify troubleshooting concentrate on the sensors and components that adapt fuel metering for a particular condition. For example, if the engine will not start when cold, the components responsible for cold start enrichment are most likely at fault, and should be tested first.

Table a lists symptoms of Bosch pulsed fuel injection problems, their probable causes, and suggested corrective actions. The boldface numbers in the corrective action column indicate headings in this chapter of the book where the test or repair procedures can be found.

Table a. Troubleshooting Bosch Pulsed Fuel Injection Systems

Symptom	Probable cause	Corrective action
1. Cold start-Engine starts hard or fails to start when cold	**a.** Cold start system faulty **b.** Fuel pump not running **c.** Engine temperature sensor faulty **d.** Fuel pressure incorrect **e.** Air-flow meter faulty **f.** Fuel injectors faulty or clogged	**a.** Test cold start valve and thermo-time switch, or cold start system. **4.** **b.** Check fuel pump fuse and pump relay. **2.3** **c.** Test sensor. **4.** **d.** Test fuel pressure. **3.2** **e.** Test air-flow meter. **5.1, 5.2** **f.** Test fuel injectors. **3.4**
2. Hot start-Engine starts hard or fails to start when warm	**a.** Insufficient residual fuel pressure **b.** Cold start valve leaking or operating continuously **c.** Fuel pressure incorrect **d.** Air-flow meter faulty	**a.** Test residual pressure. **3.2** **b.** Test cold start valve and thermo-time switch. **4.1** **c.** Test fuel pressure. **3.2** **d.** Test air-flow meter. **5.1, 5.2**
3. Engine stalls or idles roughly (cold or warm)	**a.** Vacuum (intake air) leak **b.** Idle system faulty	**a.** Check for leaks. **2.4** **b.** Test auxiliary air regulator or idle speed stabilizer. **6.1, 6.2**

continued on next page

Table a. Troubleshooting Bosch Pulsed Fuel Injection Systems

Symptom	Probable cause	Corrective action
3. Engine stalls or idles roughly (cold or warm) cont'd.	**c.** Engine temperature sensor faulty **d.** Air-flow meter faulty **e.** Cold start system faulty **f.** Fuel injectors faulty or clogged **g.** Insufficient fuel pressure or fuel delivery	**c.** Test sensor. **4.** **d.** Test air-flow meter. **5.1, 5.2** **e.** Test cold start valve and thermo-time switch, or cold start system. **4.** **f.** Test fuel injectors. **3.4** **g.** Test fuel pressure and delivery. **3.2, 3.3**
4. Engine idles too fast	**a.** Vacuum (intake air) leak **b.** Idle system faulty	**a.** Check for leaks. **2.4** **b.** Test auxiliary air regulator or idle speed stabilizer. **6.1, 6.2**
5. Engine misses, hesitates, or stalls under load	**a.** Throttle switch faulty or misadjusted **b.** Fuel injectors faulty or clogged **c.** Vacuum (intake air) leak **d.** Air-flow meter faulty **e.** Insufficient fuel pressure or fuel delivery	**a.** Test throttle switch. **5.3** **b.** Test fuel injectors. **3.4** **c.** Check for leaks. **2.4** **d.** Test air-flow meter. **5.1, 5.2** **e.** Test fuel pressure and delivery. **3.2, 3.3**
6. Low power	**a.** Throttle valve not opening fully **b.** Insufficient fuel delivery **c.** Throttle switch faulty or misadjusted **d.** Air-flow meter faulty **e.** Ignition system faulty (Motronic systems only)	**a.** Check throttle valve adjustment. **5.3** **b.** Test fuel delivery. **3.3** **c.** Test throttle switch. **5.3** **d.** Test air-flow meter. **5.1, 5.2** **e.** Check ignition timing and knock sensor control. **8.**
7. Failed emissions text	**a.** Faulty lambda sensor or electronic control unit **b.** Engine running rich	**a.** Test lambda control system. **7.2** **b.** Check air-fuel mixture. **7.1**

2.2 Engine Condition

The fuel injection system is set to operate on an engine that is in good operating condition. Because the fuel injection system is often the "new item", some people waste time checking it when the trouble may be with basic engine operation. It is a good idea to use the car manufacturer's shop manual to perform a tune-up, and to check the following systems before tackling the fuel injection system.

1. Ignition system. Check timing, including advance and retard control, ignition components and spark quality.

2. Electrical system. Check battery condition and connections, and alternator and voltage regulator.

   NOTE ━━

   Voltage spikes caused by bad alternator diodes or a faulty regulator can fool the fuel injection control unit and cause an engine miss. Check this by disconnecting the wire from the alternator to the battery with engine stopped, and then start the engine. If the missing stops, then perform a complete charging system test.

3. Air intake system. Check the air filter, PCV and crankcase connections, and the evaporative emission connections. A loose oil-filler cap or dipstick can lean the mixture by admitting extra air to the intake manifold through the PCV valve. The vapor canister can enrich the mixture by admitting fuel vapors into the intake manifold through the canister purge valve.

4. Fuel system. A clogged filter may reduce fuel flow. A faulty fuel filler cap or tank vent valve may create gas tank vacuum and reduce fuel flow. Be sure the car has fuel in the tank.

5. Mechanical operation. Check grade and condition of crankcase oil, compression, valve timing, and the exhaust system.

2.3 Electrical System

Whenever working on the wiring, take care to avoid bending any pins or connectors. Use flat pin probes if possible. Inserting the probes of a volt or ohmmeter too far into a wiring connector may spread the contacts and create a new problem.

Relay Set and System Power

A faulty relay set may prevent the fuel pump from operating, or prevent power from reaching the control unit. Remove the connector from the relay set, and with the ignition on, check for voltage at the terminals of the connector as shown in Fig. 2-1. If there is voltage, then the relay set is probably faulty, but check the continuity of the wiring to the pump and control unit as described below just to be sure.

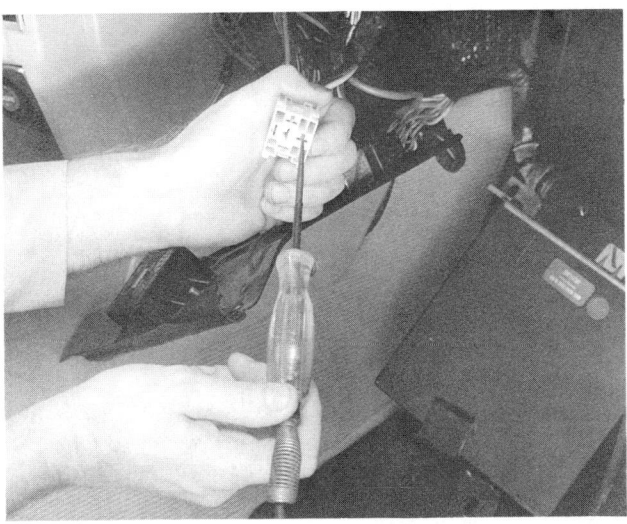

Fig. 2-1. Power supply at relay set wiring connector being checked. Check your shop manual for relay set location and terminals.

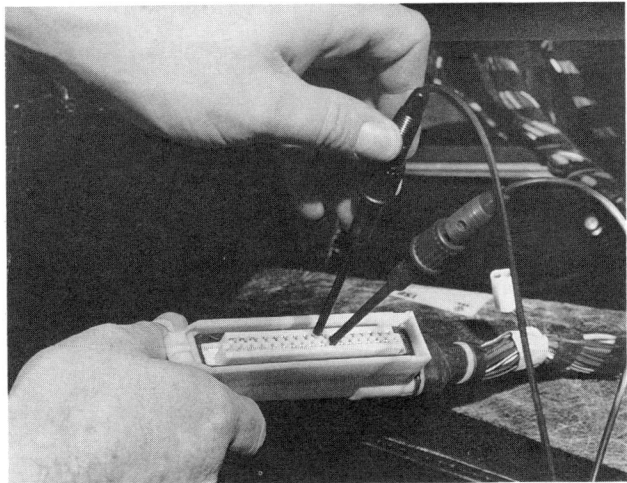

Fig. 2-2. Making electrical resistance tests of system components at the control unit connector checks wiring continuity as well as components. Be sure to use flat probes shown to avoid bending connector terminals.

Wiring Harness, Connections and Grounds

Strange as it may seem, the components of pulsed systems usually give less trouble than the wiring harnesses and connectors that link them. Even small amounts of corrosion or oxidation at the connector terminals can interfere with the small milliamp currents that signal the system to operate. This problem is compounded by the several ground paths provided to insure reliable operation. More than one owner of a "bad" control unit or component has paid for replacement when the problem was in the wiring. In many cases, cleaning connectors and grounds may solve fuel injection problems.

Identify all wiring connectors and ground locations using the shop manual. The ground locations should be secure and free from corrosion or grease and oil. With the ignition off, disconnect the wiring connectors, including the control unit connector, and also check them for corrosion or dirt. Simply disconnecting and reconnecting the connectors will clean up the contacts, but you can also use a contact cleaner designed for electronic components.

Don't forget to check for breaks or shorts in the wiring harness. A fuel injector or temperature sensor may be good, but the wiring to the control unit may be faulty. You can check this using an ohmmeter. With the ignition off, disconnect the control unit connector and test for continuity between the component terminals and the corresponding terminals at the control unit connector. A reading of zero ohms or very close to it indicates that the wiring is fine.

Some electrical tests of the components can be combined with tests of the wiring harness by removing the control unit connector and then testing between the two pins that lead to a component as shown in Fig. 2-2.

2.4 Intake Air Leakage

Another likely cause of trouble in pulsed systems is air leaks between the air-flow sensor and the intake valves. These leaks are often called "false air" because it is air that has not been measured by the air-flow sensor. As a result, the control unit may not provide fuel to burn with the excess air, leading to a lean condition and driveability or emission problems. These are often indicated by hesitation when the engine is cold, or surging at idle. Note that on systems with adaptive control, false air is rarely a problem unless the leak is very large.

False Air Checks

If you suspect an air intake leak, there are many possible sources, as shown in Fig. 2-3. The soft rubber ductwork can crack and split with age and underhood heat. Check clamps for tightness. Don't forget the vacuum hoses to the brake booster, the fuel pressure regulator, the evaporative fuel control system, and other places, such as vacuum diaphragms in the heater system inside the car. Check the intake manifold connection to the cylinder head, fuel injector seals, the auxiliary air valve, and the EGR valve, if fitted. Check openings to the crankcase, such as the dipstick, PCV valve, and oil-filler cap for air-tight fits. Check anything downstream of the air-flow sensor that could leak air into the system.

Check for leaks by pressurizing the intake system with air and then spraying a leak detector on the suspected area. An air hose inserted into the fitting for the auxiliary air valve or the idle speed stabilizer can be used to apply low pressure; only about 0.3 bar (5 psi) is needed. A spray bottle of soapy water can serve as a leak detector. Block the throttle so that it's open. Any bubbles will indicate a leak. Also listen for the sound of escaping air. See Fig. 2-4. You may have to plug the air-flow sensor intake and the exhaust tailpipe to hold enough pressure in the system.

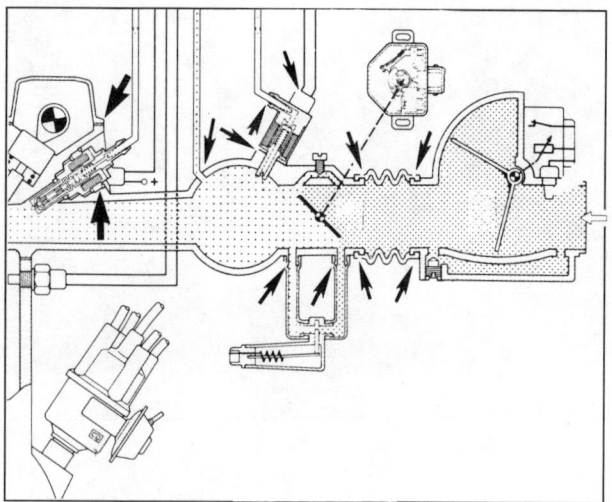

Fig. 2-3. Air intake leaks or "false air" can come from many sources (**arrows**) and cause driveability problems. In addition to rubber ductwork and hoses, check other areas such as fuel injector seals and engine valve cover.

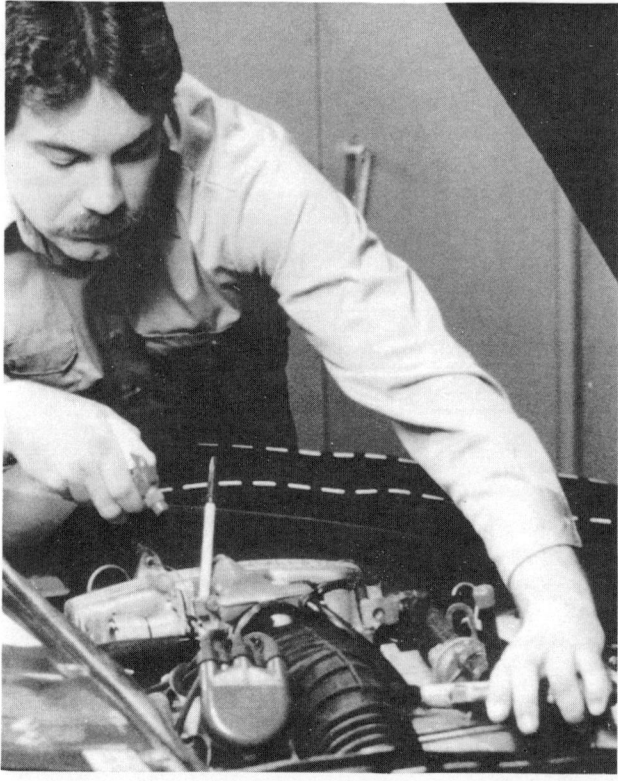

Fig. 2-4. For simple test for vacuum leaks use spray bottle of soapy water and air pressure hose. Spray suspected area with solution. Bubbles indicate a leak.

NOTE ——

On systems with an air-mass sensor, it is necessary to block the sensor inlet to hold any pressure in the system. A styrofoam coffee cup clamped in the air-mass sensor inlet makes a handy temporary plug.

An alternate leak-detection method involves squirting solvent around suspected leak locations with the engine idling. An rpm increase indicates a leak. Be sure to use an approved solvent with a high-temperature flashpoint. You can also use propane, which tends to be drawn in better than solvent.

Remember though, if the engine has an idle-speed stabilizer, it will mask any rpm increase. If you can sample CO in the exhaust ahead of the catalytic converter, look for CO increase some seconds after applying the solvent as an indication of an air leak.

To correct a leak, start by tightening clamps; you might have to replace the hose or ducting, or replace a gasket. Remember, it's the small intake air leaks that cause trouble, where the engine runs, but poorly. If there's a big air leak, the engine probably won't run at all.

3. FUEL SUPPLY

Incorrect fuel pressure or poor fuel delivery can cause:

- hard starting
- rough idle
- poor mileage
- limited maximum rpm
- incorrect idle CO
- and other emission problems.

If problems with the fuel system are suspected, begin testing by making sure that the pump runs when the ignition is turned on. If not, check the wiring and relay set as described in **2.3 Electrical System** and make sure the relay is receiving the signal indicating the engine is turning over. On some models this comes from the air-flow sensor, on others it comes from the ignition system. Continue to check basic system pressures, fuel delivery, and operation of the fuel injectors.

3.1 General Information

All of the pressure tests and many other tests of the fuel injection system, as well as component service and replacement, require either one or all of the following procedures: the relieving of fuel pressure, the installation of a pressure gauge, and the operation of the fuel pump without running the engine.

Relieving Fuel Pressure

The fuel system is under pressure even when the engine isn't running. The fuel pump check valve holds pressure in the system for many hours after the engine is turned off. To prevent fuel from spraying over yourself and the engine when opening a fuel line, there are three good ways to relieve pressure in the lines.

- Neat: pull the fuel pump fuse, then crank/run the engine until it dies

- Quick: wrap the fuel fitting in a shop cloth and loosen it, but it can be tricky and messy manipulating two wrenches, the fuel line, and the cloth

- Fancy: connect a hand vacuum-pump at the pressure regulator; when you pump vacuum, the regulator will dump pressure to the fuel tank

Always remember to remove the fuel-tank cap to relieve pressure in the tank. Pressure in the tank can be enough to squirt fuel from the lines after they're open.

Fig. 3-1. Always remove fuel tank cap before opening fuel line. Tank pressure can force fuel out open line.

Installing Fuel Pressure Gauge

After fuel pressure is relieved, install the fuel gauge. The best place to install the gauge is at the special service port on the fuel rail, as shown in Fig. 3-2. If the system doesn't have a service port, the gauge can be connected to the cold start valve supply line or, with a T-adaptor, to any other place in the high pressure side of the fuel supply, such as the pressure regulator line.

Operating Fuel Pump for Tests

There are two basic methods to jump the circuit and operate the pump, depending on the type of air-flow sensor installed in the system. Use the wiring diagram in the car shop manual to identify the pump circuit and the correct terminals for your model.

The first method is for air-flow sensors with a 6 or 7-pin connector plug. This is generally on 1974-1978 models. Jump the safety circuit at the air-flow sensor as shown in Fig. 3-3, using a fused jumper wire and a switch. Turn on the switch to operate the pump.

Fig. 3-2. Fuel pressure gauge connected to service port on Porsche 944 fuel rail. Relieve system fuel pressure before disconnecting lines or opening service port.

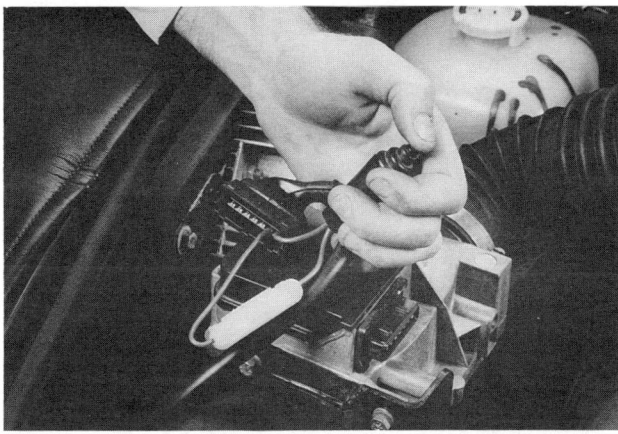

Fig. 3-3. Operating fuel pump without running engine on models with 6-or 7-pin air-flow sensor connector plug. Consult car shop manual for correct terminals.

The second method applies to cars with a 5-pin air-flow sensor connector plug. These are generally post-1978 cars. Jump the safety circuit with a fused wire at the relay set as shown in Fig. 3-4. Consult the wiring diagram in the car shop manual for the correct terminals. Turn the ignition to ON to run the pump.

There is a third method, where pushing the air-flow sensor air flap open with the ignition ON will run the pump, but that only works on early L-Jetronic systems.

4

Fig. 3-4. Operating fuel pump without running engine on models with 5-pin air-flow sensor connector plug. Consult car shop manual for correct terminals.

3.2 Pressure Tests

Fuel pressure which is too high may richen the air-fuel mixture, while fuel pressure too low may lean the mixture. The following procedures test for basic system pressures, as well as for causes of incorrect pressures. For the tests, relieve system pressure and install a pressure gauge as described in **3.1 General Information**.

System Pressure

There are two parts to testing system pressure in order to check pressure regulator function. In the first part, with the engine idling, the gauge should typically read about 2 bar (29 psi) as shown in Fig. 3-5. Check your shop manual for the correct specifications for your model. Many specifications may read 2.5 bar (36 psi), with a tolerance from 2.3 to 2.7—but remember, that's when testing with the pump running with the engine off. Here, the test checks regulation of fuel pressure by the fuel pressure regulator.

Fig. 3-5. Normal operating fuel pressure with engine idling and fuel pressure regulator connected is 2 bar (29 psi). Check your shop manual for correct specifications.

For the second part of the test, with the engine still idling disconnect the vacuum line to the intake manifold and close it off. Because the pressure regulator senses higher barometric air pressure than manifold pressure, fuel pressure should rise to about 2.5 bar (36 psi) as shown in Fig. 3-6. This most likely is what your shop manual specifies.

Fig. 3-6. Normal operating fuel pressure with engine idling and fuel pressure regulator disconnected is 2.5 bar (36 psi).

Pressures that are too low or too high indicate a problem either with the pressure regulator or with fuel delivery. Check the pressure regulator first as described below, before checking fuel delivery.

Checking Fuel Pressure Regulator

If system pressure tests do not show the 0.5 bar pressure drop with the vacuum line connected, check the vacuum line for leaks. If the line is sound, the regulator is faulty.

If system pressure is low in both parts of the test, the regulator could be returning too much fuel or the fuel pump may not be delivering properly. You can pinpoint the problem by pinching off the return line as shown in Fig. 3-7. Do this slowly to avoid a sudden pressure surge that could ruin the gauge. If the gauge shows 4 bar (59 psi) or above, the fuel-pump relief valve is working, so the regulator is faulty. If the pressure did not rise, there is a problem with fuel delivery, which should be checked as described in **3.3 Fuel Delivery**.

If system pressures are too high, temporarily remove the fuel pressure regulator return line. Attach a short hose to the pressure regulator outlet and direct it into an unbreakable container. Run the system pressure test again. If pressures are now correct, then the fuel return line is blocked. If pressures are still high, the regulator is faulty.

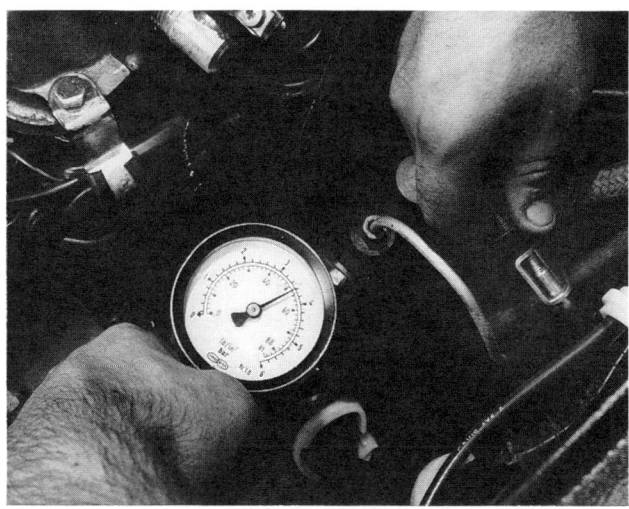

Fig. 3-7. Fuel pressure regulator return line being clamped to check regulator function.

Residual Pressure Test

With the fuel pressure gauge installed, run the engine or fuel pump briefly to build up system pressure, then shut off the engine. After 20 minutes, pressure in the system should not have fallen below 1 bar (14.5 psi) as shown in Fig. 3-8.

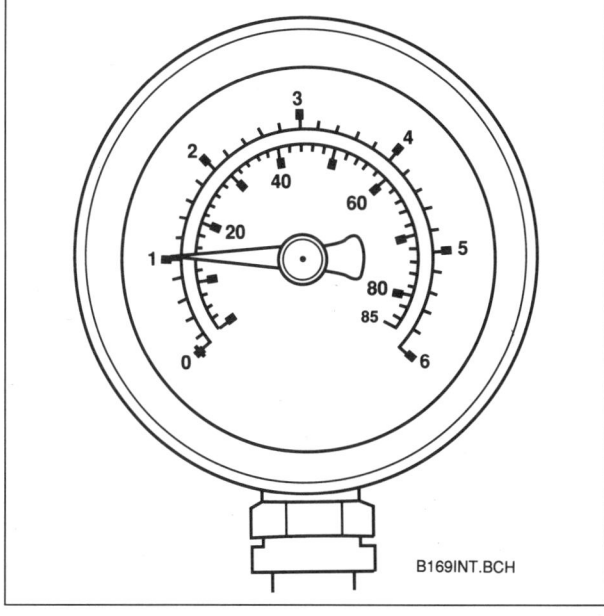

B169INT.BCH

Fig. 3-8. Residual pressure should not drop below 1 bar (14.5 psi).

If it has, there are many possible sources of leaks that could reduce pressure. Check all fuel line connections for leaks. Make sure the fuel injectors and cold start valve are not leaking. Test the fuel pump check valve by running the engine and then shutting it off. Immediately clamp shut the supply line from the fuel pump. If the pressure drops below specification, then the pressure regulator is faulty. If the pressure is now within specification, then the fuel pump check valve is faulty.

3.3 Fuel Delivery

For the fuel system to deliver sufficient fuel to the injectors, the fuel lines must be clear of blockages and the fuel pump must be able to pump a specified volume of fuel. The fuel delivery tests assume that the fuel pressure regulator has been tested and is working properly.

Checking Fuel Delivery

To check fuel delivery, disconnect the return line from the fuel pressure regulator. Run the pump without running the engine as described in **3.1 General Information** and catch fuel in an unbreakable container as shown in Fig. 3-9. Be sure the container is large enough. Most manufacturers specify a delivery of one liter in 30 seconds or less, but check your manual. You can either deliver fuel to the specified measure, shut off the pump and check the time, or run the pump for the specified time period and see how much has been delivered.

Fig. 3-9. Checking fuel-pump delivery volume.

If delivery is not to specification, replace the fuel filter, and blow compressed air through the supply line to be sure it is open. Check the fuel pump as described below. Whatever you repair or replace, recheck fuel pump delivery, even if a new pump is installed.

NOTE ▬

Wet the interior of the fuel filter with gasoline before installation. This will help prevent possible tearing of the filter element when the pump is first started.

Checking Fuel Pump

You can check the fuel pump for voltage and ground as shown in Fig. 3-10. With the pump running, you should see close to 12 volts at the positive terminal. The negative terminal should show a good ground, zero resistance. If not, clean the

Fuel Supply

terminals and check the wiring. If voltage supply and ground are fine, but delivery volume is still low, the pump must be replaced.

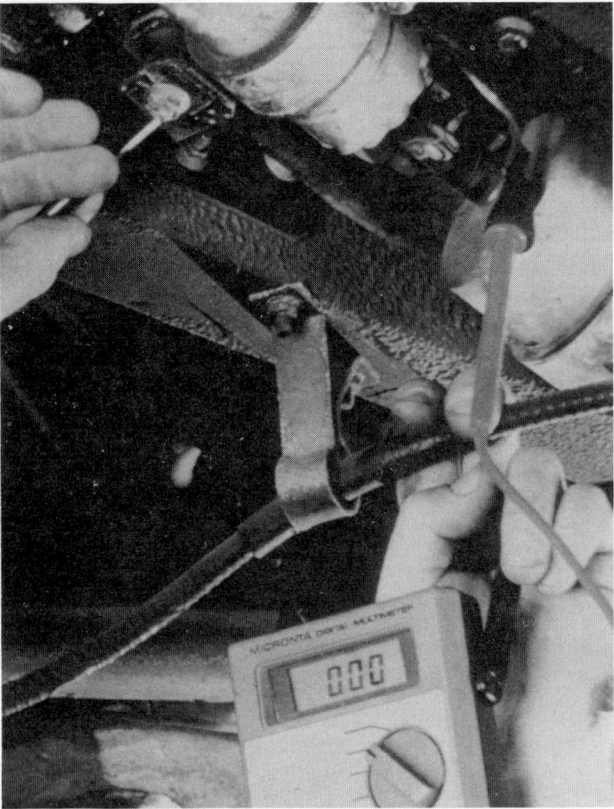

Fig. 3-10. Fuel pump being checked for voltage and ground at the pump. Peel back protective wiring boot for access to pump terminals.

3.4 Fuel Injectors

There are a number of methods to test check the operation of the fuel injectors. In addition to those listed here, don't forget to check the wiring to the control unit. Also note that while the injectors may appear to be operating correctly, even a small amount of injector clogging can affect engine performance.

> *CAUTION* —
>
> *Do not apply voltage to the fuel injectors in an attempt to test them. Excessive voltage will burn out the injectors.*

Vibration Test

A quick method is to check for injector vibration—indicating that they are opening and closing—while the engine is idling. If they're too hot to touch with your fingertips, use a mechanic's stethoscope or a screwdriver as shown in Fig. 3-11. You should hear a buzzing or clicking sound. No vibration, or a different pitch of vibration in one versus the others, indicates a bad injector or harness connection. Interchange connectors with an

injector on the same circuit from the control unit (check your wiring diagrams). If the same injector is still faulty, then replace the injector; if the injector now works, check the wiring.

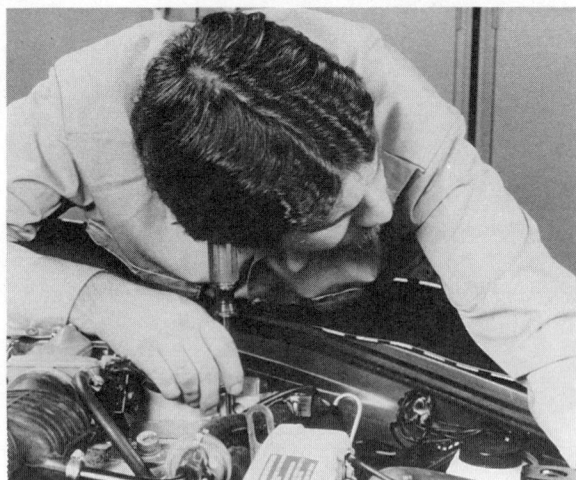

Fig. 3-11. Pulsed fuel injector operation can be checked by placing screwdriver tip against injector body and listening for clicking sound.

RPM Drop Test

With the engine idling, check rpm drop as you disconnect the injector connectors, one at a time. For example, if you read 860 rpm with all four injectors operating then you disconnect an injector, look for a drop to about 770 rpm, and a return to normal when you reconnect it. If there's no rpm drop, either the injector or its wiring is faulty. If possible, replace the injector first to narrow the cause.

If the engine has an idle speed stabilizer, the rpm drop may not be noticeable unless you freeze the idle speed at a fixed rpm before the test. Depending on the car, this can be done by grounding the appropriate test connector, as shown in Fig. 3-12, or on other cars by simply pulling the idle speed stabilizer wiring connector. Check your shop manual for more information.

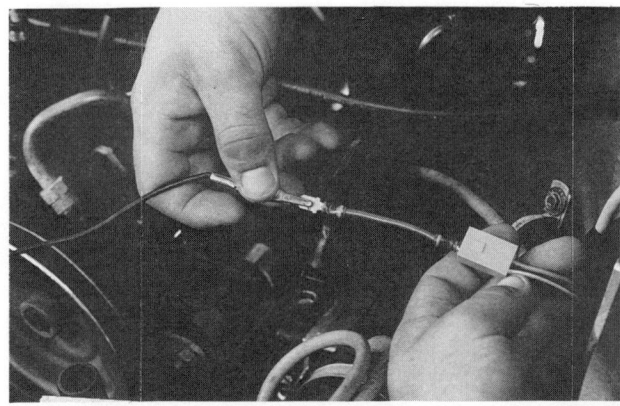

Fig. 3-12. Test connector being grounded with jumper wire to stabilize idle on cars with idle speed stabilizer. Check your shop manual for correct connector.

Injector Leak Tests

Fuel injector leaks can occur at the seams around the body of the fuel injector and bleed off fuel pressure, causing hot-start problems. Clean off the injector and look closely for seepage. The injectors usually leak most when they are cold.

The injector pintles should also be checked for leakage. Remove the injectors and the fuel rail with the injectors still attached. Run the fuel pump without running the engine to build up fuel pressure. If any of the injectors leak at the rate of more than two drops per minute, the injector may be clogged. Clean the injectors as described below. If that does not help, replace the injector.

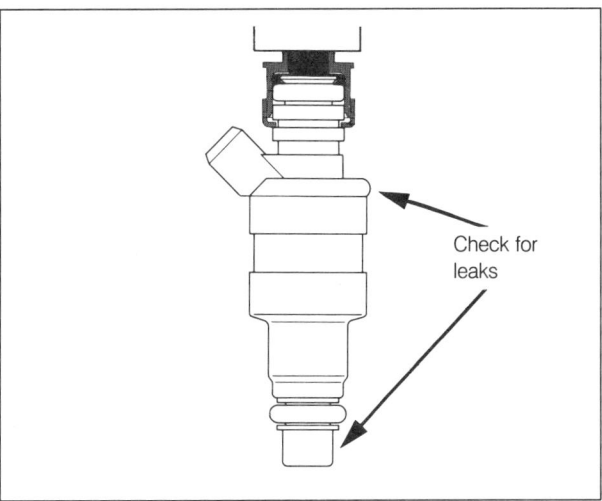

Check for leaks

Fig. 3-13. Check injector for any leakage.

Electrical Tests

Remove the harness connector from each injector as shown in Fig. 3-14 and use an ohmmeter to test the resistance across the injector terminals. The resistance depends on whether there is a separate set of series resistors in the system or not. Check your manual for the correct specifications. If the resistance is incorrect, replace the injector.

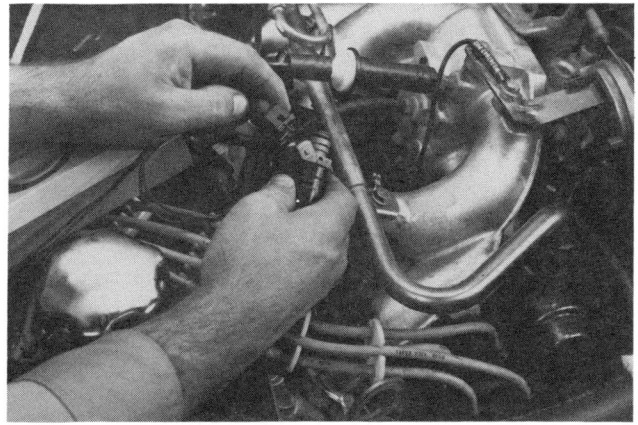

Fig. 3-14. Harness connector being removed from injector.

On models with an additional set of series resistors in the fuel injection wiring harness, the resistance of each series resistor should be from 5 to 7 ohms. With the ignition on, there should be from 11 to 12.5 volts present from terminals 43/1 and 43/2 of the resistors to ground on the chassis. Again, check your manual.

Fuel Injector Clogging

A clogged fuel injector can be indicated by rough idle, a stumble or hesitation during acceleration, or a failed emissions test. Fuel injector clogging is caused by a buildup of carbon and other deposits on the injector pintle. This reduces the flow of gasoline through the injectors and results in a poor spray pattern. See Fig. 3-15.

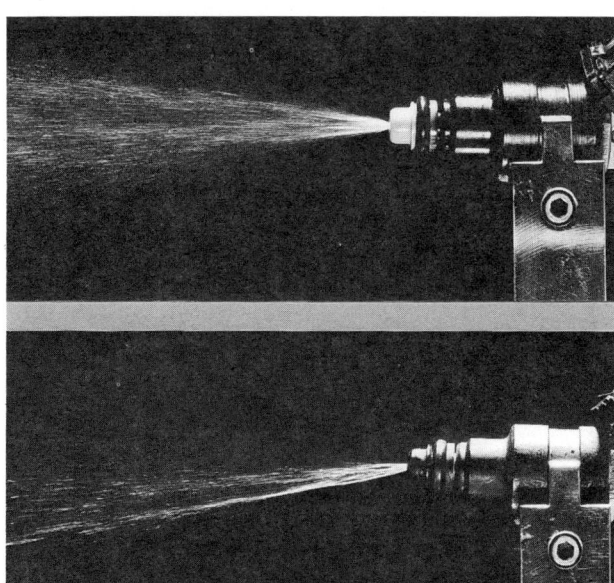

Fig. 3-15. Effect of fuel injector clogging. While new injector (top) has good spray pattern, clogged injector (bottom) has poor pattern and flow rate of 50% less.

Causes of Clogging

Clogging is the result of the combination of a number of factors:

● high underhood temperatures on smaller cars

● fuel being metered at the tip of the injector

● short driving cycles followed by hot-soak periods

● low-detergent fuels with a high carbon content and low hydrogen content

The worst clogging seems to occur with driving cycles where the car is driven for at least 15 minutes, ensuring full warm-up, then parking for about 45 minutes or more. While the

4

Fuel Supply

engine runs, the injector tips are cooled by the fuel flow. After shut-down the engine acts as a heat sink and temperatures climb, particularly at the valves and manifolds. Injector-tip temperature climbs equally high, and the small amount of fuel that is in the tip of the injector breaks down and causes the deposit. Considering the small quantities of fuel that the injector meters, and the tiny orifice of the injector tip, it doesn't take much to restrict the flow, as shown in Fig. 3-16.

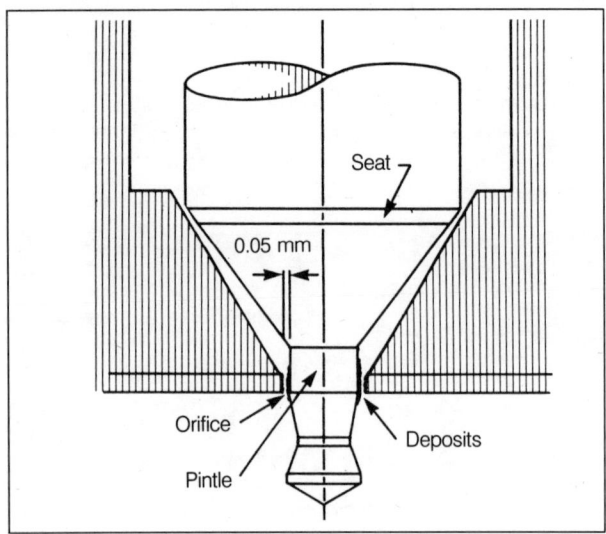

Fig. 3-16. Enlarged view of fuel injector tip. Small injector tip orifice means that very small amounts of deposits will affect engine performance.

Normally, one injector clogs before the others, reducing its delivery so that the cylinder runs lean. The lambda sensor compensates by enriching the mixture for all cylinders, which is in turn too rich for the cylinders with unclogged injectors. The result is a rough idle. The engine will most likely fail an emission test, and will send you to the gas pump more often, because it can lose as much as 25% fuel economy.

Solving the Clogging Problem

The first step towards solving the problem of fuel injector clogging is to determine whether one or more injectors is indeed clogged. The injector leak test described above gives one indication of a possible clogged injector. There are also a number of tests that any good repair shop can perform. One is called a pressure-balance test, where each injector is triggered with a fixed pulse and the pressure drop in the system is measured. If one has a greater pressure drop, it's most likely clogged.

The tendency to form injector deposits varies considerably depending on the fuel. Many cases of injector clogging can be cured by using premium fuels where manufacturers advertise more detergent additive. Most regular unleaded fuels probably have enough detergent to keep unclogged injectors clean, but they won't dissolve deposits on clogged injectors.

There are a number of special kits used by repair shops to flush out deposits. Also, pouring a separate additive in your gas tank can often clean injectors in a short time as shown in Fig. 3-17, but in some cases this may free other deposits that can clog the fuel system. Be aware that some gasoline additives that cure clogging can cause carbon deposits on the intake valves. These fluffy deposits absorb fuel and cause rough idle and hesitation, especially in cold running conditions. Check with your car manufacturer for a recommended gasoline or additive.

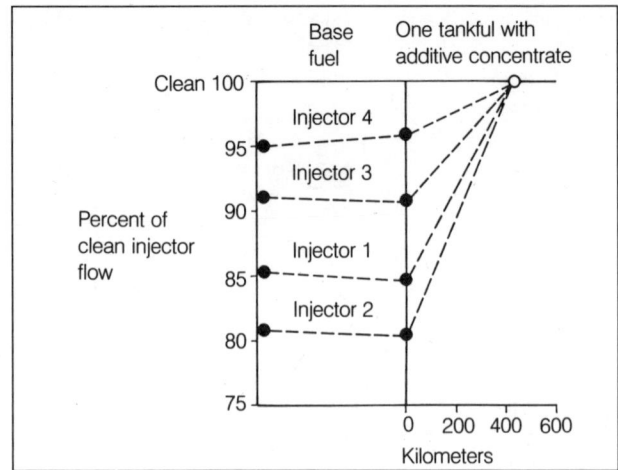

Fig. 3-17. Graph showing effect of adding fuel-injector-cleaning additive to one tank of gas.

Change your driving cycle if you still get clogged injectors with all fuels. Avoid the hot-soak after driving by adding a special fan that mounts in the engine compartment and cools the injectors after the engine is shut off. Both Audi and Nissan have such fan kits as stock parts. If you can't get good fuel, or if you want the best protection, it may be possible to install redesigned injectors. New-style Bosch injectors are designed with a chimney to help carry heat away from the injector tips as shown in Fig. 3-18. They are available as aftermarket replacement on some cars, so check with your car manufacturer's parts department for more information.

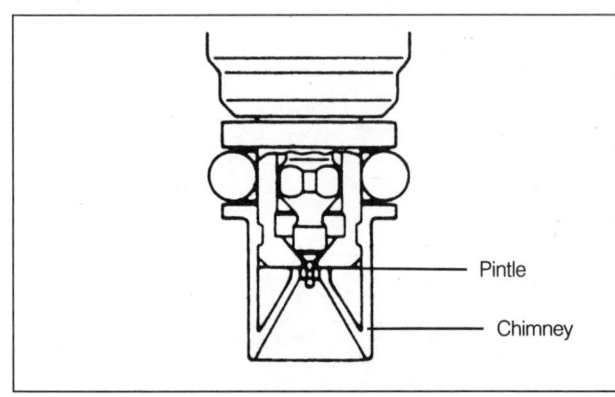

Fig. 3-18. New-style Bosch pulsed fuel injector showing chimney that reduces injector tip temperature.

Many cars may need to have their injectors cleaned on a regular basis, but with widespread distribution of fuels of increased quality, the great injector-clogging problem may pass into history.

4. START AND WARM-UP TESTING

Two types of starting systems are employed on Bosch pulsed fuel injection systems. The more common uses a thermo-time switch and cold start injector to supply additional fuel during cold starting. A second type, employed on some systems, does not have a cold-start injector or thermo-time switch. Instead, it has a start control system, which increases port injector pulse time for additional fuel.

Engine Temperature Sensor

A faulty engine temperature sensor can affect pulse-time signals to the injectors in all pulsed systems, leading to lean running when cold, or rich running when warm. Make resistance checks at the temperature sensor as shown in Fig. 4-1. See **Table b** for temperature sensor resistance values based on sensor temperature. For a more accurate test, remove the switch and cool or warm it in water as necessary.

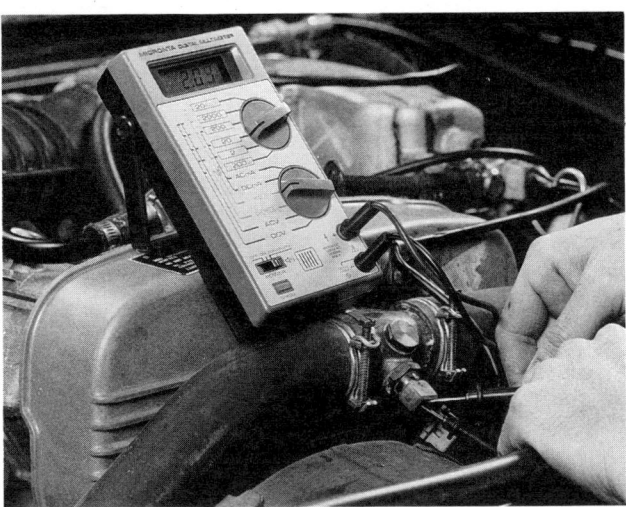

Fig. 4-1. Engine temperature sensor resistance values being checked. See **Table b** for correct values based on sensor temperature.

Table b. Engine Temperature Sensor Resistance Checks

Temperature	Resistance
68°F (20°C)	2000–3000 ohms
176°F (80°C)	250–400 ohms

Cold-Start Injector Tests

Locate the cold start injector on the intake manifold. Inspect it for any external leaks. Remove it and disconnect the wiring connector, but leave the fuel line connected. Use a spare wiring connector to fabricate a jumper wire. Attach the wiring connector to the cold-start injector, then attach one wire from the connector to ground and connect the other to the plus terminal of the ignition coil. Use an approved connector to minimize the chance of sparks. This wiring bypasses the thermo-time switch and the relay set.

WARNING ——
Performing these tests will spray gasoline and that's a fire hazard. Always keep a fire extinguisher handy.

Run the fuel pump without running the engine, as described in **3.1 General Information**. When the ignition is turned on to apply voltage, the cold-start injector should spray as shown in Fig. 4-2. If the spray is uneven, or if there is no spray, the injector is faulty and should be replaced. A few seconds of spray is enough. Turn off the ignition when done and stop the pump.

to coil + terminal

to ground on engine

Fig. 4-2. Cold-start injector being tested. When fuel system is pressurized, injector should spray when voltage is applied to terminals.

When the ignition is shut off, wipe the tip of the cold-start injector dry. Run the pump briefly again to build up fuel pressure. Without electrical power, the injector should not spray or drip. If it does, replace the cold-start injector.

Thermo-time Switch Test

The thermo-time switch can be checked for operation at the correct temperature, and for resistance values at the terminals. The coolant temperature at which the switch opens and the time it is open is stamped on the body of the switch.

Check thermo-time switch opening time with a test light installed in series with the wire to the thermo-time switch. After disabling the ignition according to your shop manual, operate the starter for a few seconds. With coolant at shop temperature, the test light should go on for about two seconds, then off. If the coolant were around freezing, it might go on for about five seconds. Checking the switch heats its bimetal strip, so do the test right the first time or else you'll have to wait for the switch to cool.

Check the resistance values between the two switch terminals and ground, and between the two terminals as shown in Fig. 4-3. The resistance should vary with thermo-time switch temperature. Your manual should have the correct values. The switch can be cooled by a blast of refrigerant, and warmed by a hair dryer, or it can be removed from the engine block and then heated and cooled in a pan of water for more precision.

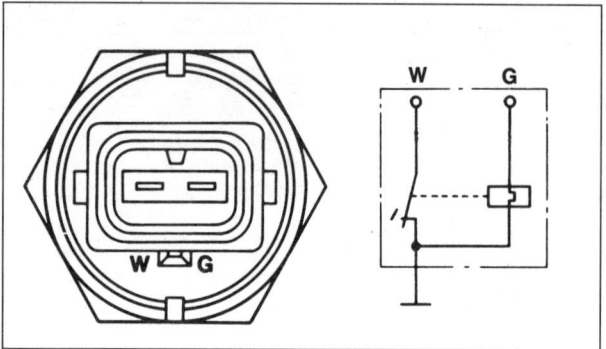

Fig. 4-3. Thermo-time switch, showing identification of wiring terminals and schematic of circuit.

Start Control System Test

The start control system is tested by measuring voltage at a port injector with the engine cranking. Begin by removing a wiring connector from any injector. Install a test lead so that you can measure voltage while the injector operates. See Fig. 4-4. To prevent raw fuel from being delivered into the cylinders, remove the fuses for the fuel pump, as well as for the pre-supply pump if the car has one. Also disable the ignition system as described in the car shop manual. It's best to do this test with the engine cold, but if the engine is warm, it's necessary to disconnect the engine-temperature sensor and install a special 10K-ohm temperature sensor to the connector to fool the system.

With the voltmeter connected across the injector leads, crank the engine. At a "cold" start, voltage drop should begin at about −2 volts, then after 10 seconds or so cut back to minus one-half volt. Check your vehicle specs, because this varies according to engine. Depending on engine temperature, start control may increase injection pulse time in two steps for cold-starting. You'll have to wait about a minute before repeating the test, since the start system won't reactivate immediately to prevent flooding. If the voltage doesn't change, and the injector wiring is sound, then you may have a bad control unit.

Fig. 4-4. Test lead being connected between fuel injector and wiring connector. Test lead allows voltage to be measured while control unit operates injector.

5. AIR-FLOW MEASUREMENT

These simple checks cover both types of air-flow measurement, the vane-type air-flow sensor and the hot-wire air-mass sensor.

5.1 Air-Flow Sensor

The air-flow sensor is not serviceable. If the vane movement is not smooth, or if any of the electrical test values are incorrect, the sensor must be replaced.

Checking Vane Movement

Remove the air intake duct from the air-flow sensor. Move the air vane with your finger as shown in Fig. 5-1. The vane should open smoothly, with just a light touch, and close smoothly and completely as you release pressure. Your fingertip will be able to sense if it's binding.

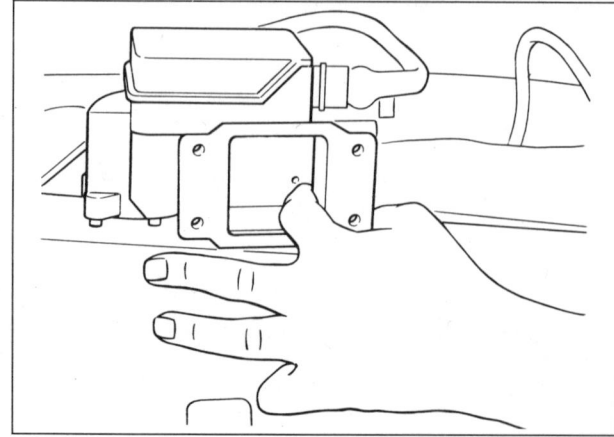

Fig. 5-1. Air-flow sensor vane movement being checked.

Electrical Tests

Test the air-flow sensor insulation. With the ignition off, disconnect the air-flow sensor wiring connector. Connect one lead of an ohmmeter to ground on the chassis and touch each air-flow sensor terminal with the other lead as shown in Fig. 5-2. A reading other than infinite ohms (no continuity) indicates a faulty air-flow sensor.

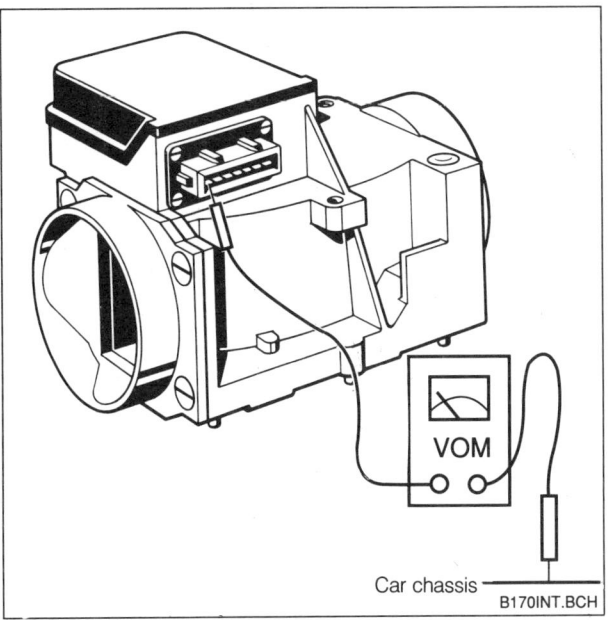

Car chassis
B170INT.BCH

Fig. 5-2. Air-flow sensor being checked for continuity.

The resistance of the air temperature sensor (usually checked across terminals 6 and 22 or 6 and 27) should be within the guidelines given in **Table b** above.

On some early models, the opening of the air vane closes a contact and signals the fuel pump to operate. Use your manual to identify the correct terminals for the contact and connect an ohmmeter across them. The reading should be infinite ohms, changing to zero ohms when the air vane is pushed open.

Check the sensor resistance track. Identify the correct terminals for resistance track input and output, and connect an ohmmeter across them. The resistance should change as the air vane opens. The actual values are less important than the fact that the values change smoothly, without any sudden drop outs.

5.2 Air-Mass Sensor

The air-mass sensor is not serviceable. If it fails any of the tests below, check for a faulty wiring connector or wiring before replacing the sensor.

Voltage Input

Peel back the protective boot on the air-mass sensor wiring connector. Check the voltage input to the sensor at the proper pin of the connector as shown in Fig. 5-3. Check your shop

manual for the correct terminal. With the igni... main relay terminal grounded, your meter sh...

Fig. 5-3. Inserting probe for air-mass sensor voltage checks. Peel protective boot from wiring connector, and insert probe from back of connector to prevent bending of terminals.

Voltage Output

Place the probe on the correct terminal for voltage output. With the engine running at normal operating temperature, the voltmeter typically should read 2.12 volts at 760 rpm, and increase to 2.83 volts at 3520 rpm. Changes are more important than the actual numbers.

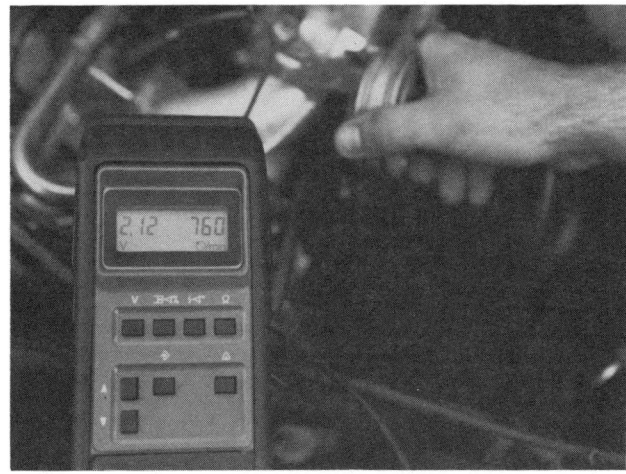

Fig. 5-4. Air-mass sensor voltage output checks being made. At 760 rpm (shown) voltmeter should read about 2.12 volts; at 3520 rpm, meter should read about 2.83 volts.

Air-Flow Measurement

Burn-off Sequence

With the voltmeter still checking output voltage, rev the engine up to 3000 rpm, then let it return to idle and turn off the engine. After approximately four seconds, the control system should send the one-second burn-off signal, visible as a voltage reading on the voltmeter. See Fig. 5-5.

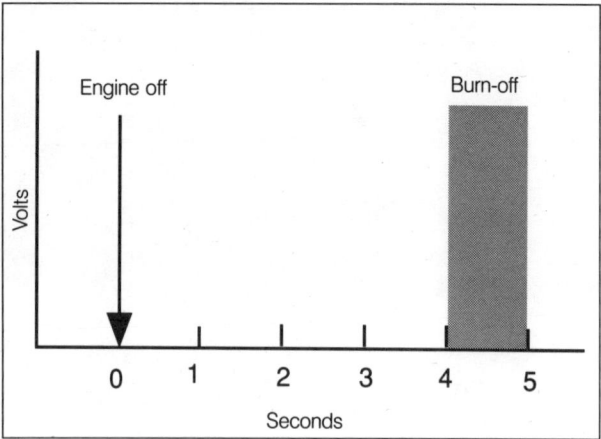

Fig. 5-5. Chart showing burn-off signal.

Visual Inspection

If the voltage readings are incorrect, check the wiring connector. Make sure that it's secure, and that the plug is not twisted and putting stress on the wiring harness.

If there's still a problem, remove the sensor and check the wire screens. Replace any broken screens, and check the hot wire for breakage as shown in Fig. 5-6. When you first look inside, you may think the air-mass sensor doesn't have any hot wire. But move it around until the wire is visible against the background. The fine wire is not easy to see.

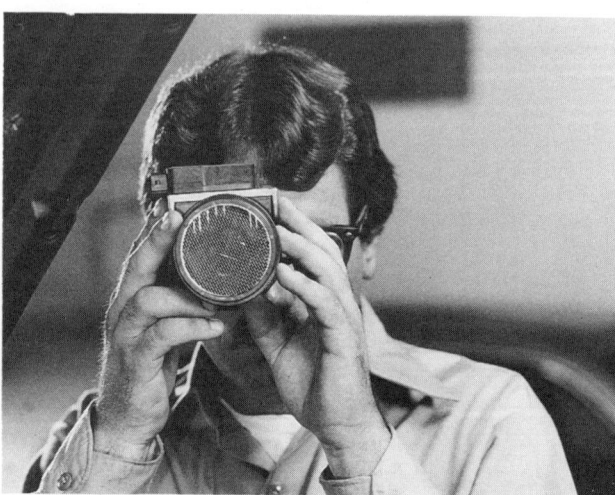

Fig. 5-6. Checking air-mass sensor for breakage of hot wire. The hot wire is very fine, so look closely.

5.3 Throttle Switches

Switches attached to the throttle valve signal the control unit for closed throttle and wide-open throttle conditions. Faulty or misadjusted switches can cause hesitation and rich or lean running. Some cars use a combination throttle switch, others use a pair of microswitches.

Throttle Valve Basic Adjustment

Before checking or adjusting the throttle switches, check the adjustment of the throttle valve. With the engine idling, try to close the throttle valve further with your hand. If engine speed drops, check to see that the throttle cable is not binding or pulling. Adjust the throttle valve by first unscrewing the adjustment screw until the throttle valve is fully closed. Screw in the adjustment screw until it just touches the stop, then screw it in an additional ¼ to ½ turn. Tighten the locknut.

NOTE ——
Do not use the throttle valve adjustment for setting the idle speed.

Checking and Adjusting Throttle Switches

Using an ohmmeter, check the throttle switch or microswitches as shown in Fig. 5-7 or Fig. 5-8. The closed-throttle function should read zero ohms at closed-throttle, and infinite ohms when the throttle just begins to open. The wide-open throttle function should read zero ohms at full throttle. Check your car manual for the correct terminals to test. Adjust the switch by loosening the mounting bolts and rotating the switch slightly.

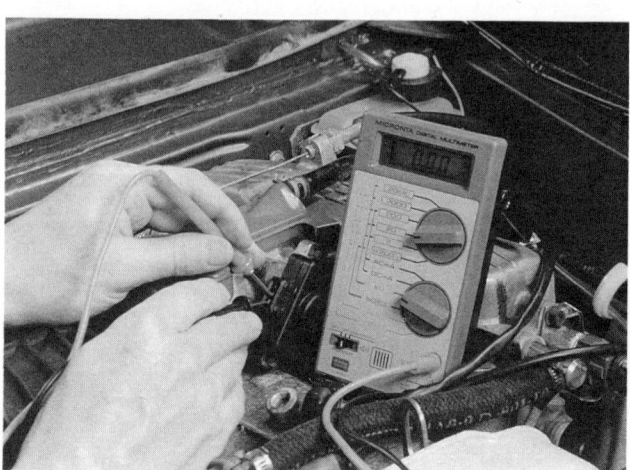

Fig. 5-7. Throttle switch adjustment being checked.

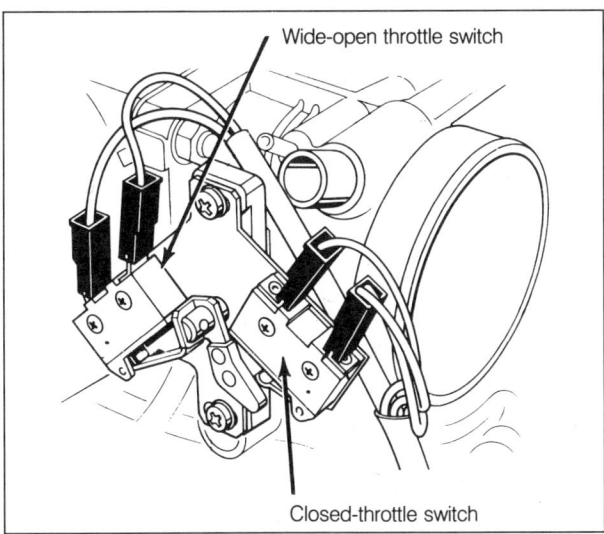

Fig. 5-8. Typical installation of microswitches, showing closed-throttle and wide-open throttle switches.

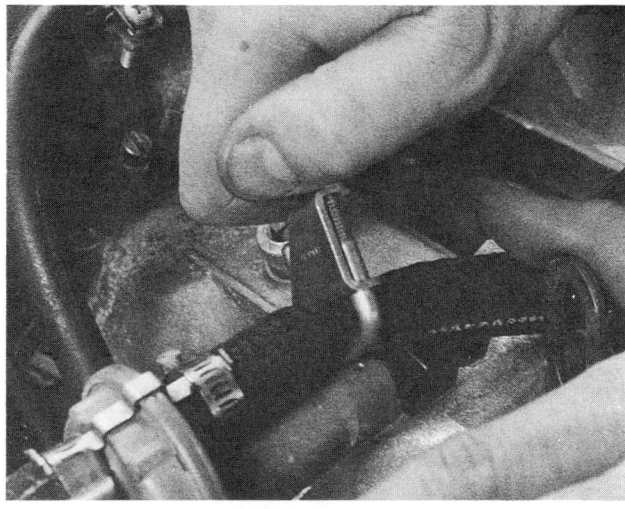

Fig. 6-1. Air hose being clamped shut to test air-flow through auxiliary air valve.

6. IDLE SPEED

Idle speed is set by an air bypass screw or knob on the throttle body which changes the amount of air that bypasses the throttle valve. The need for additional air during cold running is handled either by the auxiliary air valve, or by the idle speed stabilizer. If the engine has an idle problem, be sure that the ignition timing is correct, and check idle compensation first before adjusting idle. Idle speed and mixture are closely related. Always check mixture and adjust if necessary after adjusting idle rpm.

Auxiliary Air Valve Test

To test the air-flow through the auxiliary air valve, clamp the air hose as shown in Fig. 6-1. Perform the test with the engine cold and with the engine at normal operating temperature. With the hose clamped shut, cold-engine idle speed should drop; with a warm engine, nothing should happen to idle speed. If either result is incorrect, troubleshoot the valve as described below.

If cold-engine idle speed does not drop, remove the auxiliary air valve hoses and look through the valve. Sometimes a mirror will help. When cold (about freezing) the valve should be open as shown in Fig. 6-2. If not, replace the valve.

If warm-engine idle speed drops, remove the auxiliary air valve hoses and look through the valve. Use a mirror if necessary. The valve should be closed as shown in Fig. 6-3. If not, remove the wiring connector to the valve, and check for voltage across the connector. With the ignition on, there should be approximately 12 volts. If there is, replace the auxiliary air valve. If not, check the wiring and repair any breaks, then retest the auxiliary air valve.

Fig. 6-2. Visual inspection of cold auxiliary air valve. Valve should be open as shown (**arrow**).

Idle Speed Stabilizer Tests

Test the idle speed stabilizer using a dwell meter as shown in Fig. 6-4. Use a wiring adapter that allows dwell to be read while the stabilizer is operating. Check your car manual for the correct procedure and specifications since idle speed stabilizers differ.

With the engine idling at normal operating temperature, and with all electrical loads off (including the electrical radiator cooling fan if fitted), typical dwell is about 30°.

Raise rpm slightly to open the throttle-switch. Typical off-idle dwell is 32°, or at least a small increase from the normal idle dwell.

Idle Speed

Fig. 6-3. Visual inspection of warm auxiliary air valve. Valve should be closed as shown (**arrow**).

Fig. 6-4. Attaching dwell meter for tests of idle stabilizer valve. Check your car manual for correct test terminals and dwell specifications.

Check cold start operation by pulling the pump fuse to disable the fuel pump, and disconnect the engine-temperature sensor to simulate an engine-cold condition. During cranking, the dwell typically should show about 60°. Reinstall the pump fuse and start the engine. The engine should now idle at a higher rpm than when the tests began. Turn off the engine and reconnect the engine temperature sensor.

Run the engine at idle and turn on any electrical consumer, such as the fresh air fan or lights. Dwell should increase, and rpm should hold fairly steady. When the air conditioner is turned on, dwell should increase, and so should rpm to spin the compressor faster for better cooling.

If the idle speed stabilizer fails any tests, check the wiring between components and the control unit. If the wiring is fine, then replace the idle speed stabilizer.

Adjusting Idle RPM

Adjust the idle using the air bypass knob or screw as shown in Fig. 6-5. Check your manual for the correct procedure and specifications. On cars with an idle speed stabilizer, it is necessary to ground the stabilizer before adjusting it, as shown earlier in Fig. 3-10. This will either freeze the idle so that it can be adjusted, or drop the idle to something called the Basic Idle. Check your car shop manual for more information.

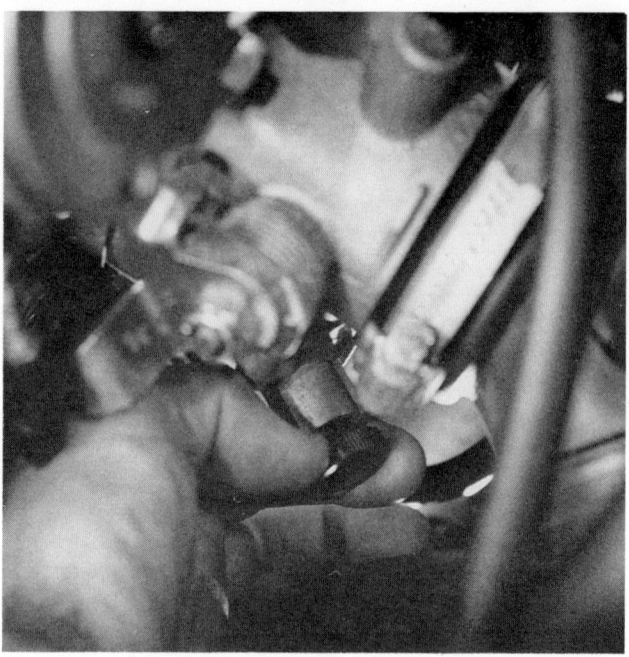

Fig. 6-5. Basic idle being adjusted using bypass air screw on throttle body.

7. MIXTURE (CO) AND LAMBDA CONTROL

7.1 Mixture Adjustment

The procedures for adjusting mixture depend greatly on the type of pulsed system in your car. Since about 1980, mixture control is usually automatic with closed-loop control from the electronic control unit and the oxygen sensor. If you pry out the seal and twist the screw to enrich the mixture, the oxygen sensor will sense the change and the control unit will change pulse time to bring the mixture back to the ideal air-fuel ratio. Beginning about 1987, pulsed systems with adaptive control do not need adjustments and none are provided. So be careful when fooling with adjustment; you may do more harm than good.

A mixture that is too rich may be caused by a faulty engine temperature sensor, a faulty air temperature sensor, a faulty cold-start system, misadjusted throttle switches, or problems with the fuel system. Lean mixtures may be caused by air intake leaks, a misadjusted throttle valve, or problems with the fuel system. Check these areas before assuming that mixture needs to be adjusted.

Adjusting Air-fuel Mixture

In general, mixture (CO) is adjusted by means of a screw on the air-flow sensor or air-mass sensor, as shown in Fig. 7-1. On air-mass sensors, the screw is to the side of the wiring harness connector. The screw may be covered by a metal plug which needs to be removed for adjustment.

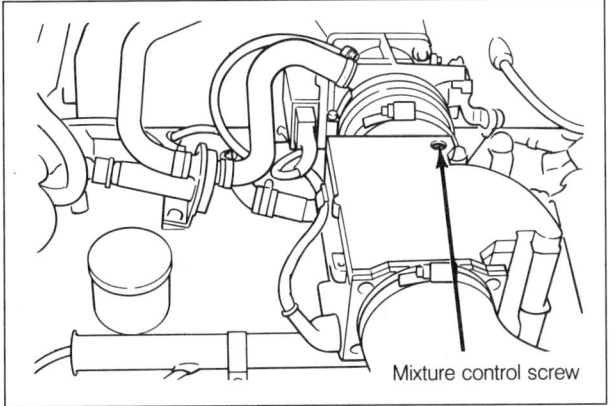

Fig. 7-1. L-Jetronic air-flow sensor mixture screw. On air-mass sensors, screw is to side of wiring harness connector. Screw may be covered by metal plug which needs to be removed for adjustment.

NOTE ——

The US government and some states consider it "illegal tampering" for a car to leave a repair shop without the plug properly installed.

To adjust the mixture, you'll need access to an exhaust-gas analyzer. While an analyzer is obviously not a run-of-the-mill tool, a local service shop or inspection station may be willing to rent some time on their machine, so check around. Mixture on some systems can be set without an analyzer, but this can only give a rough approximation, and driveability and emissions may suffer.

For the proper mixture settings, check the decal under your car hood. The engine should be warm, and the idle should be within specification. If possible, also change the oil and filter. Blowby from the crankcase may affect CO readings and adjustment.

L-Jetronic Systems without Lambda Control

If the CO on the meter is too high, the engine is running too rich, so unscrew the mixture screw to admit more air, as shown in Fig. 7-2. If the CO is too low, screw in the mixture screw.

If a CO meter is not available, adjust the mixture for the leanest setting that provides a smooth idle, then enrich the mixture an additional ¼ turn. Test drive and check for surging at part throttle. If so, enrich ¼ turn at a time until the surging disappears.

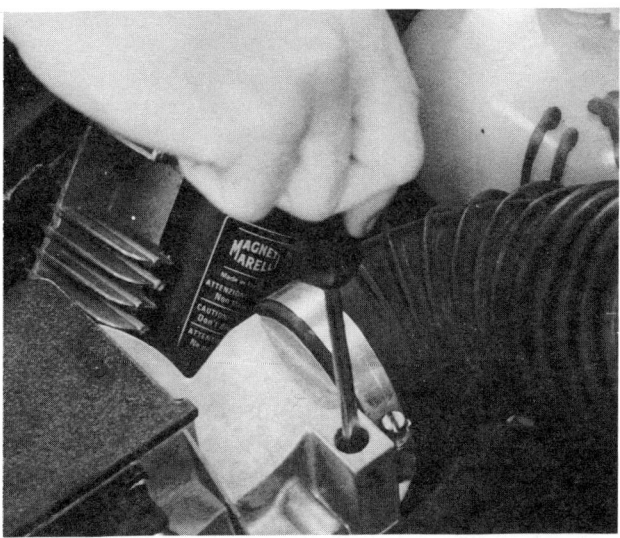

Fig. 7-2. Mixture (CO) being adjusted. On either air-flow sensors or air-mass sensors, turn screw clockwise to richen mixture, and counterclockwise to lean mixture.

L-Jetronic Systems with Lambda Control, Motronics

When checking CO on engines with a lambda sensor, remember the conditions for closed-loop operation: the engine must be at normal operating temperature and the lambda sensor must be hot. Unless the engine has a heated lambda sensor, run the engine at 3000 rpm for about 30 seconds just before the test to make sure the sensor is hot enough. If the test results are incorrect, or CO cannot be brought into specification, check the lambda system as described in **7.2 Lambda Control**.

Most cars with Bosch lambda control provide for sampling engine-out exhaust gas before the converter through an exhaust gas tap as shown in Fig. 7-3. At idle, closed loop, 0.5% volume is a typical CO specification. If you sample at the tailpipe, CO readings taken downstream of the converter will be specified at a lower level than those engine-out readings taken from the exhaust tap.

If you have the test pipe, compare closed-loop CO with open-loop CO. When you disconnect the oxygen sensor lead as shown in Fig. 7-4, allowing the system to run open loop, the CO may likely change, say up to 0.9%. If it does, lean or enrich the mixture until the open-loop CO is the same as closed-loop CO with the sensor connected. This is the correct mixture adjustment. Generally, the change is so small at the tailpipe that you may not see it if sampling CO there.

If a CO meter is not available, use the lambda sensor voltage to help set mixture to the ideal air-fuel ratio. Hook up a voltmeter to the sensor output as shown in Fig. 7-5. Turn the mixture screw toward lean until you see the: lowest voltage point; turn it toward rich until you see the highest voltage point. Turn back

Mixture (CO) and Lambda Control

and forth until it is midway between the two points. In general, this point is about 500 mV, the mid-voltage point could vary depending on the age of the sensor, as shown in Fig. 7-6. If there is no voltage, or if the voltage does not vary, replace the lambda sensor.

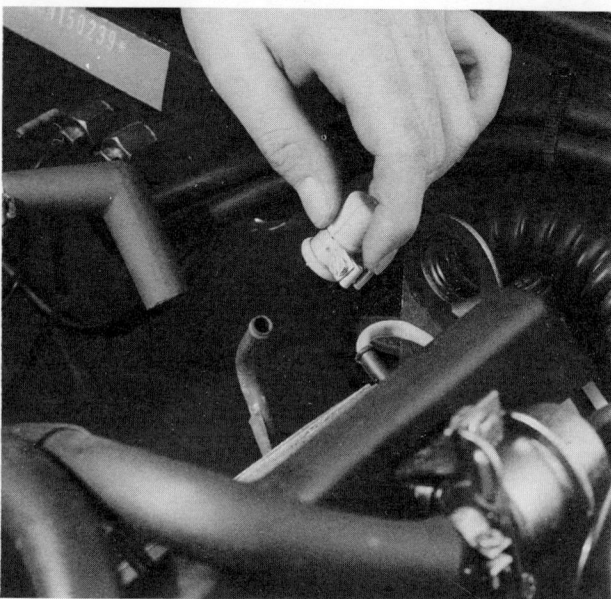

Fig. 7-3. Most engines with Bosch systems have an exhaust test pipe for sampling CO before the catalytic converter. Remove cap, usually blue, to sample.

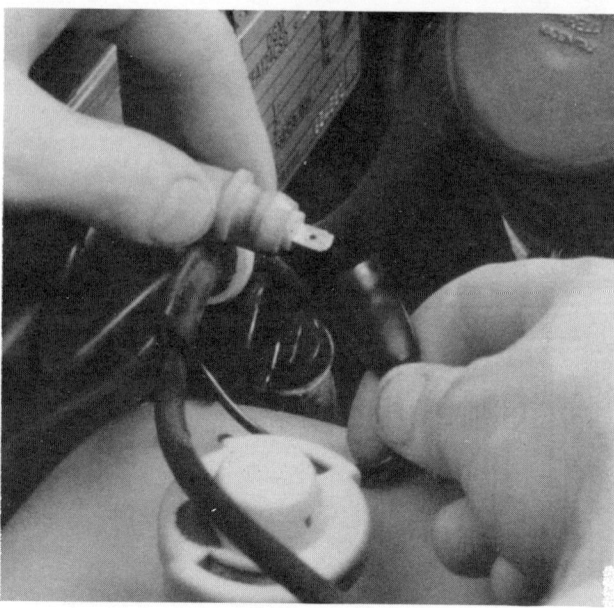

Fig. 7-4. To adjust mixture, disconnect lambda sensor and compare open-loop CO with closed-loop CO.

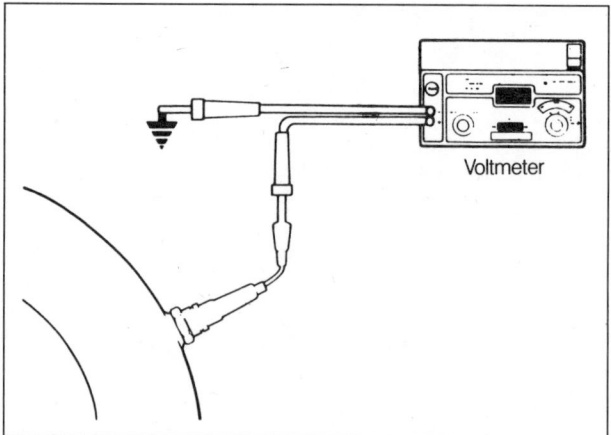

Fig. 7-5. Voltmeter attached to Lambda sensor to measure voltage output. Attach positive lead to sensor output wire, and negative lead to ground on chassis.

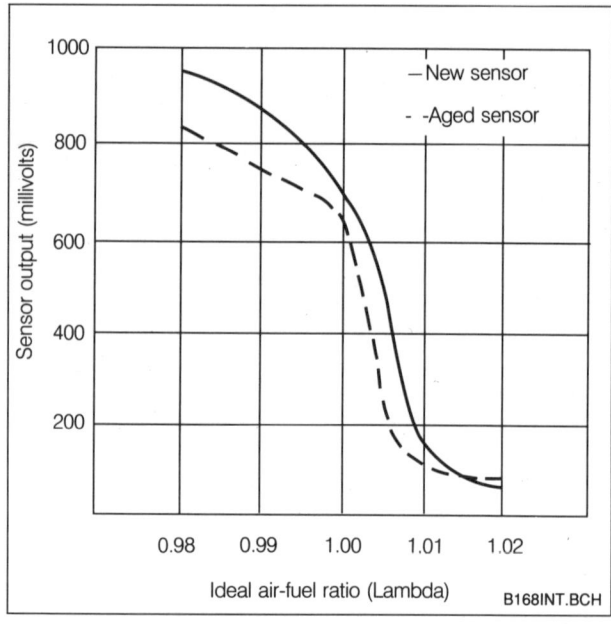

Fig. 7-6. Graph showing how lambda sensor voltage output varies with age.

7.2 Lambda Control

There are two parts to a test of the operation of the lambda sensor and the electronic control unit. To perform the tests, you must be able to sample engine-out CO at the test pipe with an exhaust-gas analyzer as shown above in Fig. 7-3.

Rich Stop Test

With the engine at normal operating temperature, disconnect the lambda sensor wire so that the system is open-loop. While watching CO, pull the vacuum hose off the fuel-pressure regulator and close it off with your thumb, as shown in Fig. 7-7. This should cause the regulator to increase fuel pressure, and thereby enrich the mixture. CO should increase, typically from 0.6% to 2%.

Fig. 7-7. Fuel pressure regulator vacuum hose disconnected and blocked, and lambda sensor wire disconnected for rich stop test of lambda control.

Rev the engine to 3000 rpm for at least 30 seconds, then let the engine idle. Leave the vacuum hose disconnected and blocked and reconnect the lambda sensor wire. CO should return to the original value, indicating that the lambda sensor and control unit can handle a rich mixture. Reconnect the vacuum hose when finished.

Lean Stop Test

Disconnect the lambda sensor wire to let the system run open-loop and pull the engine dipstick as shown in Fig. 7-8. The CO reading should drop – typically from 0.6% to 0.3% – indicating that the false air entering the engine through the crankcase has leaned the mixture.

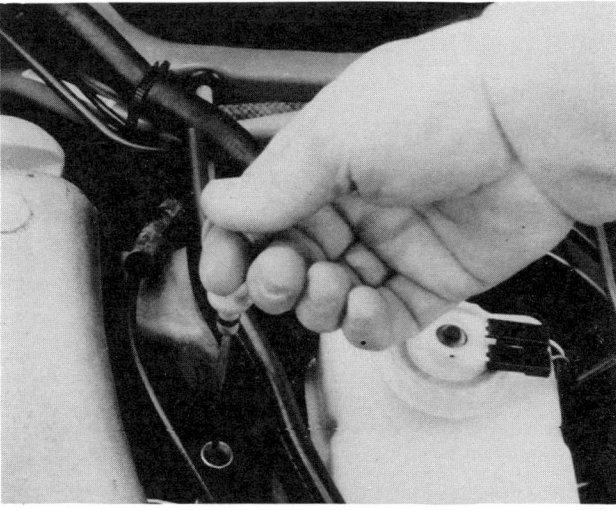

Fig. 7-8. Dipstick being withdrawn for lean stop test of lambda control.

Rev the engine to 3000 rpm for at least 30 seconds, then let the engine idle. Leave the dipstick out. Reconnect the lambda sensor wire. CO should return to the original value, indicating that the sensor and control unit can handle a lean mixture.

If CO does not return to the original value in either case, check the fuel pressure regulator and system wiring. If the regulator and wiring are not faulty, check the lambda sensor by measuring its voltage output with the engine running at 3000 rpm. If there is no voltage fluctuation (in milliamps), replace the sensor. If the sensor is working, then the control unit may be faulty.

8. MOTRONIC TIMING AND DWELL

Timing and dwell on Motronic systems are not adjustable, but you can check how they change according to specification with changing engine conditions, and identify possible faulty components.

Timing

Engine timing changes with engine temperature, so it cannot be checked with the engine cold or warming up. On the other hand, some engines cannot be too hot, so it may be necessary to run a fan over the engine while checking timing to keep the intake air cool enough.

Check timing at the specified curb-idle rpm as specified in your shop manual. On some engines, a TDC (top dead center) sensor can be connected to an engine analyzer for direct readout of timing or dwell as well as rpm.

Use a timing light to read the pointer, or mark on the flywheel. You'll see small timing changes, as Motronic makes small changes with each revolution to compensate for small changes in idle rpm. See Fig. 8-1. Average these by eyeball. Open the throttle partially and check to see that timing advances at the higher rpm.

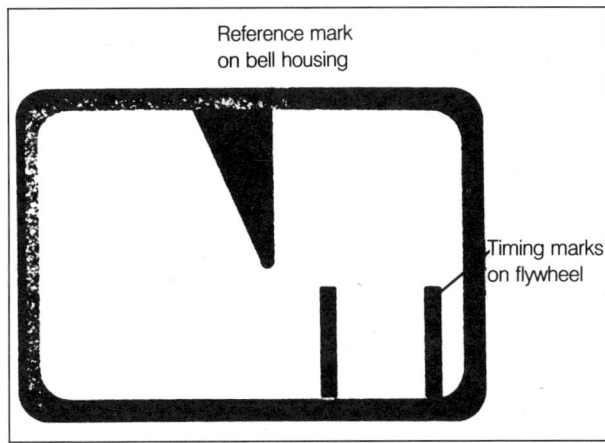

Reference mark on bell housing

Timing marks on flywheel

Fig. 8-1. Typical timing marks showing timing retard as knock sensor is tapped.

If idle timing is off, refer to your shop manual to check all input components, as well as the electrical connections and grounds as described in **2.3 Electrical System**—everything that could affect idle-timing signals. If all is OK, then the control unit may be faulty.

Knock Sensor Control

If your engine has one, locate the knock sensor, another input controlling timing, on the cylinder head, or the block. Rev the engine at about 2500 rpm, and lightly tap the knock sensor to simulate the vibrations that come from engine knocking. The timing marks should retard, or even move right out of the window. It shouldn't take much of a tap. If the timing does not change, and idle timing and advance is correct, then the sensor is most likely faulty.

Dwell

With the engine off, hook up the multimeter to the test connector to check dwell. At idle, you should see a basic value, for example 16.6%, which will increase as rpm increases. If dwell is incorrect or does not increase, check the engine speed sensor and the wiring from the control unit to the coil.

Motronic Timing and Dwell

Chapter 5

Continuous Injection – Theory

Contents

5

2   CONTINUOUS INJECTION — THEORY

1. GENERAL DESCRIPTION

This chapter covers the operating theory of Bosch continuous fuel-injection systems. Continuous Injection System (CIS) is the term widely used to refer to this branch of the Bosch fuel injection family. Do not confuse this with the Volvo term CIS (Constant Idle Stabilization), which is their term for the Idle-Speed Stabilizer, an unfortunate overlap of terminology. The continuous systems I'll discuss in this chapter include:

- K-Jetronic (I'll call it K-basic)

- K-Jetronic with lambda control (I'll call it K-lambda)

- KE-Jetronic and its variations: KE3-Jetronic, and KE-Motronic (I'll call them KE systems)

I'll begin this chapter with a general description of how Bosch continuous systems meter fuel to match a given air flow. I'll then examine in detail how the different subsystems of CIS — air flow measurement and fuel delivery — interact mechanically to create this basic fuel metering function. Finally, I'll explain how K-basic, K-lambda, and KE control systems fine-tune this basic fuel metering to compensate for different operating conditions.

For more information on the general principles of fuel-injection, see chapter 1 and chapter 2. To determine which Bosch system is installed on your car, see **1.2 Applications - Identifying Features**, or the detailed applications table at the end of chapter 1.

1.1 Continuous Injection System (CIS)

As I described in chapter 1 and chapter 2, the aim of the fuel-injection system is to measure the amount of air the engine is taking in and to meter a precise amount of pressurized fuel to match that air and create the correct air-fuel mixture. All Bosch continuous systems provide this basic air-measuring and fuel-metering function in the same way: in the mixture-control unit.

Mixture-Control Unit

The mixture-control unit is the heart of continuous injection. As Fig. 1-1 shows, it is where the air-flow measurement system and the fuel delivery system interact. The mixture-control unit measures the engine's intake air, and then meters fuel in proportion to that air flow.

The mixture-control unit is actually the combination of two separate components: the air-flow sensor, and the fuel distributor. See Fig. 1-2. The air-flow sensor measures the air entering the engine. The fuel distributor, in turn, delivers a proportional amount of pressurized fuel to the injectors.

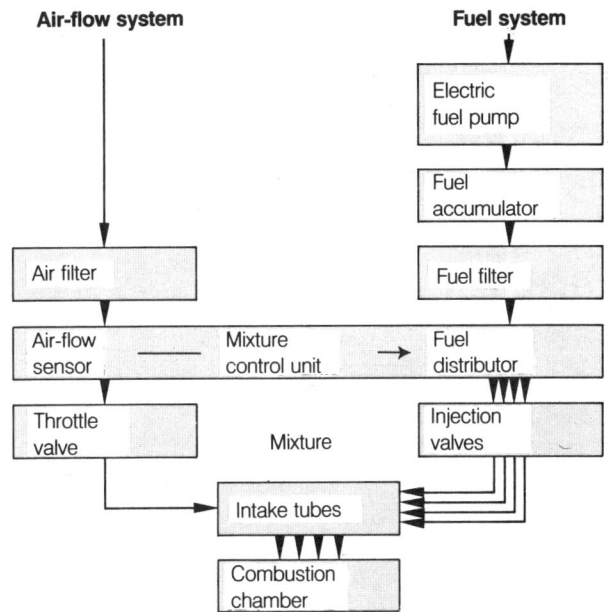

Fig. 1-1. Air-flow measurement system and fuel delivery system interact in mixture control unit to deliver the basic air-fuel mixture.

Fig. 1-2. The mixture-control unit is the combination of the air-flow sensor (**1**) and the fuel distributor (**2**). Circular sensor plate (**a**) is lifted by flow of engine intake air.

Air Flow Measurement and Fuel Metering

Fig. 1-3 is a schematic view of the operation of the mixture-control unit. The circular air-flow sensor plate is positioned in the intake tract so that all air entering the engine flows past it. The plate is attached to a lever which pivots, allowing the plate to move up and down. Intake air flowing through the housing raises the sensor plate. The movement of the sensor plate and lever is in direct proportion to the volume of the incoming air.

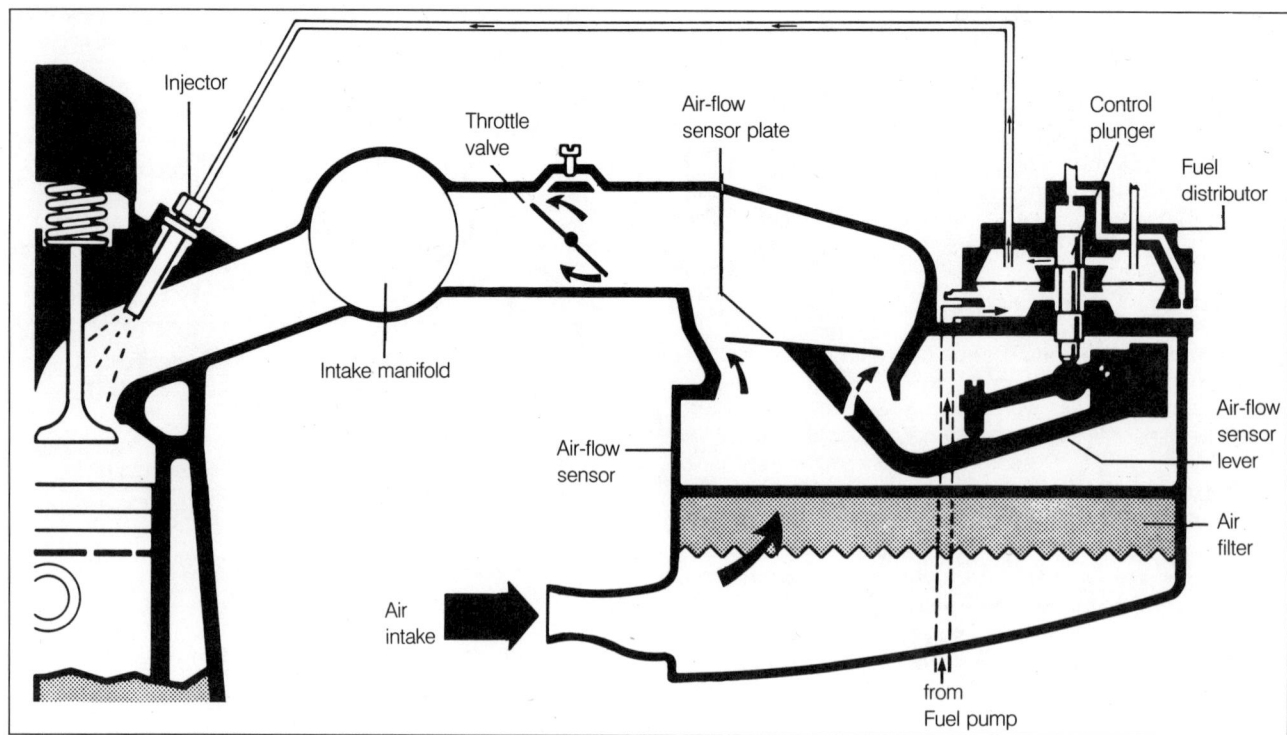

Fig. 1-3. Schematic of continuous injection air-flow measurement and fuel metering. Air entering intake manifold raises air-flow sensor plate. Rise of sensor plate mechanically raises control plunger. This basic fuel metering function applies to all systems in this chapter.

This air measurement is turned into an injection quantity by the control plunger in the fuel distributor. The plunger rests on the air-flow sensor lever, and it rises and falls at the same rate as the sensor plate. The position of the plunger controls fuel flow to the injectors. When air flow into the engine increases and raises the air-flow sensor and plunger, the rise of the plunger increases fuel flow proportionally. This maintains the correct air-fuel ratio. See Fig. 1-4. This mechanical lift of the control plunger by the air-flow sensor is how all continuous injection systems perform basic fuel metering.

In continuous injection all fuel metering takes place in the fuel distributor. The fuel injectors flow fuel continuously while the engine is running; their only function is atomization of the fuel. This is in contrast to the pulsed systems described in previous chapters, where fuel metering is controlled by opening and closing of the injectors.

Now you can see why continuous systems are often referred to as "mechanical" fuel injection, because basic fuel metering is controlled by the mechanical relationship between the air-flow sensor and the control plunger in the fuel distributor. As you'll see later in this chapter, continuous systems are also referred to as "hydraulic" fuel injection. This is because their control systems alter this basic air-fuel mixture for different operating conditions by changing fuel pressures in various parts of the system.

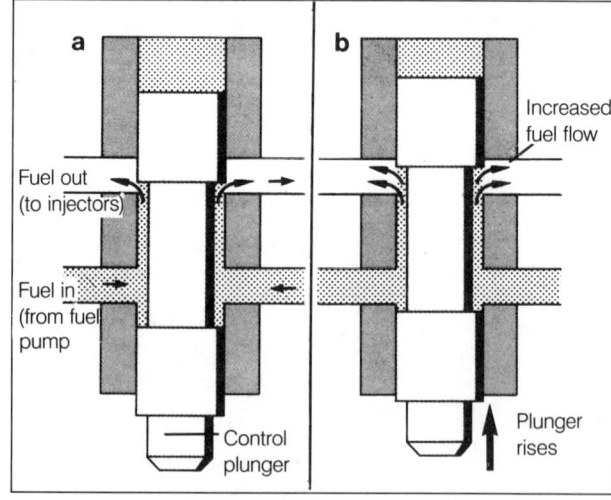

Fig. 1-4. Fuel metering by control plunger in the fuel distributor: At low air flow (**a**), rise of air-flow sensor and control plunger is low so less fuel flows to injectors. At increased air flow (**b**), when air-flow sensor rises higher, control plunger in turn rises, so more fuel flows to injectors.

1.2 Applications - Identifying Features

Bosch continuous injection systems are installed only by European car manufacturers. As of 1989, no Japanese or U.S. manufacturer is installing K-Jetronic fuel injection or its derivatives. With a large installed base on Volkswagen models, several million continuous systems are on the road in the U.S. Other makers using Bosch continuous systems include: Audi, BMW, Ferrari, Lotus, Mercedes-Benz, Peugeot, Porsche, Rolls-Royce, Saab, and Volvo. The 12-cylinder Ferrari Testarossa and Mercedes-Benz engines both use twin KE-Jetronic mixture control units, one for each 6 cylinders. Clearly, Bosch continuous systems are capable of performing well in many different applications.

To identify continuous systems, look for the separate mixture-control unit. It is usually mounted on a fender panel with flexible air duct leading to the throttle body and intake manifold. Also, continuous systems have separate injector fuel lines delivering fuel to each injector, instead of the common fuel rail found on pulsed systems. The fuel lines are usually flexible lines enclosed in protective metal braiding. See Fig. 1-5.

Fig. 1-5. Separate continuous mixture-control unit (**1**) between air cleaner and intake manifold. Each injector gets fuel through separate fuel line (**2**). No fuel rail, as in pulsed injection.

There are exceptions: On Mercedes cars, and on Peugeot, Renault, and Volvo cars using the PRV V-6 engine, the mixture-control unit is mounted directly on the intake manifold, as shown in Fig. 1-6. It usually delivers fuel through rigid metal lines to the injectors. The large air cleaner covers most of this so it may look like a carbureted engine at first glance.

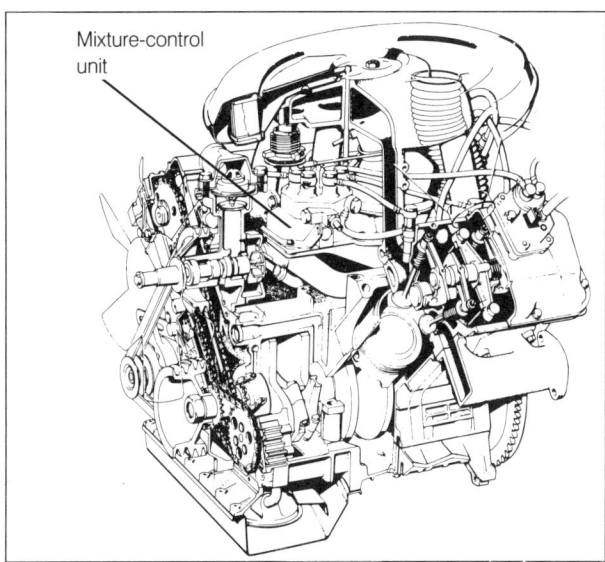

Fig. 1-6. On the PRV V-6 engine used by Peugeot, Renault, Volvo, the mixture-control unit hides under air cleaner. Location on Mercedes 4-cyl., 6-cyl. and 8-cyl. engines is similar.

How can you tell the systems apart? You can distinguish between each by component and function.

K-Jetronic and K-Jetronic with Lambda Control
(K-Basic and K-Lambda)

K-basic and K-lambda fuel distributors are made of cast iron and usually painted black. K-lambda systems are further distinguished from K-basic by the solenoid-type lambda valve next to the fuel distributor; it has an electrical connector and looks almost like a pulsed injector. See Fig. 1-7.

Table a. Bosch Continuous Fuel Injection Systems

Name	First use	Fuel pressure	Computer/pins	Lambda control
K-basic	1974 Porsche 911T	4.8–5.2 bar (70–75 psi)	no	no
K-lambda	1980 almost all mfrs.	4.8 bar (70 psi)	analog/25	yes
KE-Jet 2	1984 Mercedes 190E	5.4 bar (78 psi)	digital/25	yes
KE-Jet 3	1986 Mercedes 300/560	5.4–5.8 bar (78–84 psi)	digital/25	yes
KE-Motronic	1988 Audi 4-cylinder	6.2 bar (90 psi)	digital/35	yes

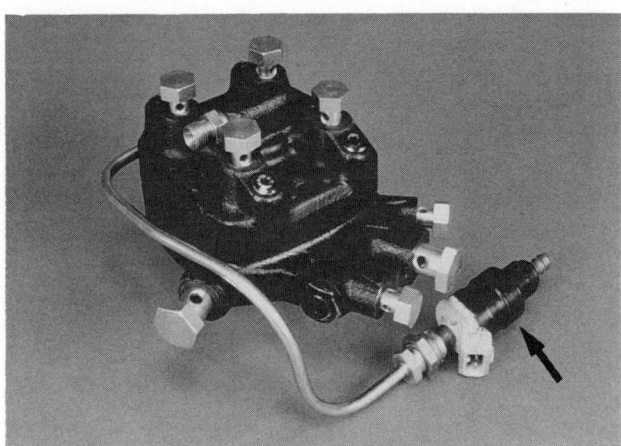

Fig. 1-7. You can tell both K-basic and K-lambda by the black color of the fuel distributor. Only K-lambda has the lambda control valve (**arrow**), which looks something like a pulsed injector.

KE-Jetronic, KE3-Jetronic, and KE-Motronic

(KE Systems)

All KE systems have a pressure actuator fastened to the fuel distributor as shown in Fig. 1-8. Also, KE fuel distributors are usually unpainted aluminum. Further differences which distinguish the three different KE systems are mostly in the electronics; not immediately apparent. For example, KE-Motronic systems lack vacuum lines to the distributor, but some KE-Jetronics also lack such vacuum lines (1988 Audis with separate electronic ignition-timing control, for example). Check your manual to identify which KE system you have.

Fig. 1-8. You can identify KE systems by the attached pressure actuator. Some are gray, some are black; all have electrical connector for control-unit wiring connection.

2. AIR-FLOW SYSTEM

The most basic function of a fuel-injection system is to measure the amount of air drawn in by the engine and to meter a proportional amount of fuel. The air-flow system performs this air measurement function. The air flow system also includes the throttle valve which controls engine speed by regulating engine air intake. See Fig. 2-1.

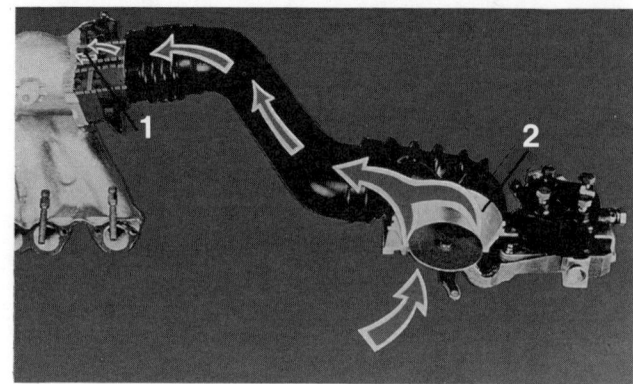

Fig. 2-1. Continuous injection air-flow system. Throttle valve (**1**) regulates air flow into intake manifold and engine. Air-flow sensor (**2**) measures volume of intake air.

The air-flow system includes the air-flow sensor (part of the mixture-control unit), the ducting to the throttle body, the throttle valve, and the intake manifold. The throttle valve controls the amount of air entering the engine. The air-flow sensor is deflected by that air, measuring its volume. The movement of the sensor in turn controls the basic fuel metering in the fuel distributor. The air-flow system also provides adjustment for engine idle speed.

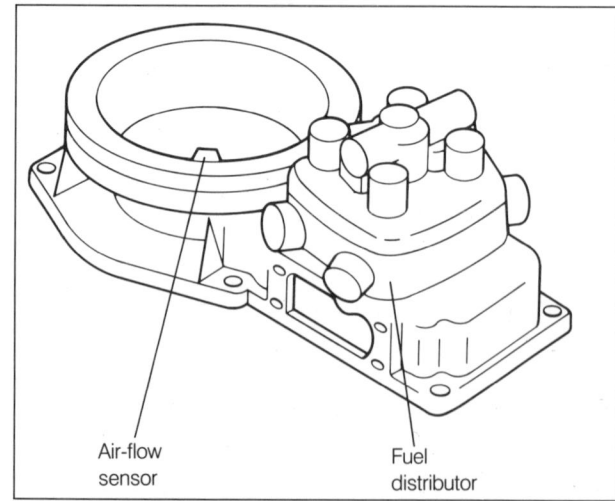

Air-flow sensor

Fuel distributor

Fig. 2-2. The air-flow system measures engine air intake at the air-flow sensor, and turns that measurement into a proportional amount of fuel in the fuel distributor. Together, the fuel distributor and air-flow sensor are known as the mixture-control unit.

2.1 Air-Flow Sensor

The key to air-flow sensor operation is a circular sensor plate that is deflected by engine intake air moving through the sensor plate housing. The movement of the plate in turn lifts the control plunger in the fuel distributor.

Sensor Plate and Control Plunger

When the engine is off, the sensor plate rests on a spring support, as shown in Fig. 2-3. This is known as the Zero Position. The plate is carried by a pivoting main lever. This allows the sensor plate to move in the intake air flow. The fuel distributor's control plunger rests directly on the same lever that carries the sensor plate.

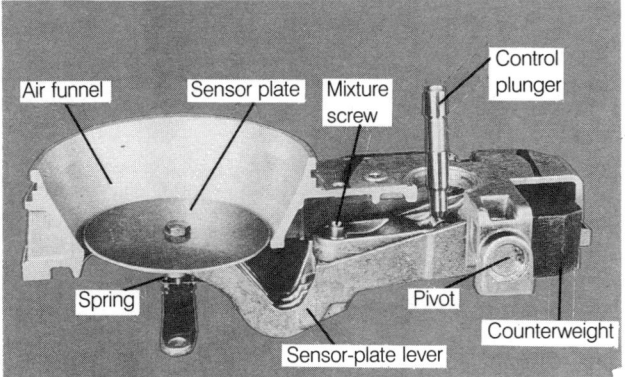

Fig. 2-3. Cut-away view of air-flow sensor shows sensor plate carried by a lever on a pivot, balanced by a counterweight. Shown with fuel distributor removed.

As the engine takes in air, the sensor plate rises as shown in Fig. 2-4. The more air flow, the more the plate and lever rise. The control plunger, because of its position on the lever, rises at the same rate as the sensor plate. This is a linear relationship: a doubling of air flow doubles sensor-plate lift, which in turn doubles the lift of the control plunger. The height of the plunger determines fuel flow to the injectors, so the higher it rises the greater the fuel flow. It is this mechanical relationship between the air flow sensor and the control plunger in the fuel distributor that determines basic fuel metering.

There are actually two types of air-flow sensors that are used in continuous injection, updraft and downdraft. In updraft sensors, as you've just seen, intake air flows from below and lifts the sensor plate **up**. In downdraft sensors, air flows from above and pushes the sensor plate **down**. Although the direction of air flow through the two sensors is different, the operating principles are identical: as the sensor plate is deflected by the intake air, the control plunger **rises**, increasing fuel flow. See Fig. 2-5.

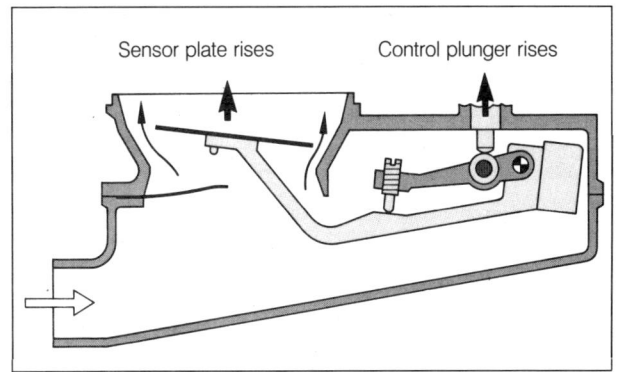

Fig. 2-4. As air is drawn into the engine, the sensor plate lifts, floating on the column of air. The movement of the sensor plate also lifts the control plunger.

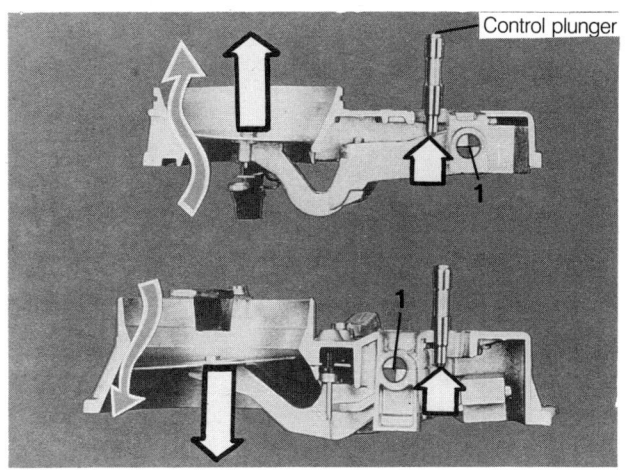

Fig. 2-5. When an updraft sensor plate rises, it lifts the plunger — on the same side of the pivot (**1**). When a downdraft sensor plate is forced down, it also lifts the plunger — on the opposite side of the pivot.

In both types of air-flow sensors, the weight of the sensor plate and lever is balanced by a counterweight. This means that the plate acts as though it were weightless in the intake air stream. The force of the air flow easily lifts the sensor plate, so much so that a counter-acting force is necessary to stabilize the movement of the plate. As you'll see later, fuel pressure force is used, applied to the top of the control plunger.

In later continuous systems, the counterweight is replaced by a balance spring; the spring has less inertia than the weight. This permits the sensor plate to adjust more quickly to changing air flow, so fuel metering and throttle response are faster.

Funnel Angles

The operation of the air-flow sensor is based on something called the "floating body principle." This principle states that under the right conditions a column of pressurized air will support and move an object, and that the movement of this object will be in direct proportion to the volume of air flowing past.

Air-Flow System

In the air-flow sensor, all intake air flows through the sensor air funnel to create a column of air. Because the counter-weighted sensor plate is positioned in the middle of the air funnel, it floats on the column of air. As the engine uses more air, the flow through the funnel is greater and the sensor plate rises in the column of air.

The shape of the sides of the funnel is tailored so that for a given air flow, the strength of the column of air is greater or less, and the lift of the sensor plate will be greater or less. See Fig. 2-6. The funnel is designed for each system on which it is installed, according to the power characteristics of the particular engine design.

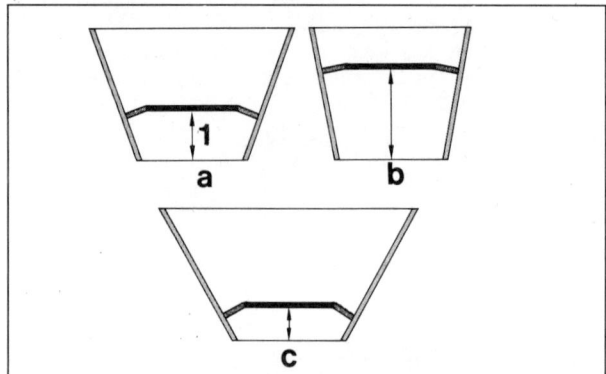

Fig. 2-6. Air-flow sensor funnel shape affects lift of sensor plate. For a medium cone angle (**a**), the sensor plate lifts to height **1** for a given air flow around the sensor plate. If the funnel walls are steep (**b**), the plate would have to lift much higher to pass the same amount of air. If the funnel wall angle is shallow (**c**), the same amount of air pushes with much less lift.

If you remember that sensor-plate lift affects control-plunger lift and basic fuel metering, you can see that funnel angles can be changed to affect the air-fuel ratio for a given air flow.

For example, in K-basic systems, the funnel is usually shaped with different angles according to engine needs for different operating conditions, as shown in Fig. 2-7. A steeper wall section near the top of the funnel causes more sensor-plate and plunger lift for the air flow, giving a richer mixture for full load. A shallower wall section in the middle causes less lift for a leaner mixture at part-load. At the bottom of the funnel, another steeper wall section gives more lift and a richer mixture for idle/off-idle conditions. In K-lambda and KE systems, the funnel shape is usually conical, so the proportion of the basic fuel metering is constant for the entire range of sensor plate lift.

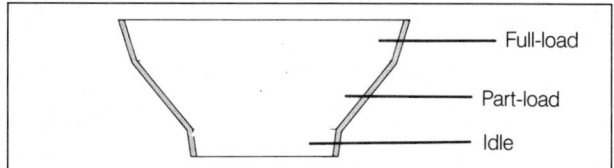

Fig. 2-7. In K-basic, funnel shape is designed to fit specific engines for enrichment at different air flows. Funnel shape is matched to each engine and model; do not interchange funnels.

Control-Plunger Counterforce

Some additional force is needed to balance the air-flow force that raises the sensor plate and control plunger. Otherwise, the movement of the plate would be too quick, and when the engine is started the plunger would rise to the top of its travel and stay there.

Fuel pressure supplies this counterforce, applied to the top of the control plunger in the fuel distributor. This pushes down against the upward force of the air flow sensor and the upward movement of the control plunger, regulating their movement and allowing the sensor plate to float in the air stream. The counterforce pushes the plunger back down for less fuel flow when there is less air flow, and returns the control plunger to its rest position when the engine stops running.

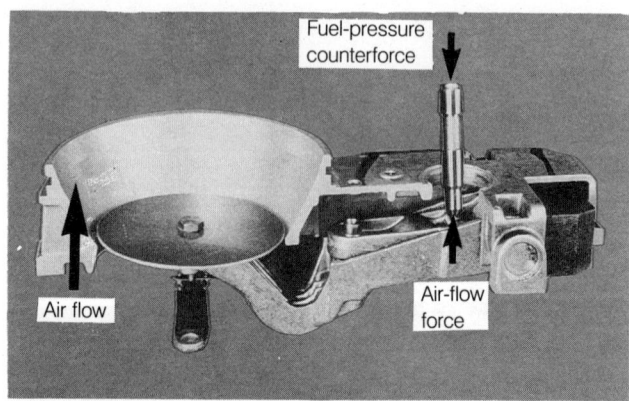

Fig. 2-8. Air-flow force lifts sensor plate and control plunger. Fuel-pressure counterforce presses down on the plunger. Balance of these two forces determines amount of plunger lift for a given air flow.

As you'll see later, this fuel-pressure counterforce can be varied to affect control-plunger lift for a given air flow, changing the proportion of fuel injected, and therefore changing the basic air-fuel ratio. In K-basic and K-lambda, the counterforce is known as the control pressure. I'll explain control pressure in **5. Control Systems**.

Beginning with 1983 models, a return spring atop the control plunger ensures that the plunger follows the movement of the air-flow sensor during starting and engine deceleration. This improves starting and reduces hydrocarbon emissions during coasting. See Fig. 2-9.

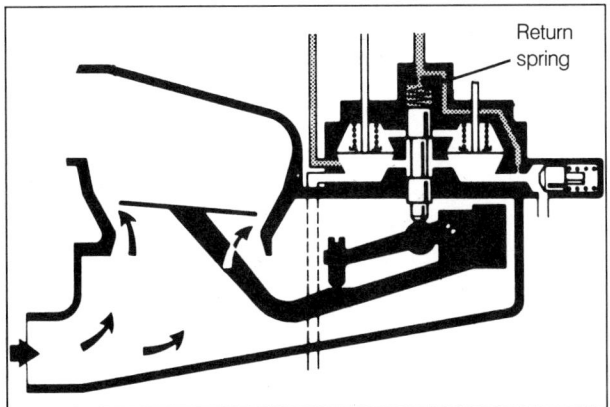

Fig. 2-9. Beginning in 1983, a small spring presses down to ensure control plunger follows sensor lever more closely.

Backfire Protection

The clearance between the edge of the sensor plate and the small part of the air-flow sensor funnel is very small, only about .10 mm. Because the plate effectively blocks the intake tract, any backfire in the intake manifold could build up reverse pressure and blow the ductwork, bend the sensor plate, or damage vacuum-operated systems.

To prevent damage in the event of a backfire, the air-flow sensor is designed with a secondary relief area in the funnel. See Fig. 2-10. Any backfire explosion in the manifold drives the sensor plate against its spring support and rubber bumper so that the pressure is vented around the edge of the plate. The sensor plate is driven past the zero position, down against its stop. (In downdraft sensors the plate is driven up against its stop).

KE Sensor-Plate Positions

Up until now, I've been talking about principles that apply to all continuous air-flow sensors. KE sensors operate in the same basic way as K-basic and K-lambda sensors: intake air lifts the air-flow sensor plate, which lifts the control plunger to meter fuel. But in KE, you'll find a big difference in the rest position of the control plunger due to some design changes. Because this difference affects restarting, it's important to know for chapter 6 when you'll be making checks of the air-flow sensor.

In K-basic and K-lambda, the control plunger rests on the lever when the engine is off (zero position); the sensor plate's rest position is determined by its stop. In contrast, KE is designed so that, at zero position, the plunger rests on an O-ring seal in the fuel distributor; there is a small amount of clearance between the plunger and the lever so the plunger seats firmly on the O-ring. See Fig. 2-11. This seals off fuel pressure more completely when the engine is off, and helps prevent vapor lock in the fuel lines.

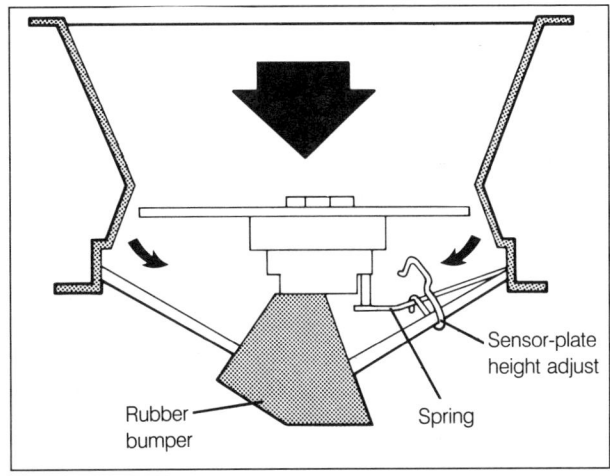

Fig. 2-10. In case of backfire, plate is pressed into second part of funnel relieving pressure in manifold.

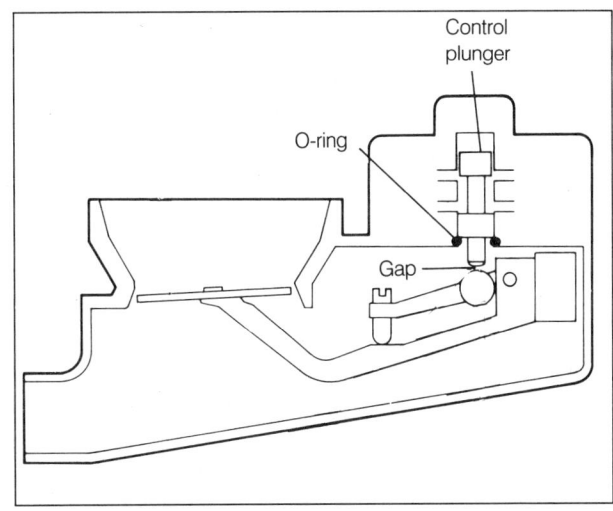

Fig. 2-11. KE control plunger rests on O-ring that seals in fuel pressure at base of plunger.

This difference in KE means that there are actually two important KE sensor-plate positions. The first is the zero position as described above, where a wire-clip supports the sensor plate when the engine is off and there is clearance between the control plunger and the lever. The second is the basic position (and that's only in KE), the position of the plate in the air funnel when it is lifted so the lever just touches the control plunger. See Fig. 2-12.

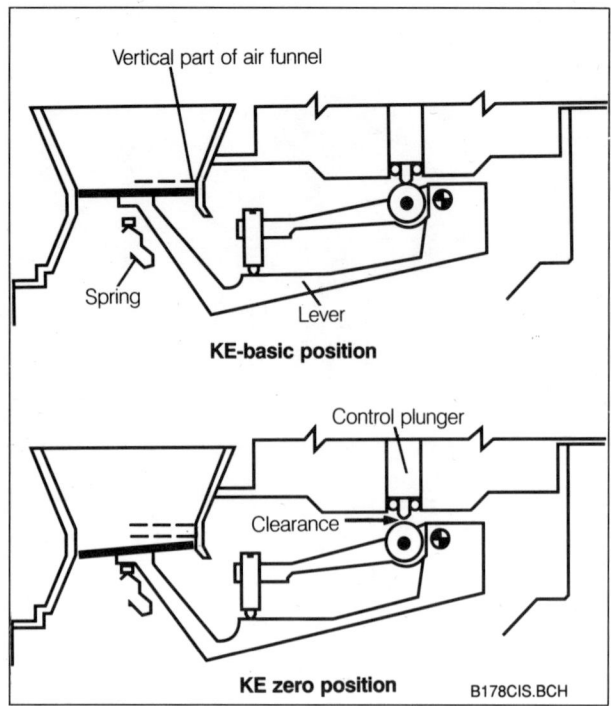

Fig. 2-12. In updraft KE basic position (**top**), sensor plate is at top of vertical part of air funnel (**dotted line**) and lever just touches control plunger. In zero position (**bottom**), plate rests on spring, and there is a gap between plunger and lever. You should be able to see all of the vertical part of funnel (**dotted lines**) around the sensor plate.

Incorrect clearance between the sensor lever and the control plunger will cause problems. If there is no clearance, the plunger will not seal against the O-ring, and residual fuel may dribble down into the mixture-control unit when the engine is off. This can cause flooding-type restart problems and, in a worst case, a potential fire hazard.

In contrast, if there is too much clearance between the sensor plate lever and the plunger you may experience a different start problem. During cranking, the air flow must lift the sensor plate a certain amount before the lever contacts the plunger to lift it and increase fuel flow. If there is excess clearance the lever will not lift the plunger soon enough for the air flow, upsetting the basic air-flow and fuel-metering function.

In chapter 6 you'll see how to check these important KE sensor plate positions. Also, remember the distinction between how updraft sensors and downdraft sensors operate: In updraft KE air-flow sensors, the sensor plate is lifted up from the zero position to the basic position; in downdraft air-flow sensors, the plate is pushed down from the zero position to the basic position.

Fig. 2-13. KE-Jetronic air-flow sensors; downdraft (top) and updraft (bottom). Note that on downdraft sensors, sensor plate is angled.

Mixture (CO)

To ensure that the basic air-fuel mixture is in the best range for emission control, all continuous systems have a provision for mixture adjustment. Although this adjustment is made at idle, it affects the basic sensor-plate and control-plunger relationship over the entire range of operating conditions. For a given lift of the sensor plate, the lift of the control plunger is adjusted to change the amount of injected fuel, and therefore the air-fuel mixture. This adjustment is also known as the CO adjustment because the change in the air-fuel mixture changes the amount of CO (carbon monoxide) in the exhaust. The accuracy of the mixture adjustment is commonly checked by measuring exhaust CO.

You adjust mixture with the mixture-control screw, which is shown in Fig. 2-14. You can see that the air-flow sensor lever is actually two parts, and that turning the mixture screw in or out moves the mixture lever only. This moves the control plunger without changing the position of the sensor-plate lever or sensor plate.

For example, for a constant air flow at idle, turning the mixture screw clockwise pushes the mixture lever up. This lifts the control plunger further and more fuel is injected for the same air flow, enriching the mixture. If the screw is turned counterclockwise, the control plunger is lowered and the mixture is leaned. As you'll see in chapter 6, because the mixture screw is located in the air-flow sensor you'll need a special tool to reach it.

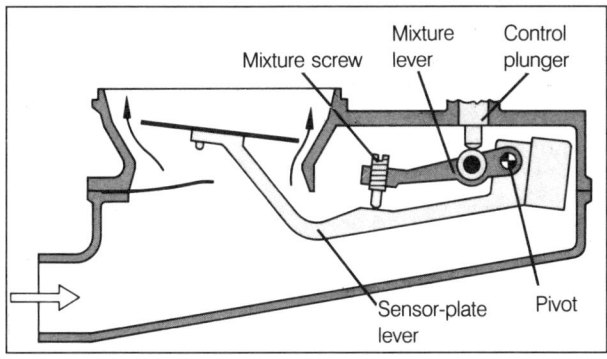

Fig. 2-14. Schematic view of air-flow sensor showing location of mixture-control screw.

2.2 Throttle Valve

The throttle valve regulates air flow into the engine. The more it is open, the greater the air flow into the engine. The throttle valve is downstream of the mixture-control unit, usually mounted on the intake manifold, and is controlled by the accelerator cable. The throttle valve does not control idle speed. This is done by a separate idle screw as described below.

Idle Speed

You adjust idle speed on continuous systems not by fine adjustment of the throttle valve, but by controlling a small amount of air that is allowed to bypass the throttle. This is more reliable since it's independent of accelerator-cable stretch or wear in the throttle valve mechanism.

The idle air bypass is regulated by the idle screw, as shown schematically in Fig. 2-15. The screw is usually on the throttle-valve housing. Turning the screw in or out changes the amount of bypass air. The screw is downstream of the air-flow sensor. Any change in bypass air results in a corresponding change in the amount of fuel injected, so idle speed adjustments do not affect the mixture adjustment.

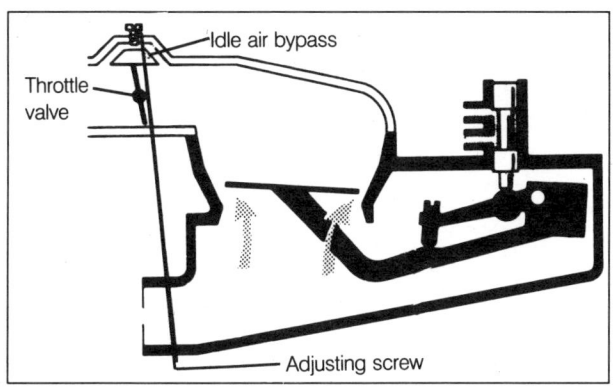

Fig. 2-15. Schematic view of air-flow sensor and throttle valve shows idle air bypass and adjusting screw.

3. Fuel System

In continuous systems control of fuel pressure is extremely important, even more so than in pulsed systems. As you'll see, fuel pressure opens the fuel injectors, and the control systems manipulate fuel pressure to alter the basic air-fuel mixture when compensating for different operating conditions. It's vital that the fuel system not only supply adequate fuel for the amount of air the engine is using, but also that the fuel be under pressure, maintained and controlled within a narrow range.

The fuel system includes the fuel tank to store the fuel, the electric fuel pump to pressurize the fuel, the fuel accumulator to damp pressure surges in the system, the filter, the fuel distributor to meter and distribute the fuel, the pressure regulator to maintain primary system pressure, and the fuel injectors to atomize the fuel. See Fig. 3-1. The electric pump and the filter operate similarly to those used in pulsed systems. For a discussion of how they operate see chapter 3. Some cars have an additional electric low-pressure in-tank pump, called the pre-supply or transfer pump. It ensures delivery to the main pump without vapor lock.

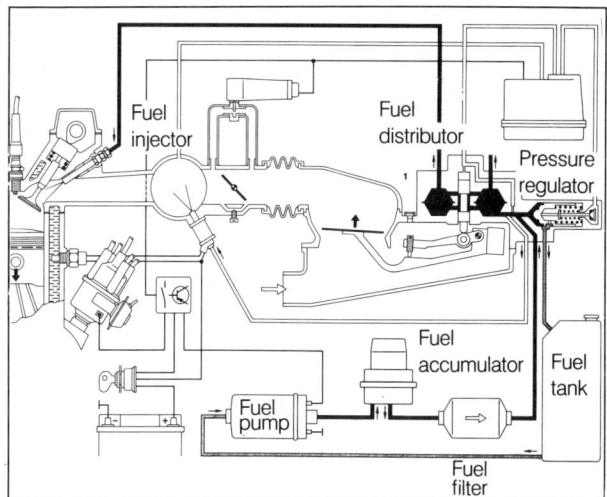

Fig. 3-1. Continuous injection fuel system. Supply components (fuel pump, filter) are generally similar to those of pulsed systems, with addition of accumulator. K-basic shown, K-lambda and KE systems similar.

3.1 Fuel Accumulator

Most continuous systems include a fuel accumulator, shown in Fig. 3-2, which damps fuel-pressure surges and holds residual pressure in the system when the engine is shut off. The accumulator's damping action protects the fuel distributor from the rapid build-up of fuel-pressure during startup, and reduces fuel pump noise when the engine is running. When the engine stops, the accumulator holds fuel pressure to help prevent vapor lock and improve hot restarts.

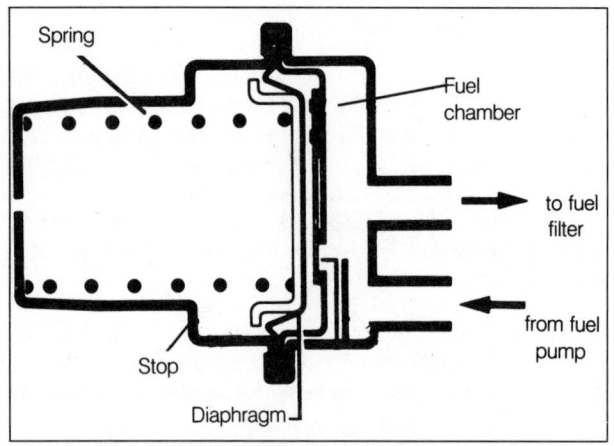

Fig. 3-2. Fuel accumulator. When engine starts, fuel chamber fills with fuel. Pump noise is dampened by the diaphragm. With engine off, accumulator spring maintains pressure in lines to help hot starts.

The main components of the accumulator are a spring and a diaphragm that separates a spring chamber and a fuel chamber. When the pump begins running, pressure builds in about 1 second as fuel fills the fuel chamber and presses the diaphragm against the stop. At pump shut-off, the pressure drops rapidly to about 2 bar (30 psi). See Fig. 3-3. This residual pressure is maintained by the force of the spring pressing against the diaphragm and the fuel in the fuel chamber. The accumulator holds pressure for a half hour or more to prevent vapor lock caused by fuel vaporization in the fuel lines of a hot engine.

3.2 Fuel Distributor

The fuel distributor is the part of the mixture-control unit that meters the fuel and then distributes it to the individual injectors. See Fig. 3-4. All fuel metering takes place in the center of the fuel distributor at the control plunger as it rises and falls. Secondary mixture control by the control systems to fine-tune the basic air-fuel mixture takes place through the manipulation of fuel pressures in the fuel distributor. The fuel distributor's role in fuel metering is discussed in **4. Fuel Metering**; the action of the control systems to further manipulate fuel pressures is discussed in **5. Control Systems**.

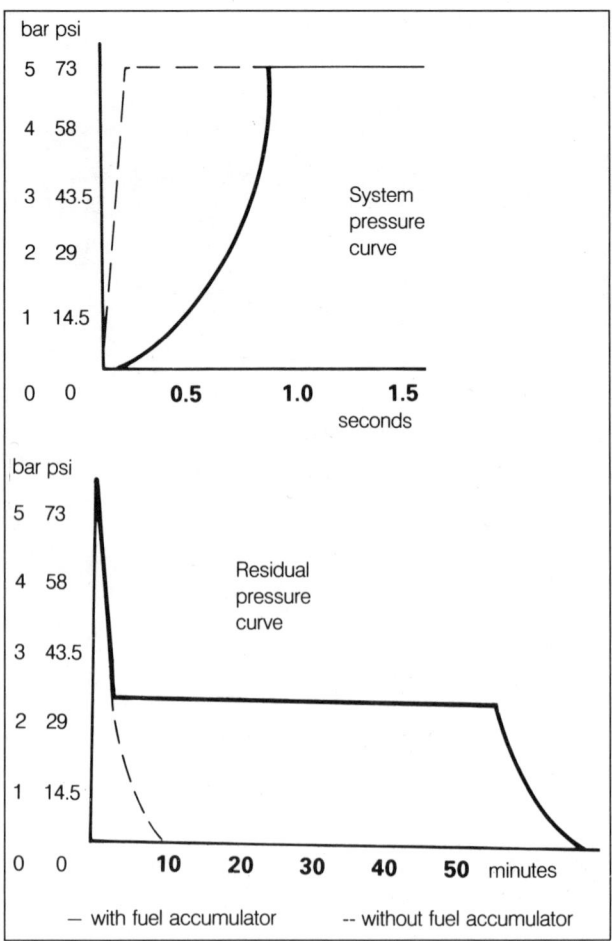

Fig. 3-3. Fuel pressure builds rapidly during cranking, damped by accumulator (**top**). After shut-off, pressure falls rapidly to close injectors, preventing fuel dribble at injector tips (**bottom**). Then accumulator maintains pressure in lines to aid hot starts. As engine cools over time, pressure falls — rate depends on minor leaks in lines and injectors.

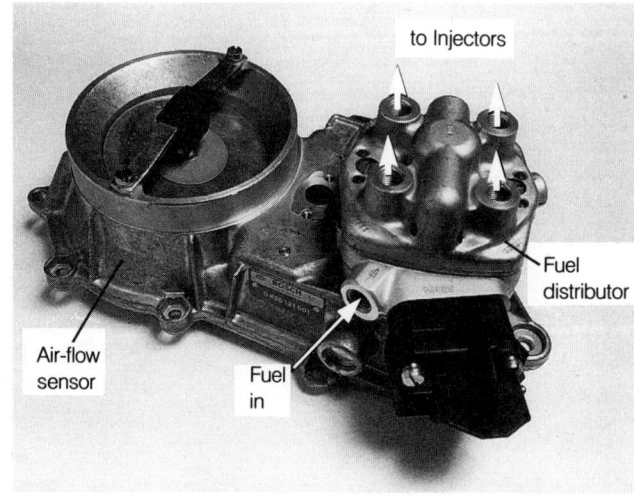

Fig. 3-4. Fuel distributor is part of mixture control unit; bolted to air-flow sensor. It meters and distributes fuel to the injectors.

3.3 System-Pressure Regulator

Continuous injection depends on fuel pressure to manipulate the air-fuel mixture, so that pressure must be closely regulated. The system-pressure regulator maintains system pressure (sometimes called primary pressure) in the fuel system at the specified level. There are two different types of pressure regulators, depending on the system. K-basic and K-lambda use a pressure relief valve located in the fuel distributor housing; KE systems use a separate diaphragm pressure regulator. Both regulate pressure by recirculating excess fuel back to the fuel tank.

K-Basic, K-Lambda

In K-basic and K-lambda, the system-pressure regulator is built into the fuel distributor. System pressure is set by spring force against the regulator relief valve. When the pump is turned on, fuel pressure builds until it overcomes spring pressure and opens the relief valve, returning excess fuel to the tank. See Fig. 3-5.

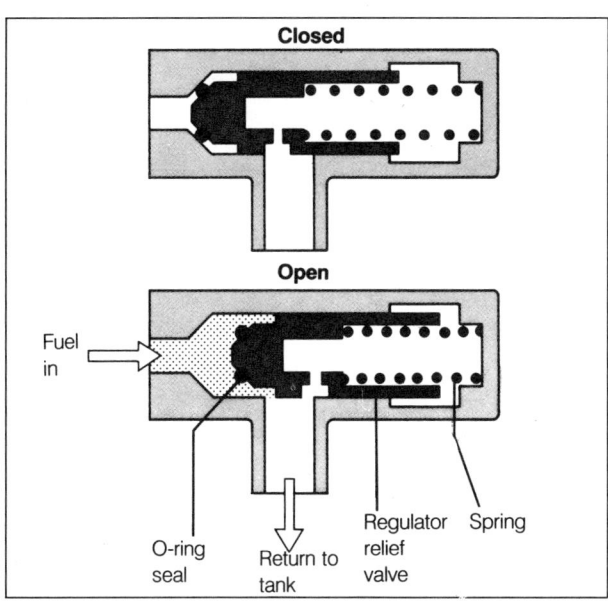

Fig. 3-5. Basic operation of K-basic and K-lambda system-pressure regulator. Fuel-pressure force overcomes spring force and opens regulating valve at specified pressure, returning excess fuel to the fuel tank.

There are actually two variations of the K-basic and K-lambda system-pressure regulator, depending on when the system was manufactured. The first one, shown above, was used until about 1978 and controls only system-pressure. Beginning about 1978, a new type of system-pressure regulator performs a second pressure-control function. See Fig. 3-6.

This slightly larger type includes what is known as a push valve. In the push-valve regulator, the opening of the relief valve also opens the push valve. The push valve controls the return of fuel from something called the control-pressure regulator. You'll see more about the control-pressure regulator in **5. Control Systems**, but for now, all you need to know is that the push valve O-ring shuts off fuel return from the control-pressure regulator to better maintain residual system pressure. The difference between the two regulators will affect how you measure fuel pressure in chapter 6; you can tell the two apart by the position of the system-pressure regulator on the fuel distributor. See Fig. 3-7.

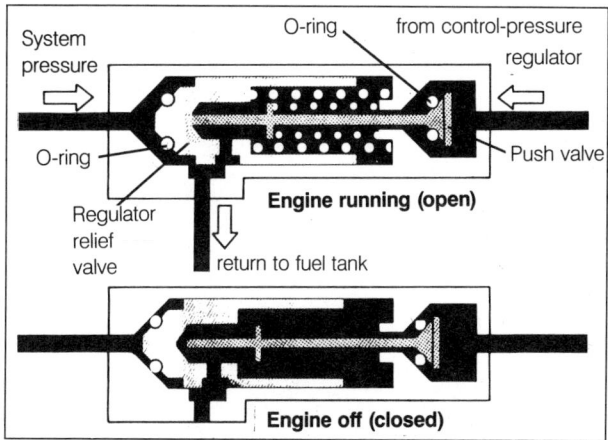

Fig. 3-6. Post-1978 K-basic and K-lambda system-pressure regulator with push valve. When engine is running, relief-valve plunger opens push valve. With engine off, larger spring closes relief plunger and smaller spring closes push valve, sealing control pressure in the system.

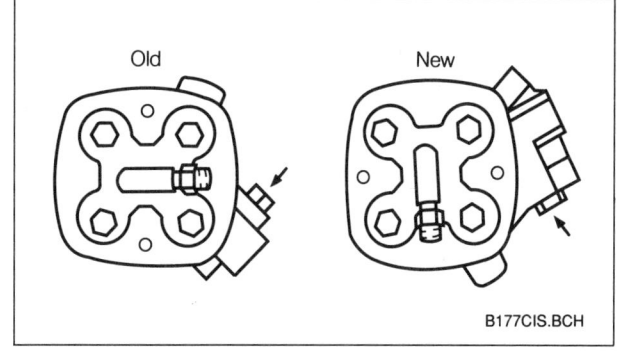

B177CIS.BCH

Fig. 3-7. K-basic and K-lambda system-pressure regulator (arrows) is built into the fuel distributor casting. Learn the difference between old (used before 1978) without push valve, and new with push valve.

5

KE Systems

The KE system-pressure regulator is mounted outside the fuel distributor as shown in Fig. 3-8. System pressure is set by fuel-pressure deflecting a diaphragm. See Fig. 3-9. When the fuel pump is running, fuel pressure builds until it overcomes spring pressure in the regulator and moves the diaphragm. This opens the regulating valve to return excess fuel to the tank and maintain pressure at the specified level. The thinner return line carries fuel from the fuel distributor to the regulator; the thicker line supplies fuel from the pump.

Fig. 3-8. KE pressure regulator looks like that of pulsed systems. It controls system pressure, and also controls return flow from the fuel distributor.

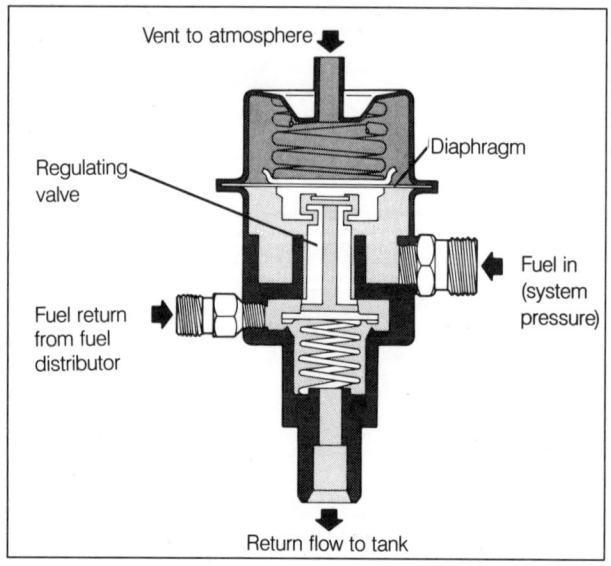

Fig. 3-9. KE fuel pressure regulator in open position.

3.4 Fuel Injectors

The fuel injectors used on continuous systems are mechanical. See Fig. 3-10. Pressure of the fuel delivered from the mixture-control unit overcomes spring pressure and opens the injector. Injector opening pressure is determined by the spring in each injector. There are no electric signals, and the injectors do not meter the fuel; they just continually inject fuel and atomize it. When an injector is working, you can hear a chatter as fuel pressure vibrates the open valve pins to atomize the fuel. Delivery stops when fuel pressure drops; the injector closes to keep fuel in the injector lines.

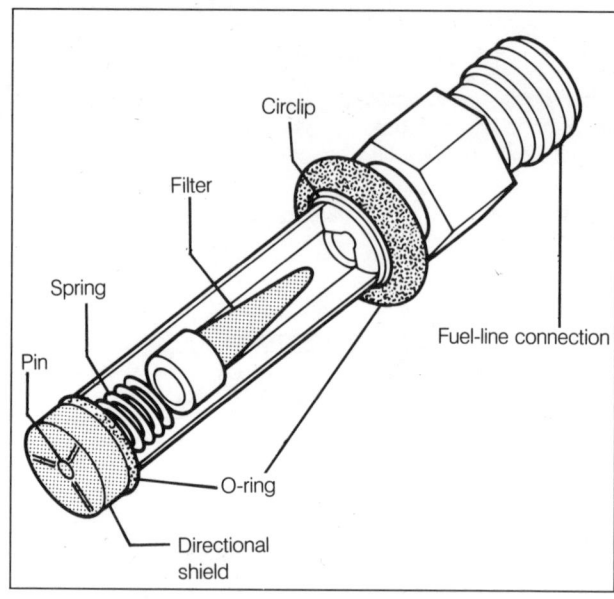

Fig. 3-10. Mechanical injectors are simple, no electrical connections. K-basic, K-lambda, and KE are all similar.

Air-Shrouded Injectors

Beginning in 1984, some KE systems have air-shrouded injectors to improve fuel atomization. Better fuel atomization means more efficient combustion. This saves fuel and reduces exhaust emissions. The injectors are mounted in a special air shroud in the intake manifold as shown in Fig. 3-11. At idle, lower manifold pressure at the injector tips induces air flow from upstream of the throttle to flow around the injected fuel stream.

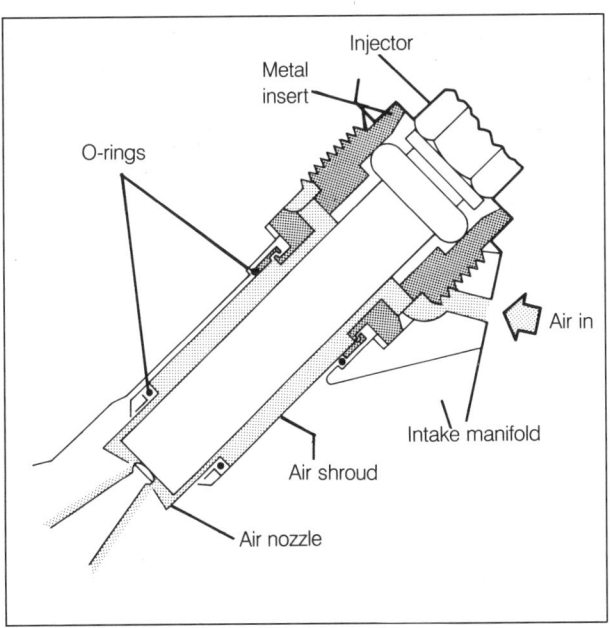

Fig. 3-11. Most KE injectors are air-shrouded to improve fuel atomization for smoother idle, better fuel economy and reduced emissions.

4. FUEL METERING

In continuous systems, all fuel metering takes place in the fuel distributor. This section describes basic fuel metering for all systems: matching a proportional amount of fuel to the air entering the engine to create the correct air-fuel mixture. This basic air-fuel mixture is then fine-tuned by the control system to compensate for different operating conditions. The control systems for each system are different; I'll describe them in detail in the next section, **5. Control Systems**.

4.1 Basic Fuel Metering

Basic fuel metering occurs when air entering the engine lifts the sensor plate, which in turn lifts the control plunger in the fuel distributor, allowing fuel to flow to the injectors. See Fig. 4-1. You've already seen how the air-flow sensor measures air flow and lifts the plunger, now I'll show you just how the control plunger and fuel distributor turn that air-flow measurement into a proportional fuel quantity.

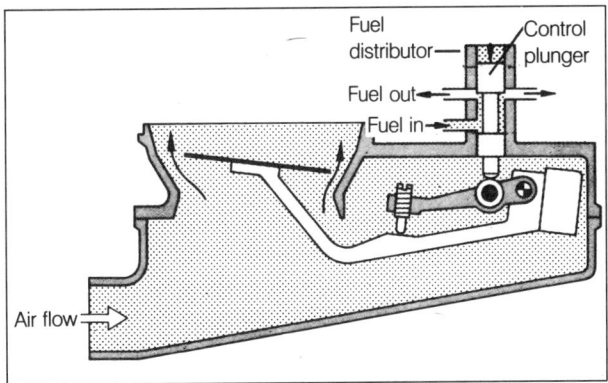

Fig. 4-1. Schematic view of mixture-control unit. Basic fuel metering begins when air-flow sensor lifts control plunger in fuel distributor, allowing fuel to flow to injectors.

Control Plunger Movement and Slit Size

The control plunger is mounted in the center of the fuel distributor, in a precision housing called the "barrel". See Fig. 4-2. The fuel pump supplies fuel at system pressure to the lower portion of the barrel. As the plunger is lifted, the fuel flows though upper slits to the injectors. See Fig. 4-3. There is one slit for each injector.

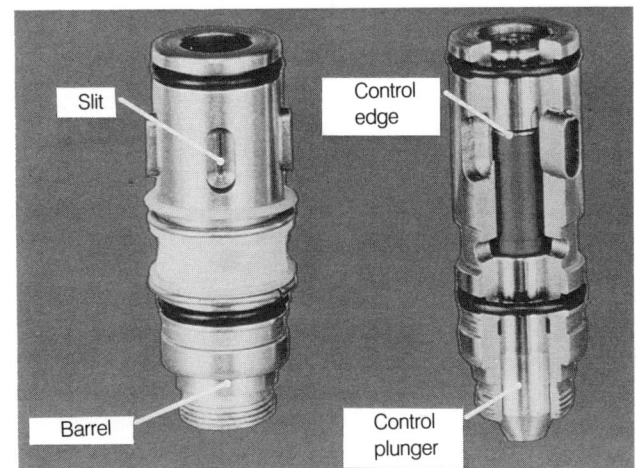

Fig. 4-2. The control-plunger and its barrel. Metering of fuel takes place at precision slit, one for each cylinder. Same for K-basic, K-lambda, KE.

5

Fuel Metering

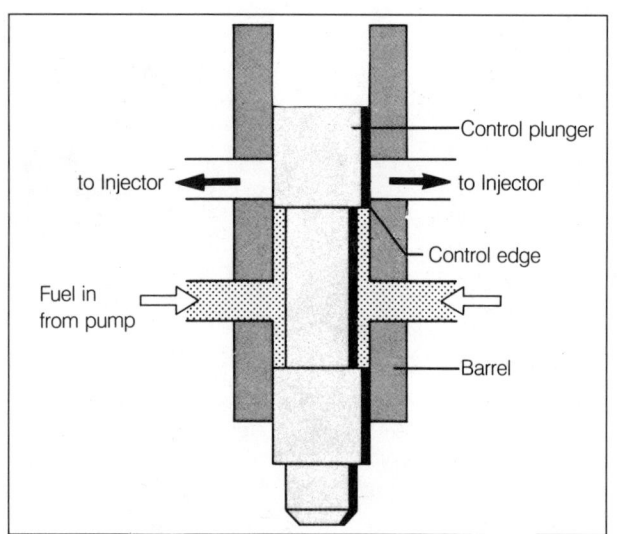

Fig. 4-3. Fuel flow through control-plunger barrel. As the plunger control edge rises, it uncovers the slits. Precisely sized slits are very narrow: approximately 0.2 mm (.008 in.).

The key to control-plunger fuel metering is the control edge of the plunger. This controls the amount that the upper slits are open, and therefore the amount of fuel that flows through the silts to the injectors. See Fig. 4-4. As the plunger is lifted in the barrel, the control edge of the plunger exposes more of each slit. Low air flow into the engine causes a small lift of the sensor plate and plunger; a small amount of each slit is exposed, delivering a small quantity of fuel through each slit. Increased air flow increases sensor-plate and plunger lift; more of each slit is exposed and fuel flow increases.

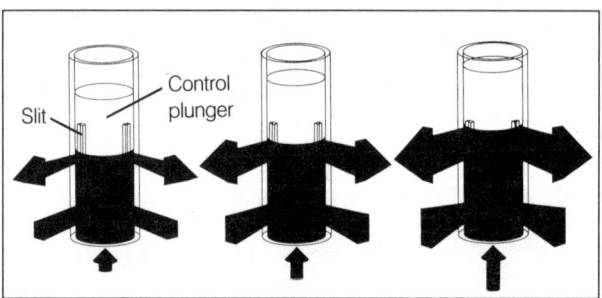

Fig. 4-4. The more the control plunger rises, the greater the flow of fuel through each slit.

Fuel Metering and Pressure Drop

Fuel flow to the injectors is influenced not only by the size of the exposed slit, but also by the pressure drop at the slit. Think of it this way: the pressure drop at the metering slits is the pressure differential between system pressure inside the slit pushing fuel out, and the lower pressure outside the slit in the fuel distributor. If the pressure drop changes for a given air flow and plunger lift, fuel flow through the slit also changes. For example, increased pressure drop forces more fuel to flow through the slit; fuel delivery increases.

Changing pressure drop at the slits causes problems with the linear relationship between air flow and fuel flow. When the control plunger rises, a more open slit changes the pressure drop and upsets the basic proportional relationship between air flow and fuel flow. So while increased air flow produces a linear increase in plunger lift, it does not necessarily provide a linear increase in fuel flow. At a certain point the change in pressure will cause the air-fuel mixture to be incorrect.

To hold constant pressure drop at the slits for the entire range of plunger lift, all continuous systems have something called Differential-Pressure Valves in the fuel distributor.

Differential-Pressure Valves and Pressure Drop

The differential-pressure valves maintain a constant pressure drop at the metering slits of the control plunger by reacting to the increased fuel flow that comes with a larger slit opening. There is one differential-pressure valve for each injector. They are mounted in chambers in the fuel distributor, and have a flexible metal diaphragm that separates the upper and lower halves of the chamber. See Fig. 4-5.

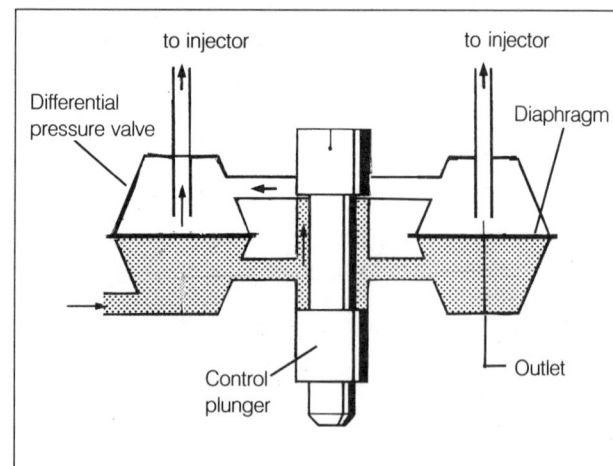

Fig. 4-5. Schematic view of fuel distributor shows location of differential-pressure valves. There is one valve for each fuel injector. Arrows indicate fuel flow through the control plunger and differential-pressure valves.

As the control plunger rises, and fuel flow into the differential-pressure valve increases, the fuel pressure deflects the diaphragm. This causes a proportional increase in the area of the fuel outlet to the injector and maintains the same pressure drop at the control-plunger slit. Even though there is more fuel flow through a more open slit, the differential between pressure inside the slit and pressure outside the slit is constant.

It may seem, because the deflection of the diaphragm enlarges the outlet to the fuel injectors, that fuel metering is a function of how close the diaphragm is to the outlet. But this is not the case. All fuel metering takes place at the control-plunger slits. The movement of the diaphragm acts only to maintain constant pressure drop—to eliminate pressure drop as a variable and maintain accurate metering to each injector, regardless of changing fuel flow rates.

As you'll see in **5. Control Systems**, the differential-pressure valves play an important part in how the control systems fine-tune the basic air-fuel mixture. The control systems manipulate fuel pressure in the differential-pressure valves to change the pressure drop at the slits for a given plunger lift.

Comparison of Differential-Pressure Valves

In each differential-pressure valve the diaphragm separates the flow of injector fuel through the upper chambers from the fuel in the lower chambers. The upper chambers are separate so that the pressure-drop regulating function is independent for each injector and its line.

In K-basic and K-lambda, there is a spring which pushes down on the diaphragm to reduce the pressure in the upper chamber by 0.1 bar (1.5 psi). See Fig. 4-6. It is this pressure drop which causes fuel to flow out through the slits to the upper chambers and from there to the injectors.

In K-basic, the lower chambers are always at system pressure so the job of each diaphragm is to flex up and down to maintain the constant 0.1 bar pressure drop regardless of flow in the upper chamber.

In K-lambda systems, the electronic control systems operate to change the lower-chamber pressure. The flexible diaphragms still move up and down to equalize the pressures in between the upper and lower chambers, except as modified by the spring in each differential-pressure valve and by the change in lower-chamber pressure.

In KE systems, the control systems also change the lower chamber pressure, but the valves are different from K-lambda. The spring presses up from the bottom as shown in Fig. 4-7. The reasons for this difference are explained in detail in **5. Control Systems.**

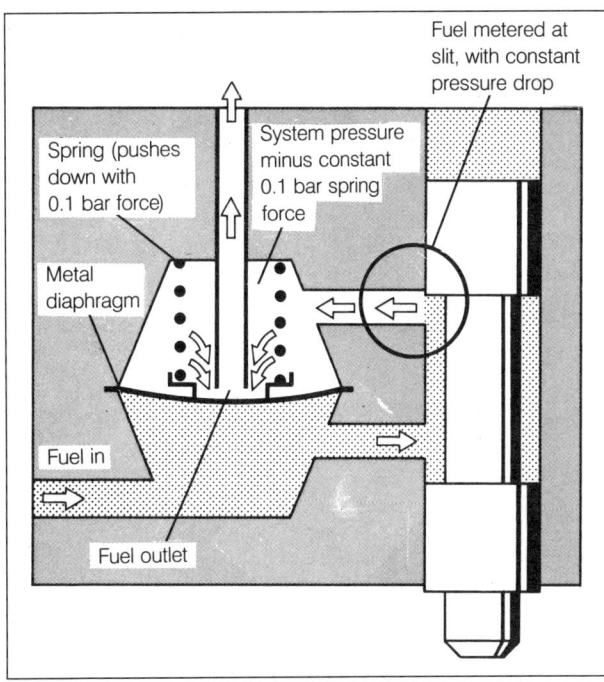

Fig. 4-6. Cross-section of K-basic differential-pressure valve in fuel distributor shows deflection of diaphragm from fuel flow. Accuracy of fuel metering is precise, based on constant pressure drop for each cylinder slit regardless of changing fuel flow rates.

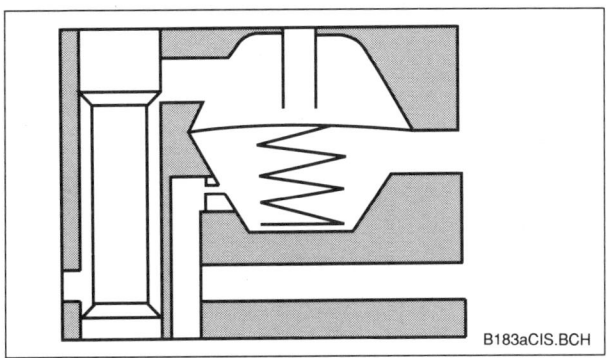

Fig. 4-7. In KE systems, spring in differential-pressure valve presses up against diaphragm. Flexible metal diaphragm still deflects with fuel flow, as modified by lower-chamber pressure.

Summary of Basic Fuel Metering

To briefly summarize basic fuel metering of continuous systems: Air flow entering the engine lifts the sensor plate and, in turn, the control plunger in the fuel distributor. Fuel-pressure counterforce applied to the top of the plunger balances the force of the air entering the engine and stabilizes plunger and sensor-plate movement. As the control plunger rises, its control edge uncovers the metering slits in the plunger barrel. As the slits open, fuel flows through the differential-pressure valves and out to the injectors. Fuel pressure opens the injectors, and the injectors atomize the fuel.

Fuel Metering

The differential-pressure valves operate to equalize pressures between each upper chamber and its lower chamber. As a result, even as the delivery quantity changes to meet changing engine operating conditions, the pressure drop at each slit remains consistent with the spring pressure in the chamber and the lower chamber pressure. That insures equal delivery to each cylinder for increased engine smoothness.

5. CONTROL SYSTEMS

In **4. Fuel Metering**, I've described the basic fuel metering function of the air-flow sensor and the fuel distributor—the basic matching of fuel and intake air in the proper proportions. As I've described in chapter 2, modern fuel injection systems must make even finer adjustments to the mixture to maintain the stoichiometric air-fuel ratio and to compensate for the demands of different operating conditions.

This section describes how control systems manipulate fuel pressures in the system to fine-tune the basic air-fuel mixture. For any given air flow the basic fuel metering system creates the basic air-fuel mixture, and the control systems enrich or lean the basic mixture for more precise control of fuel metering.

There are two methods that the control systems use to adjust the basic mixture—changing slit size, and changing pressure drop at the slits. Later, you'll see how the control systems of each of the three continuous systems, K-Basic, K-Lambda and KE, apply these methods in different ways.

Changing Slit Size

The first method of adjusting fuel metering is to change the position of the control plunger for a given air flow by changing the counterforce on the plunger. You've seen how the fuel pressure counterforce is used to oppose the lift of the control plunger (against the air-flow force). Reducing the counterforce allows the sensor plate to lift the plunger higher for the same air flow, exposing more of the slits and allowing more fuel to flow for a richer mixture. Increasing the counterforce reduces plunger lift for a given air flow, exposing less of the slits and resulting in less fuel flow for a leaner mixture. This control is a fine adjustment to the basic movement of the plunger by the air-flow sensor; I'll refer to it as "changing slit size."

Changing Pressure Drop

The second method the control systems use for mixture adjustment is the manipulation of pressures on either side of the differential-pressure valves. This changes the deflection of the diaphragm and, in turn, changes the pressure drop at the slit. Remember, pressure drop here is the difference between pressure inside the slit and pressure outside the slit. For any given air flow and plunger lift, a change in the pressure drop changes the amount of fuel flowing through the slit, enriching or leaning the mixture. This control is in addition to the normal action of the differential-pressure valves; I'll refer to it as "changing pressure drop."

5.1 K-Basic—Changing Slit Size

K-basic systems use fuel under pressure as a hydraulic control fluid to maintain counterforce on the control plunger. This is called control pressure. To fine-tune the basic fuel metering, K-basic systems vary control pressure to change the counterforce on the control plunger. This changes the lift of the plunger and the size of the metering slit for a given air flow. The K-basic control system does not affect the differential-pressure valves. They deflect in a normal way to keep the pressure drop at the control-plunger metering slits constant.

Control-Pressure

As shown in Fig. 5-2, control pressure is applied through passageways in the fuel distributor. A flow restrictor admits fuel pressure from the system-pressure side to the control-pressure side without delivering full system pressure to the top of the plunger. A damping restrictor admits control-pressure force to the top of the plunger in order to balance the air-flow force of the air-flow sensor plate. The damping restrictor is large enough to permit transfer of pressure, but small enough to dampen any possible oscillation of the sensor plate and plunger which could result from pulsating air flow.

Control-Pressure Regulator

The control-pressure regulator adjusts control pressure by returning excess fuel from the control pressure circuit to the fuel tank. See Fig. 5-3. If more fuel is bled off to the tank, control pressure is reduced. If less fuel is bled off, control pressure is increased.

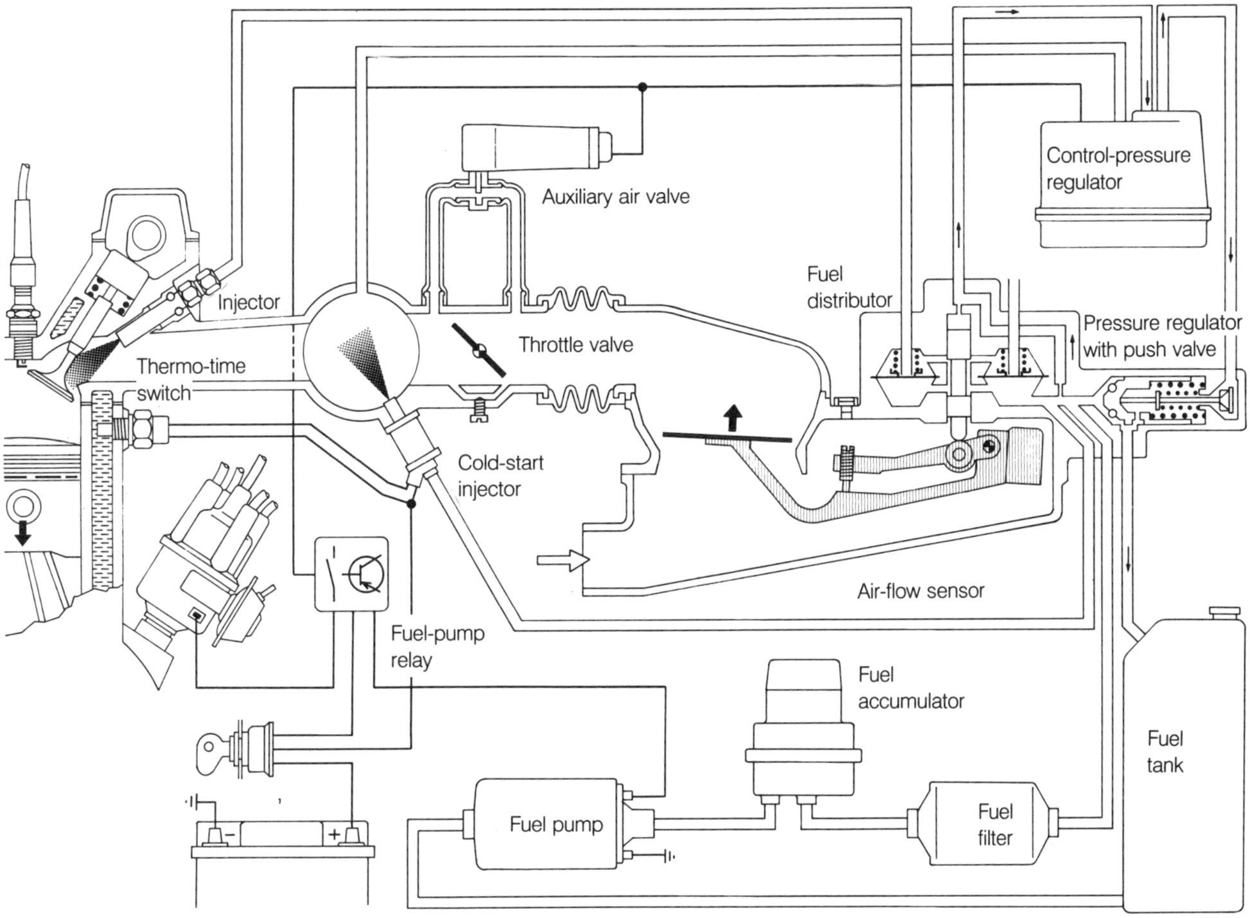

Auxiliary air valve

Injector

Throttle valve

Thermo-time
switch

Cold-start
injector

Fuel
distributor

Control-pressure
regulator

Pressure regulator
with push valve

Air-flow sensor

Fuel-pump
relay

Fuel
accumulator

Fuel
tank

Fuel pump

Fuel
filter

Fig. 5-1. Schematic of K-basic system shows all of the components. Control-pressure regulator and its compensation of the basic air-fuel mixture is described in this section.

5

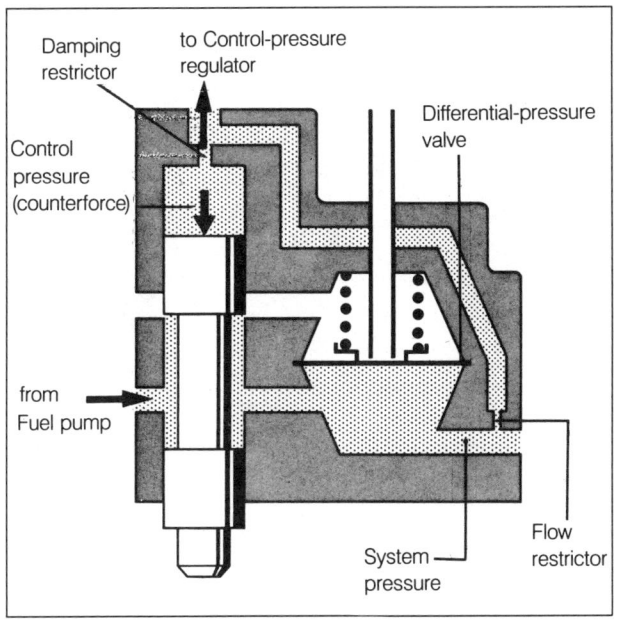

Damping
restrictor

to Control-pressure
regulator

Differential-pressure
valve

Control
pressure
(counterforce)

from
Fuel pump

System
pressure

Flow
restrictor

Fig. 5-2. Relationship of control pressure and system pressure in K-basic. System pressure is constant, control pressure varies.

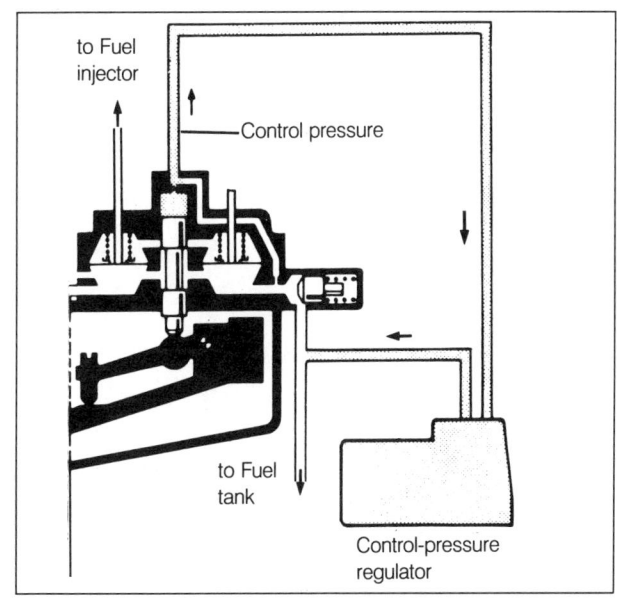

to Fuel
injector

Control pressure

to Fuel
tank

Control-pressure
regulator

Fig. 5-3. Control pressure regulator adjusts control-pressure counterforce on the plunger by bleeding off fuel to the fuel tank.

Control Systems

As shown in Fig. 5-4, the control-pressure regulator consists of a valve mounted on a bimetal arm. The arm is made up of two different metals of different expansion rates. It bends one way when it is cold and the other way when it is warm, changing control pressure in response to temperature.

When cold, the natural shape of the bimetal arm pulls the diaphragm away from the valve. As a result, the valve:

● increases fuel return to the tank, and

● reduces control pressure

As the regulator warms, the bimetal arm bends the other way. The spring closes the valve, which:

● decreases fuel return to tank, and

● increases control pressure

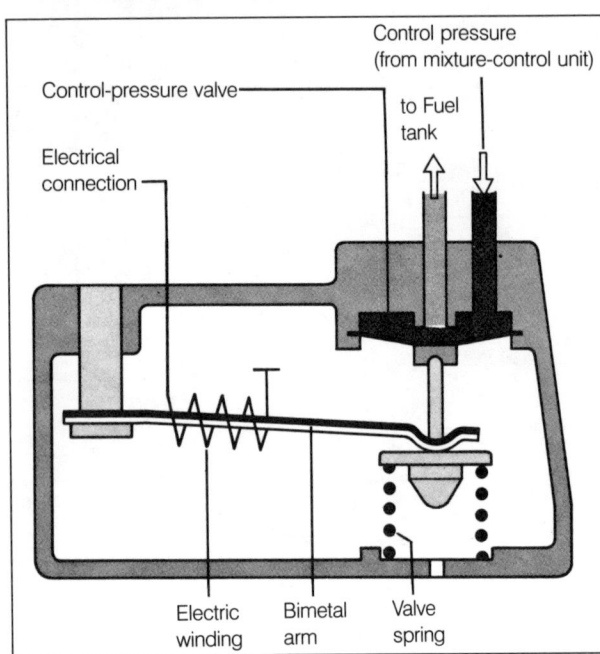

Fig. 5-4. K-basic control-pressure regulator. When cold, bimetal arm bends down. This increases fuel return and lowers control pressure, reducing counterforce on the control plunger. When warm, spring lifts diaphragm to increase control pressure and counterforce on the plunger.

Control Pressure and Mixture Adjustment

Remember that the basic action of the fuel-pressure counterforce on the plunger is to balance the force of the air flow entering the engine, so that the movement of the sensor plate and plunger is opposed. If this counterforce is changed, but the air flow stays the same, the amount that the sensor plate and plunger lift will also change.

The primary action of the control-pressure regulator is to enrich the basic mixture for cold operating conditions by reducing the control-pressure counterforce on the control plunger. As you saw above, when the control-pressure regulator is cold, control-pressure counterforce on the plunger is low. The plunger is able to rise higher for a given air-flow force, uncovering more of the metering slits so more fuel flows for the same amount of air. This enriches the mixture, improving cold driveability.

As the control-pressure regulator and its bimetal arm warm up, control-pressure increases. This increases counterforce on the plunger and it is forced lower for a given air-flow force. This reduces the size of the slits and leans the mixture. The regulator eventually warms to the point where it no longer compensates the mixture for cold conditions; only the basic sensor-plate lift of the control plunger determines the mixture.

A typical control pressure at freezing temperatures might be 1 bar (14.5 psi), and warm control pressure 3.5 bar (51 psi). As you'll see in chapter 6, measuring control pressure is the best way to check the operation of the regulator as pressures change with temperature.

The control-pressure regulator is usually located on the engine block, as shown in Fig. 5-5, so that engine-block temperature can affect the position of the bimetal arm. The regulator is often buried and hard to find. Look for two fuel lines to the fuel distributor: one is the control-pressure line to the regulator, and one returns fuel to the tank. In pre-1978 systems, return fuel flows uncontrolled through a "tee" into the return line from the system-pressure regulator. In 1978 and later, return fuel flow is controlled by the push valve, through the system-pressure regulator and back to the tank. See earlier Fig. 3-7.

Fig. 5-5. Control-pressure regulator (arrow) is often buried, located on the block to react to engine temperature. Look for two lines from the mixture-control unit.

After a cold start, fuel enrichment needs to be cut back in one or two minutes, but the engine block might take 10 minutes to warm up to the point where it affects the bimetal arm in the control-pressure regulator. To insure that the mixture is not enriched for too long, an electric heating coil warms the arm when the ignition is turned on. Warm-up needs are different for each car model, so do not interchange regulators.

Additional Control-Pressure Compensations

The control-pressure regulator was originally called the warm-up regulator (WUR), because in its early applications, warm-up compensation was its only job. In some cars, control-pressure changes are also used for other reasons; to enrich the mixture during full-load operation or when under boost from a turbocharger, or to lean the mixture for altitude compensation.

For load compensation, the special control-pressure regulator has a second diaphragm which responds to engine load by sensing intake manifold pressure. See Fig. 5-6. High manifold pressure from an open throttle or from turbocharger boost influences the diaphragm and lowers control pressure. Lower control pressure allows greater plunger lift, more open slits, and greater fuel flow for a richer mixture. You can recognize these special control-pressure regulators by the vacuum-hose connection to the intake manifold, and by the extra chamber along the bottom of the regulator.

At part-throttle, when little enrichment is needed, manifold pressure is less than atmospheric pressure in the lower chamber of the diaphragm. This raises the diaphragm to add its spring pressure to the control-pressure valve. Control pressure is high, and there is little additional lift of the control plunger to enrich the mixture.

Altitude compensation is provided by an aneroid chamber at the bottom of the regulator. This chamber expands as the air gets thinner at higher altitudes; this in turn acts on the control-pressure valve to increase control pressure and lean the mixture.

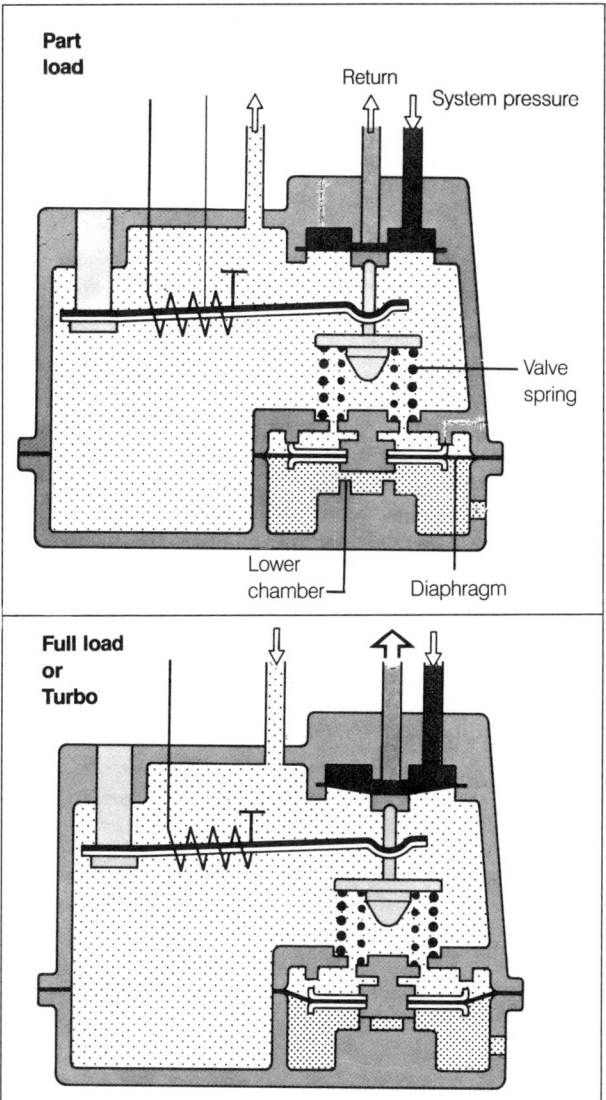

Fig. 5-6. Full-load control-pressure regulator on some K-basic and K-lambda systems has vacuum connection to intake manifold so control pressure is modified according to manifold pressure. Full-load or turbocharger-boosted manifold pressure (**bottom**) presses down on lower diaphragm to reduce control pressure and enrich mixture.

Control Systems

5.2 K-Lambda – Changing Slit Size and Pressure Drop

K-lambda control systems adjust the basic air-fuel mixture by a combination of two ways. The first is by changing slit size for a given air flow. This is done just as in K-basic, by the control-pressure regulator, primarily for cold running.

The second method of adjusting the basic air-fuel mixture is by changing the pressure drop at the control-plunger slit so that fuel flow changes for a given plunger lift. This is done by manipulating pressures in the differential-pressure valves. Pressure in the valves' lower chambers is controlled by the Lambda Control System, in response to the signals from a lambda sensor in the exhaust system.

K-Lambda Differential-Pressure Valves and Mixture Adjustment

As you saw earlier, the K-basic differential-pressure valves hold the pressure drop constant at the metering slits of the control plunger. The deflection of the valve diaphragm increases with fuel flow for this pressure-drop function. If air flow and plunger lift stay the same but the deflection of the diaphragm is changed for some other reason, the pressure drop, and therefore the amount of fuel flowing through the slits, will also change.

For example, with a given plunger lift and fuel flow through the slits the pressure drop at the slits is a constant. But if the lower-chamber pressure of the differential-pressure valves is lowered, the diaphragm will deflect further. This extra deflection of the diaphragm decreases pressure in the upper chamber and increases the pressure drop at the slits; more fuel flows for the same plunger lift, enriching the mixture.

K-lambda control systems use this method to adjust air-fuel mixture. The lower-chamber pressure of the differential-pressure valves is not system pressure, as in K-basic. The pressure in the lower chambers, and therefore the pressure drop at the slits, is controlled by a separate pressure circuit which is in turn controlled by the lambda control system. Fuel is admitted to the lower chambers through a restrictor, and flows back to the tank through the lambda valve and the fuel return lines. In that way, the pressure of all lower chambers is controlled by the lambda valve.

Lambda Control System

The lambda control system includes a lambda sensor, a lambda electronic control unit, and a lambda control valve. See Fig. 5-7. The lambda valve (sometimes called a frequency valve, or timing valve) looks similar to a pulsed fuel injector, as shown earlier in Fig. 1-5. It opens and closes in a similar manner, but its opening and closing does not meter fuel. Rather, the amount of time the valve is open controls the pressure in the lower-chambers of the differential pressure valves. This in turn changes upper-chamber pressure to change the pressure drop and change enrichment.

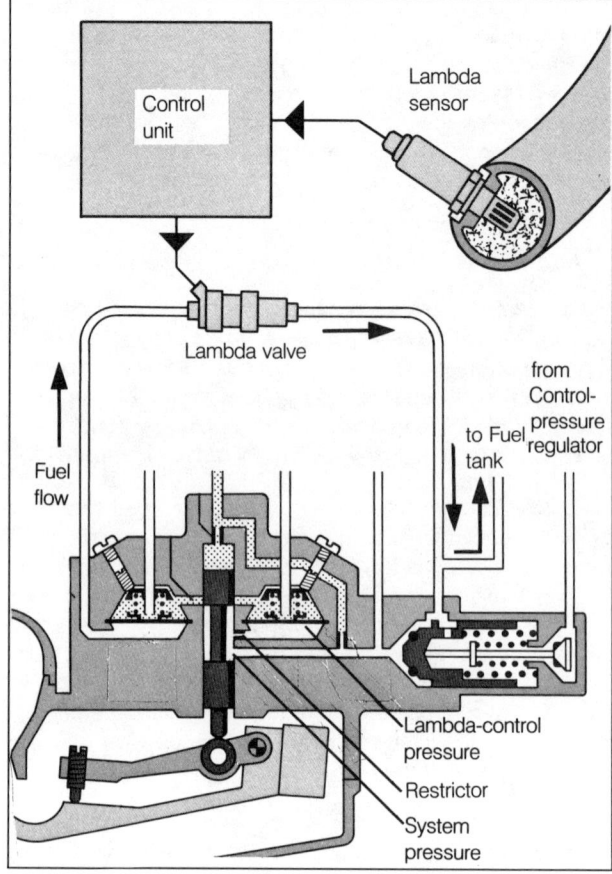

Fig. 5-7. In addition to K-basic control-pressure regulator, K-lambda control adds lambda sensor (or oxygen sensor), control unit, and lambda valve. Note additional fuel connection to pressure regulator.

The opening of the lambda valve is controlled by the electronic control unit, which monitors the signal from the lambda sensor (also known as the oxygen sensor.) The lambda sensor sends a signal to the control unit which is based on the oxygen content in the exhaust. The control unit then operates the lambda valve to control lower-chamber pressure.

The lambda valve is a solenoid valve which is either open or closed. The control unit sends on-off voltage pulses to the lambda valve. The frequency of these pulses determines the amount of time the valve is open and, therefore, the amount of fuel that is returned to the fuel tank. The on-off cycle of these pulses is known as the duty cycle. Duty cycle is measured on a scale of 0 to 100%. When the duty cycle is 50%, that means the lambda valve is open 50% of the time and closed the rest of the time.

Duty cycle can also be measured using a dwell meter scale of 0 to 90°. If you read a 50% duty cycle on a dwell meter set to the 4-cylinder scale, it would read 45° (50% of 90°).

Lambda Control and Changing Pressure Drop

When the lambda valve's duty cycle increases so that it is open longer and returning more fuel to the tank, lower-chamber pressure is reduced. The differential-pressure valve diaphragms deflect further, reducing upper-chamber pressure. This increases the pressure drop at each metering slit and more fuel flows through the slits for any given plunger lift. In short, the greater the duty cycle, the richer the mixture.

On the other hand, reduced duty cycle means less fuel bleeds off from the lower chamber; that increases chamber pressure and decreases deflection of the diaphragm. This reduces the pressure drop and fuel flow at the slits. So, less duty cycle means a leaner mixture. See Fig. 5-8.

Lambda Closed-Loop Control

The control of enrichment of the basic mixture by the lambda-control system is closed-loop, based on the signal from the lambda sensor. The control system continuously corrects the air-fuel mixture for a given plunger lift so that the mixture is held near the ideal, stoichiometric ratio necessary for catalytic converter operation.

As exhaust gas passes the lambda sensor, oxygen-content signals are sent to the lambda control unit. The control unit changes the duty cycle of the lambda-control valve to change pressure drop at the slits and adjust the mixture. The resulting change in exhaust-gas oxygen content changes the lambda-sensor signal. The control unit senses the change and continuously changes the duty cycle of the lambda valve. The cycle is continuous. See Fig. 5-9.

The oxygen content of the exhaust is continuously changing, so the lambda-control system averages to achieve the ideal mixture. It tends to correct to either side of the stoichiometric ratio, but always in a very narrow range. For more on closed-loop systems and the operation of the lambda sensor, see chapter 3.

Lambda Thermoswitch

Because lambda sensor operation depends on high temperature, the system operates open-loop before the sensor warms up. This can cause driveability problems and increased exhaust emissions due to imprecise mixture control. To correct this, many cars have a lambda thermoswitch, located in the coolant (or on the cylinder head of air-cooled Porsches) to give the control unit an additional input and insure proper fuel metering during engine warm-up.

When the engine is cold, the thermoswitch is closed. The control unit sends a fixed, slightly-rich duty-cycle signal of 60% to the lambda control valve. When the thermoswitch warms enough to open, the control unit sends a fixed middle signal, 50% duty cycle. When the lambda sensor reaches operating temperature so its signals are valid, the control unit switches to closed-loop operation. Remember, in closed-loop the duty-cycle signals are constantly changing, cycling back-and-forth between about 45% and 55%, so that the air-fuel ratio is maintained near stoichiometric; lambda (λ) = one.

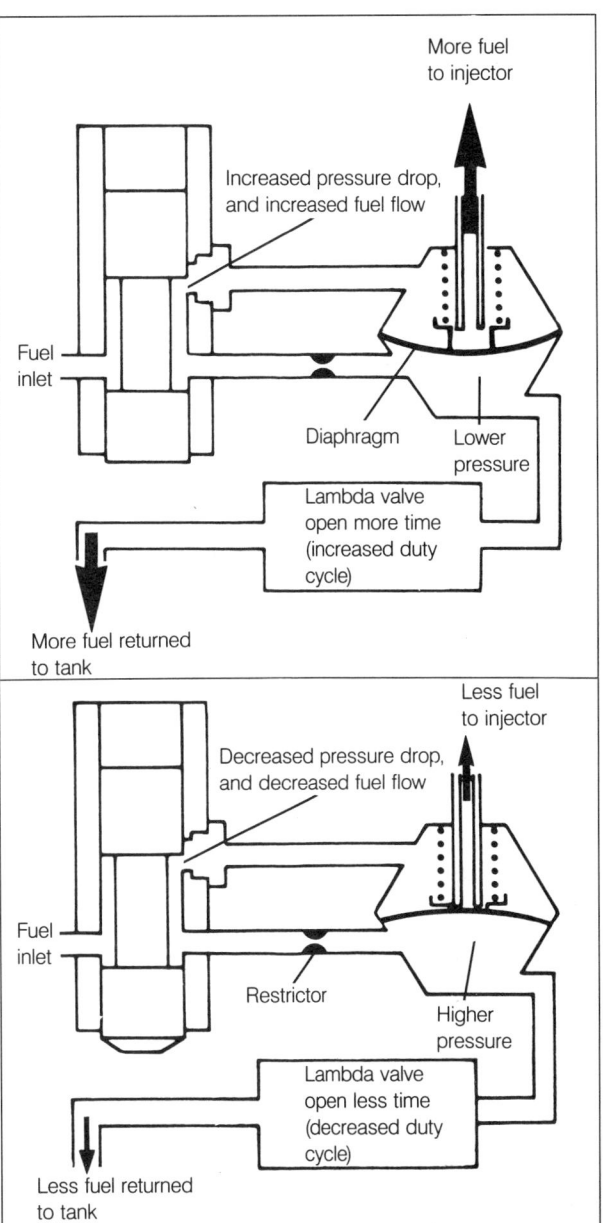

Fig. 5-8. K-lambda valve adjusts mixture by changing pressure drop at the metering slits. **Top**: More open valve returns more fuel, decreases lower and upper chamber pressure, and increases pressure drop and fuel delivery at slits, enriching the mixture. **Bottom**: Decreased opening of valve decreases fuel return, increases lower and upper chamber pressure, and decreases pressure drop and fuel delivery at slits, leaning the mixture. Note that fuel is metered at the slits, not at the diaphragm.

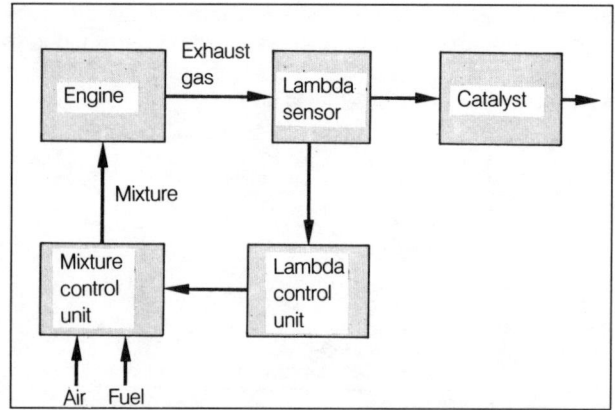

Fig. 5-9. When K-lambda operates closed-loop, the lambda sensor in the exhaust signals oxygen content to the lambda control unit. Signals to the lambda valve on the mixture-control unit change the air-fuel ratio to achieve stoichiometric mixture as burned in the engine.

5.3 KE Systems—Changing Pressure Drop

The KE control system adjusts the basic air-fuel mixture in only one way, by changing pressure drop at the control-plunger slits. KE systems do not have a control-pressure regulator or lambda-control valve. Instead, these controls are replaced by the KE control system—a series of sensors similar to those used in pulsed fuel injection, a control unit to monitor those sensors, and a pressure actuator on the fuel distributor, operated by the control unit.

For any given air flow, slit size in KE systems is governed only by the basic fuel metering—by the lift of the sensor plate. The KE control system changes pressures in the chambers of the differential-pressure valves, and therefore pressure-drop at the slits, to handle all adjustments to the basic air-fuel mixture to compensate for warm-up, emission control demands, and all other operating conditions.

On KE systems, the counterforce on the plunger that balances the air-flow force is system pressure, which is constant. So for any given air flow, plunger lift and slit size is constant. See Fig. 5-10. That's in contrast to K-basic and K-lambda, where reducing control pressure allows the plunger to rise further for a given air flow, increasing slit size to increase fuel delivery.

KE Differential-Pressure Valves and Separate Fuel Flows

In order to achieve more accurate pressure control by the KE control system, there are two fuel flow paths into the chambers of the differential-pressure valves. See Fig. 5-11. In one path, fuel enters the control-plunger barrel at system pressure. It then flows through each slit into each separate upper chamber at reduced pressure, and out to the injectors. I'll call this injector fuel. The basic rate of injector fuel flow is determined by the lift of the control plunger by the air-flow sensor, just as in the other systems.

In the other path, fuel at system pressure first flows to something called the pressure actuator. Fuel then flows at reduced pressure through the lower chambers and a restrictor and back to the fuel tank. I'll call this actuator fuel. The fixed restrictor controls the rate of flow out of the lower chambers. The variable pressure actuator controls the amount of fuel throughflow passing into the lower chambers, and therefore controls lower-chamber pressure and pressure drop at the slits. The operation of this control system is in addition to the basic fuel metering of the control plunger.

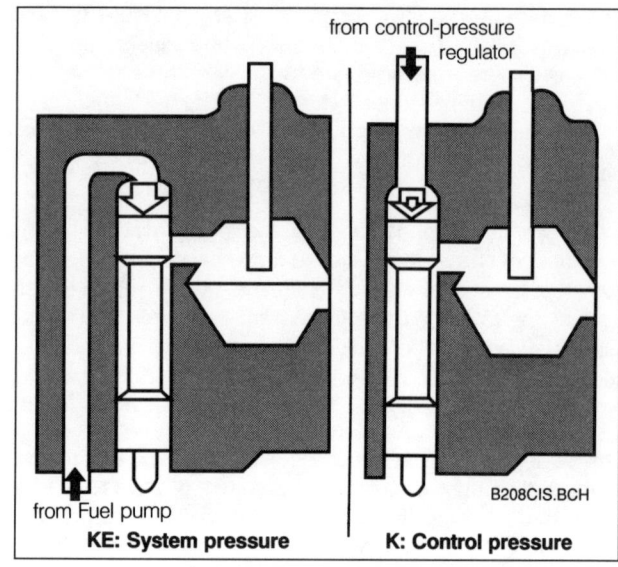

Fig. 5-10. KE meters with constant system pressure on the control plunger, so for any given air flow plunger lift is constant and slit size stays the same.

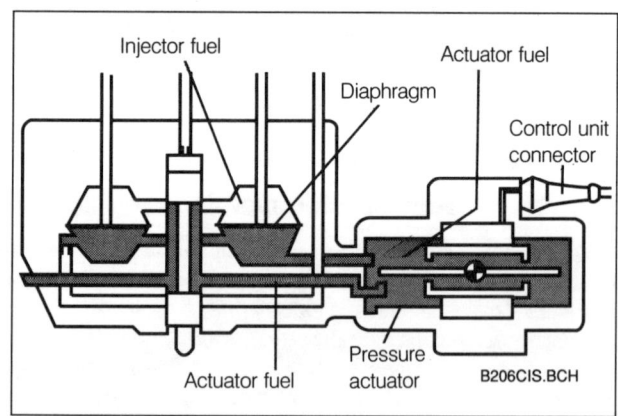

Fig. 5-11. KE injected fuel flow through each upper chamber is separated from actuator fuel flow below the diaphragm, which is spring loaded from below. KE actuator fuel flows continuously through the lower chambers. This throughflow is controlled by the pressure actuator.

KE Differential Pressure and Pressure Drop

In KE systems, the pressure drop at the slits, and therefore adjustment of the basic fuel metering, is controlled by the pressure drop at the entrance to the pressure actuator. This is also known as the Differential Pressure, the difference between actuator fuel pressure in the lower chambers and system pressure entering the pressure actuator.

The differential pressure directly affects the deflection of the differential-pressure valve diaphragms and so controls the pressure drop at the slits. Remember, by changing the pressure drop at the metering slits of the control plunger, fuel flow is changed for a given air flow and slit size.

The differential pressure is controlled by changing the flow of actuator fuel through the lower chambers. This control of throughflow is by the pressure actuator at its inlet. See Fig. 5-12.

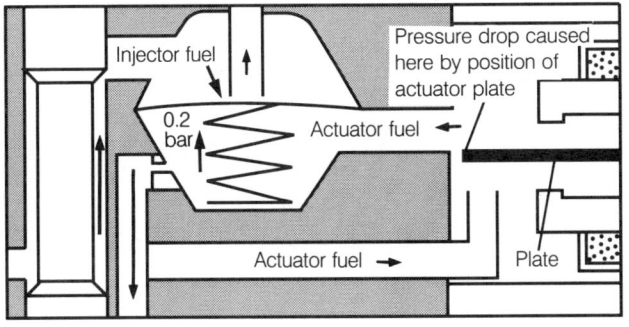

B183bCIS.BCH

Fig. 5-12. In KE systems, pressure drop at the slits is related through the diaphragm to pressure in the lower chamber. The differential pressure – the difference between system pressure and lower-chamber pressure – determines the pressure drop. Pressure drop at the slits is differential pressure minus 0.2 bar upward spring pressure.

Pressure Actuator

The pressure actuator is part of the mixture-control unit that controls actuator fuel flow through the lower chambers. For any given air flow and its related control-plunger slit opening, the actuator compensates the mixture by changing actuator fuel flow to change the differential pressure. The actuator is shown mounted on the fuel distributor in Fig. 5-13.

The key to the operation of the pressure actuator is the small plate valve that controls fuel throughflow. See Fig. 5-14. A small electromagnet causes the plate to move closer to or farther away from the fuel inlet to the actuator. For example, fuel flows to the actuator inlet at system pressure. Increased current flow through the electromagnet increases the magnetic force and

pulls the plate closer to the inlet. This reduces the flow and pressure of the actuator fuel in the lower chambers, changing the differential pressure. The actuator plate valve is controlled by the electronic control unit, as you'll see later. The valve is very small so that it can move rapidly in response to changing engine conditions.

Fig. 5-13. KE pressure actuator controls mixture by controlling flow of actuator fuel through the lower chambers.

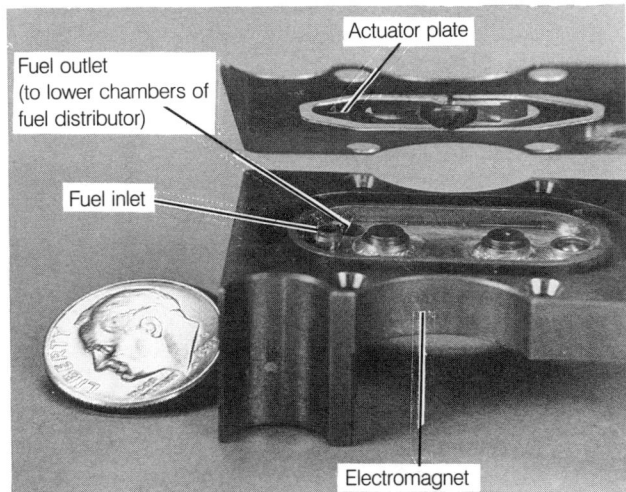

Fig. 5-14. KE pressure-actuator plate valve controls fuel flow through pressure actuator.

5

Pressure Actuator and Mixture Adjustment

Fig. 5-15 shows the relationship of differential pressure and pressure drop at the metering slits and how it affects basic fuel metering. When the plate is closer to the actuator inlet it restricts the flow of actuator fuel. Differential pressure increases and the diaphragms deflect further. This increases the pressure drop at the control-plunger slits and increases the flow of injector fuel for a given plunger lift. When the plate is further away from the inlet the differential pressure is less so the diaphragms deflect less. This decreases pressure drop at the slits and reduces the flow of injector fuel for a given plunger lift.

You can think of the pressure function of the actuator this way: Pressure drop at the metering slits is equal to system pressure minus upper-chamber pressure. Upper-chamber pressure is equal to system pressure plus the 0.2 bar spring pressure, minus the differential pressure. See Fig. 5-16. Remember, differential pressure is the difference between system pressure and pressure in the lower chambers controlled by the pressure actuator. As you'll see in chapter 6, you can't test differential pressure directly. You have to first measure system pressure and then subtract measured lower-chamber pressure to get the differential.

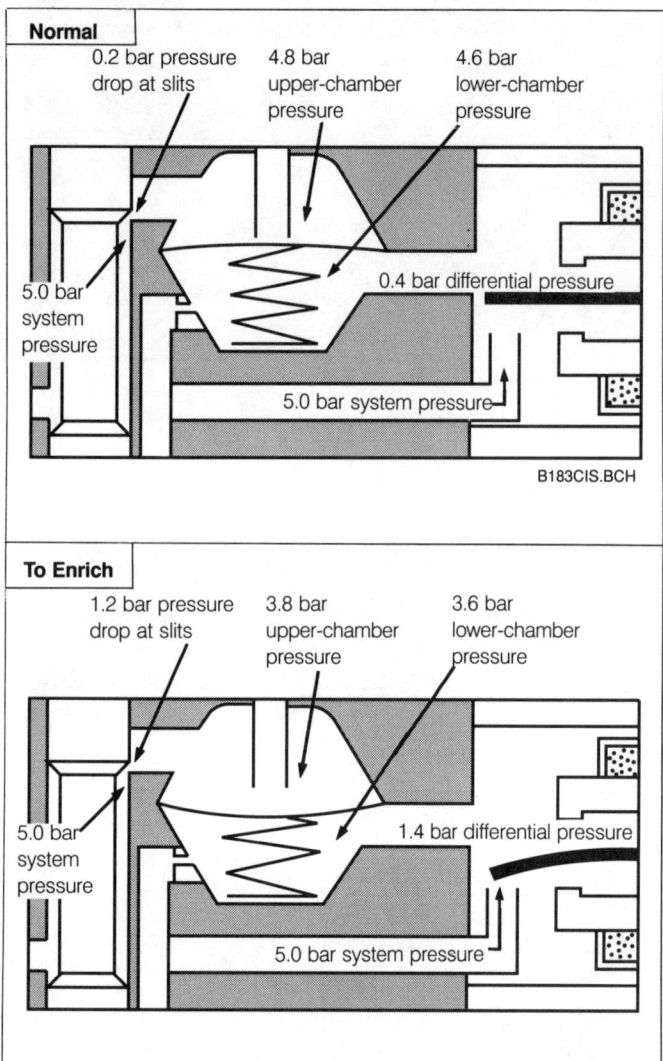

Fig. 5-15. Pressures in the KE differential-pressure valve chambers compared to differential pressure. **Top:** With little deflection of actuator plate differential pressure is low. Actuator fuel flow increases to increase upper-chamber pressure. Pressure drop at the slits drops, restricting the flow of injector fuel. **Bottom:** On the other hand, increased magnetic force increases differential pressure, increases pressure drop at slits, and increases fuel flow.

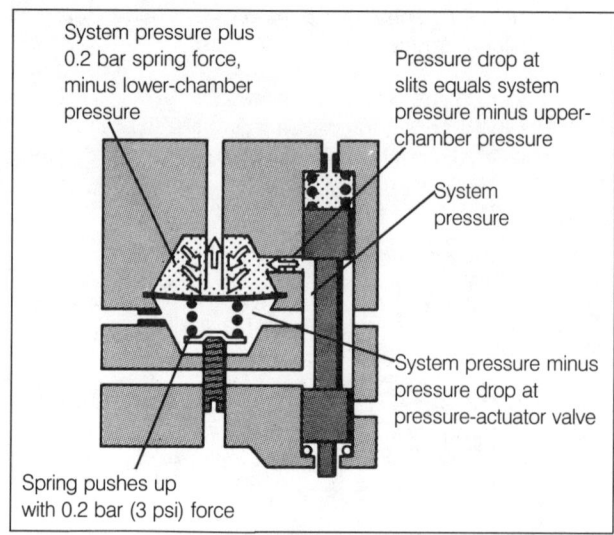

Fig. 5-16. Relationship of pressures in KE differential-pressure valves.

Control Unit and Sensors

The amount of current supplied to the pressure-actuator electromagnet for mixture control is determined by the KE control unit. To determine the correct current, the control unit depends on inputs from sensors much like those used on pulsed systems. See Fig. 5-17. These inputs include those from the ignition circuit for engine-speed, the lambda sensor, the engine-temperature sensor, the intake-air temperature sensor (on Mercedes-Benz, beginning 1986), the sensor-plate potentiometer for acceleration, and the throttle switch for closed and wide-open throttle indications. Control-unit outputs include the pressure-actuator signal and, on some models, signals to an idle-speed stabilizer.

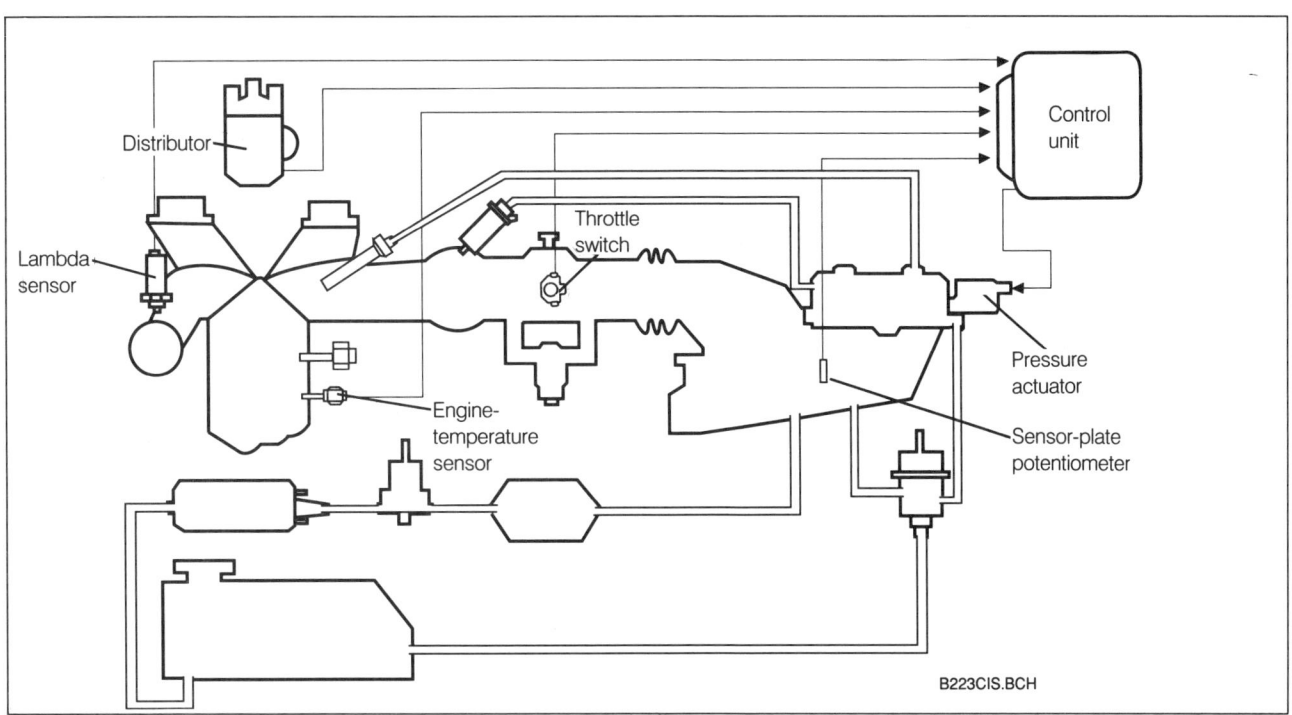

Fig. 5-17. Most KE inputs to control unit are similar to those of pulsed systems. Sensor-plate potentiometer is unique to KE, operating something like a throttle-position sensor. For more information see **6. Functions and Compensation.**

Actuator Current, Pressures, and Air-Fuel Ratios

When troubleshooting, a digital meter can be used to measure the control-unit actuator current. It's very small, measured in milliamps (mA), so be aware that small milliamp currents are affected by oxidation on connectors. The connectors must remain clean and must not be bent by probes inserted to read small currents. The following are some examples of how the control current relates to pressure-actuator mixture adjustment.

For typical warm-engine operation, current flow is about 10 mA. If you remember the discussion above, this causes a small deflection of the plate valve for a small difference between system pressure and lower-chamber pressure. The differential on a fuel gauge would be about 0.4 bar. Pressure drop and injector fuel flow at the slits is normal.

To lean the mixture, the plate is allowed to move away as the current is reduced, to perhaps only 1 mA. The differential pressure falls to only 0.2 bar, so injector fuel decreases.

To enrich the mixture, as in engine warm-up, the plate is pulled closer with more current, say 70 mA. Differential pressure increases to 1 bar, so injector fuel increases. More actuator current means more differential pressure, more deflection

of the differential-pressure valve diaphragms, and more pressure drop at the metering slits, so more fuel is injected. When you've got that, you've got the basic idea of the KE-Jetronic control system's adjustment of the basic fuel mixture.

The graph shown in Fig. 5-18 looks complicated, but a moment of study will give you a good summary of actuator current as it affects the air-fuel ratio. This typical curve does not apply to any specific car.

On the baseline, read the pressure-actuator current in milliamps (mA). On the left, one vertical scale shows the pressure drop at the slit: the difference between system pressure and upper-chamber pressure. The other vertical scale shows the specified differential-pressure, the difference between system pressure and lower-chamber pressure. These numbers are 0.2 bar higher than corresponding pressure drop numbers, the result of the 0.2 bar pressure from the spring in the lower chamber.

The right column shows the corresponding air-fuel ratios. Follow the curved line to see how changing pressure-actuator current changes the air-fuel ratio.

Control Systems

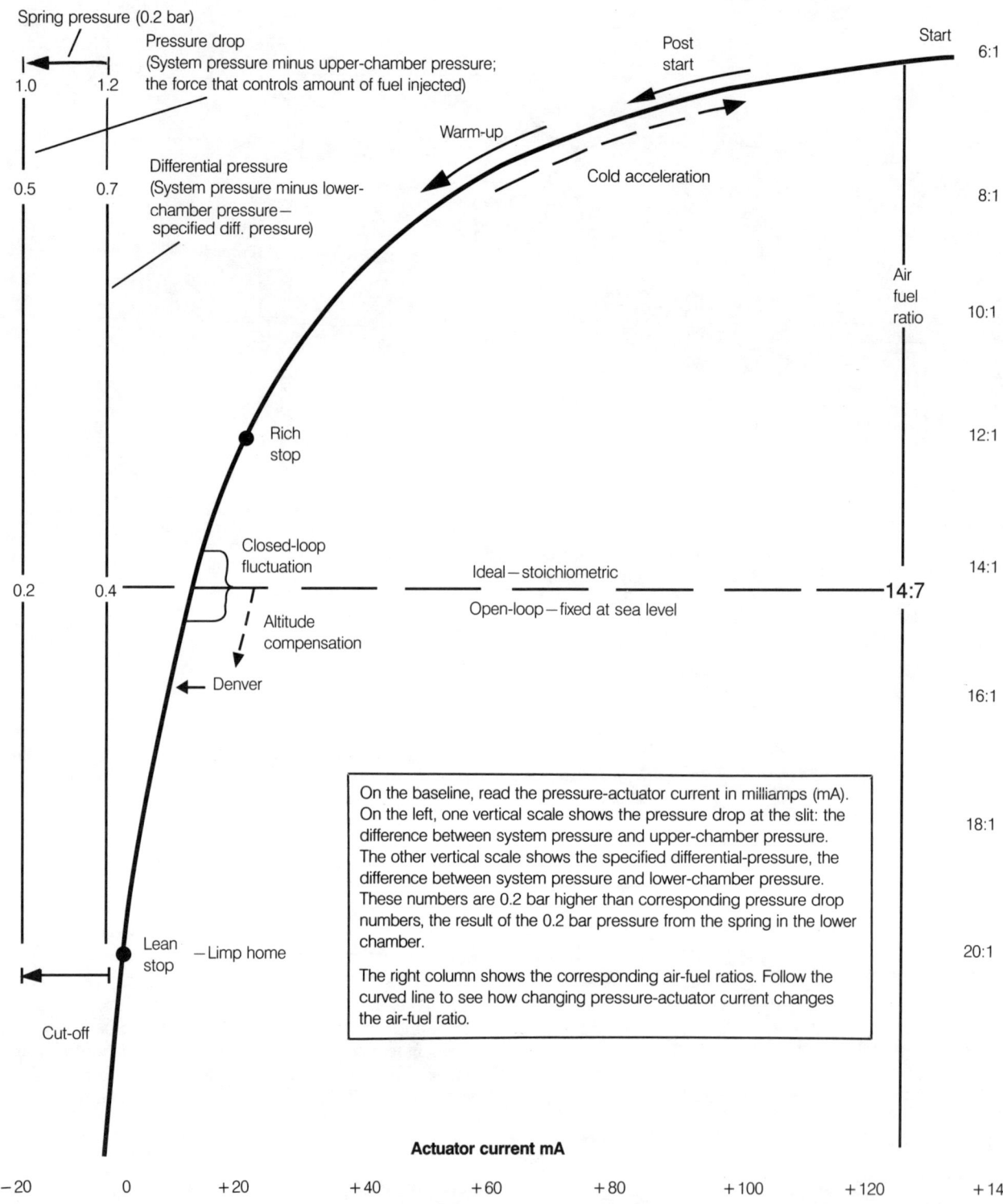

Pressure (in bar)

Spring pressure (0.2 bar)

Pressure drop
(System pressure minus upper-chamber pressure;
the force that controls amount of fuel injected)

1.0   1.2

Differential pressure
(System pressure minus lower-
chamber pressure –
specified diff. pressure)

0.5   0.7

Warm-up

Post start

Cold acceleration

Start

6:1

8:1

Air fuel ratio

10:1

Rich stop

12:1

Closed-loop fluctuation

Ideal – stoichiometric

14:1

0.2   0.4

Open-loop – fixed at sea level

14:7

Altitude compensation

Denver

16:1

On the baseline, read the pressure-actuator current in milliamps (mA).
On the left, one vertical scale shows the pressure drop at the slit: the
difference between system pressure and upper-chamber pressure.
The other vertical scale shows the specified differential-pressure, the
difference between system pressure and lower-chamber pressure.
These numbers are 0.2 bar higher than corresponding pressure drop
numbers, the result of the 0.2 bar pressure from the spring in the lower
chamber.

The right column shows the corresponding air-fuel ratios. Follow the
curved line to see how changing pressure-actuator current changes
the air-fuel ratio.

18:1

Lean stop – Limp home

20:1

Cut-off

Actuator current mA

– 20   0   + 20   + 40   + 60   + 80   + 100   + 120   + 140

B181CIS.BCH

Fig. 5-18. KE actuator current changes typical pressure
drop at the slits, and therefore the differential
pressure (system pressure minus lower-chamber
pressure) and the air-fuel ratio. Notice small
closed-loop fluctuation around stoichiometric air-
fuel ratio at 8–12 mA, and the limp-home setting at
lean stop. These values do not apply to any specific car.

Limp-Home Mode

The actuator plate is designed to rest at a "limp-home" position if the control-unit current is interrupted. The engine will continue to run, but slightly leaner than normal since the actuator current is not fine-tuning the mixture. In this circumstance, mixture is determined solely by the basic mechanical metering of the air-flow sensor and control plunger. The uncontrolled plate in the pressure actuator allows just enough pressure drop at the slits to keep a warm engine running. The engine may not start, but if it's running, it will get you home. You can simulate the limp-home condition on a running warm engine by disconnecting the pressure-actuator connector.

Logic Circuits

Beginning in 1986, some KE systems have logic circuits in the control unit which help it to recognize and ignore erroneous signals. The computer recognizes rapid changes in input from some sensors as improper and switches to pre-programmed outputs. For example, the coolant temperature signal does not normally change rapidly. If it does, the computer interprets this as a short or an open circuit, and switches to a predetermined value associated with a warm engine. Because this capability is so new and so vehicle-specific, refer to your car service manual for more information.

5.4 Summary of Control Systems: K-Basic, K-Lambda, KE

Now that I've discussed how the control systems of the three types of continuous fuel injection fine-tune the basic fuel metering of the air-flow sensor and control plunger, I'll summarize these systems according to the two important factors: changing control-plunger slit size for a given air flow, and changing pressure drop at the slits for a given plunger lift. In all cases, pressure drop at the slits is the difference between system pressure inside the slit and upper-chamber pressure in the differential-pressure valves.

In all three types of continuous systems, the air-flow sensor and the control plunger meter the basic amount of proportional fuel to match the air flow. The control systems then manipulate pressures in the system to adjust the basic air-fuel mixture to compensate for different operating conditions. Refer to Fig. 5-19 on the next page for comparison of the manipulation of pressures in the differential-pressure valves of the three systems. Note that the following control descriptions are not for any specific air flow or operating condition.

Table b illustrates these comparisons with numbers. The table uses the same system pressure of 5.0 bar (73 psi) for clarity, but be aware that system pressures do vary.

K-Basic

K-basic systems increase fuel delivery to enrich the mixture for a given air flow by reducing counterforce on the plunger, allowing it to rise higher. When the engine is cold, the control-pressure regulator opens more and increases return flow from the control-pressure circuit. This reduces control-pressure counterforce and the plunger rises higher for the same air-flow force. A larger portion of the plunger slit is exposed and fuel flow to the injectors increases.

Table b. CIS Upper-Chamber and Lower-Chamber Pressures

System	System pressure	Lower-chamber pressure	Upper-chamber pressure	Pressure drop at slit (system pressure minus upper-chamber pressure)	Enrichment method
K-Basic	5.0 bar (73 psi)	5.0 bar (73 psi)	4.9 bar (71.5 psi) (lower-chamber pressure minus 0.1 bar (1.5 psi) spring pressure	0.1 bar (1.5 psi)	Increase slit size (by reducing control pressure)
K-Lambda	5.0 bar (73 psi)	5.0 bar (73 psi) minus reduction caused by return of fuel at lambda valve	Lower-chamber pressure minus 0.1 bar (1.5 psi) spring pressure	Varies: 0.1 bar (1.5 psi) plus the reduction in lower-chamber pressure	Increase slit size (by control pressure), increase pressure drop (by lambda valve)
KE	5.0 bar (73 psi)	5.0 bar (73 psi) minus reduction caused by pressure actuator	5.0 bar (73 psi) plus 0.2 bar (3 psi) spring pressure, minus differential pressure (difference between system pressure and lower-chamber pressure)	Varies: differential pressure minus 0.2 bar (3 psi) lower spring force	Increase pressure drop (by pressure actuator)

Pressure drop at the metering slits of the control plunger is a constant 0.1 bar (1.5 psi), maintained by the spring in the upper chamber pressing down on the diaphragm of the differential-pressure valve. Each flexible diaphragm of the differential-pressure valves moves with a change in fuel flow, so upper-chamber pressure is always 0.1 bar less than system pressure in the lower chamber.

K-Lambda

K-lambda systems use control pressure to change the mixture, just as in K-basic. They also have an additional control system. K-lambda changes the pressure drop at the slit by changing the lower-chamber pressure in the differential-pressure valve.

Fuel pressure in the lower chambers is controlled by the lambda control system. To enrich as a result of lean signal from the lambda sensor, the control unit increases the duty cycle of the lambda-control valve. More fuel is returned from the lambda-control circuit so lower-chamber pressure drops. Through the movement of the differential-pressure valve diaphragm, this decreases upper-chamber pressure and increases the pressure drop at the slits. More fuel flows for a given plunger lift and slit size.

With the spring in the upper chamber pressing down, each diaphragm moves so the pressure drop at the slit is always 0.1 bar (1.5 psi) less than lower-chamber pressure. Upper-chamber pressure is always 0.1 bar less than lower-chamber pressure as lower-chamber pressure varies according to action of the lambda control valve.

KE Systems

KE systems adjust the basic mixture only by changing the pressure drop at the slit. KE slit size does not vary for a given air-flow force because the counterforce on the plunger is system pressure, which is constant. The fuel flow through the lower chambers, and therefore lower-chamber pressure, is controlled by the pressure actuator. This KE control system compensates for all operating conditions, as well as maintaining a stoichiometric air-fuel ratio for emission control.

To enrich, increased current from the control unit pulls down the pressure-actuator plate valve to reduce lower-chamber pressure. This increases differential pressure – the difference between system pressure and pressure in the lower chambers. For any given air flow, this increases diaphragm deflection in the differential-pressure valves. The pressure drop at the slits increases, increasing fuel delivery to the injectors.

With a spring pushing up on each differential-pressure valve, upper-chamber pressure is 0.2 bar (3 psi) higher than lower-chamber pressure. Lower-chamber pressure is system pressure minus differential pressure. Vehicle specifications will list differential pressures for corresponding actuator currents in milliamps (mA).

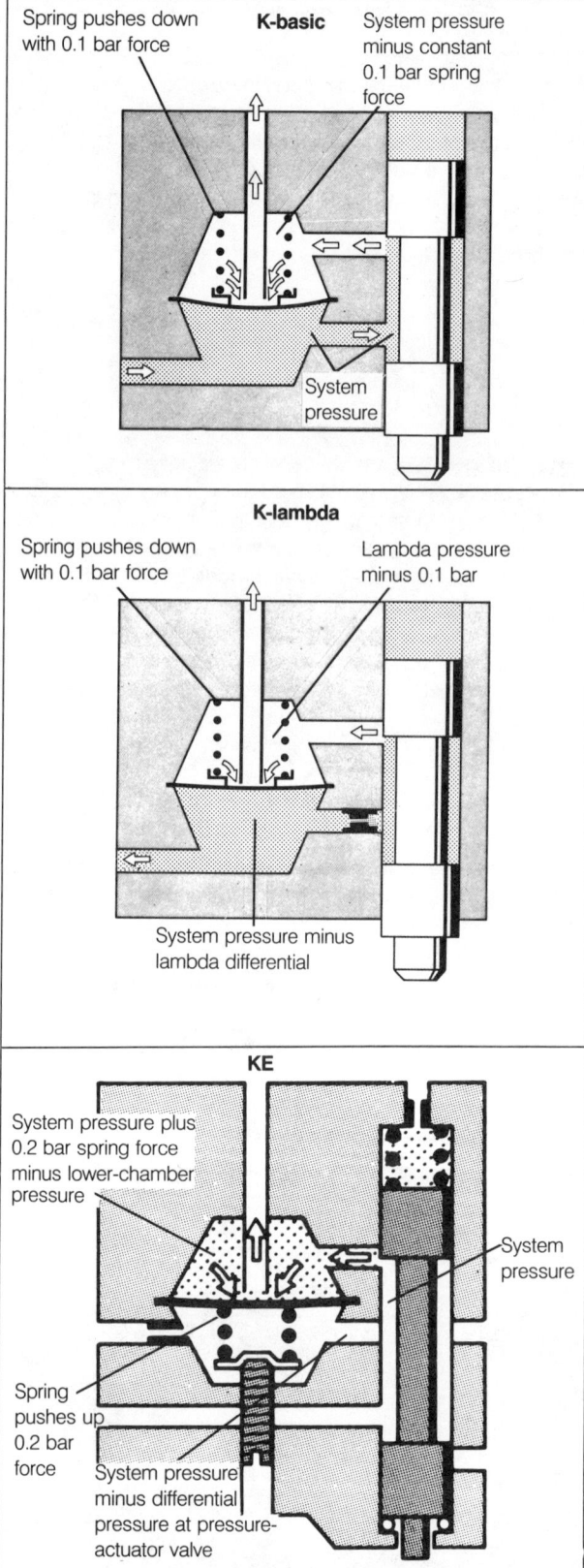

Fig. 5-19. Comparison of differential-pressure valve pressures in continuous systems. Lift of plunger by air flow determines basic fuel quantity. Control systems then manipulate pressure to change mixture. All three have system pressure inside the control-plunger slit.

Control Systems

6. FUNCTIONS AND COMPENSATIONS

You've seen how each control system adjusts the basic air-fuel mixture. Now you'll see how that control is applied for all operating conditions, and you'll also see how other components add additional enrichment as necessary.

6.1 Starting

When the key is turned to start, the fuel pump is turned on through the safety circuit to build system pressure; if the engine stops turning for more than one second (as when the engine will not start) the pump shuts off, even with the ignition key on. For starting enrichment, extra fuel is delivered to the port injectors. In K-basic and K-lambda, reduced control pressure allows more plunger lift, increasing slit size to deliver more fuel. In KE systems, the control unit sends more current to the pressure actuator for start enrichment, lasting about one second. The current is high for maximum pressure drop at the metering slits, enriching the mixture delivered to the port injectors.

Cold Starting

For cold starting, in addition to enrichment at the port injectors, continuous systems enrich the mixture with a cold-start injector system, similar to that used on pulsed systems. Also, an auxiliary-air valve or idle-speed stabilizer adds air, bypassing the throttle plate to increase idle rpm. See chapter 3 for more information. Beginning about 1985, some KE systems ground the start injector through the control unit, so the thermo-time switch is eliminated.

Hot-Start Pulse Relay

Some cars – notably VWs beginning about 1985 – have a special hot-start pulse relay. See Fig. 6-1. The relay bypasses the thermo-time switch and grounds the cold-start injector in short pulses when a warm engine is started, spraying fuel into the manifold. This provides additional fuel in the intake manifold to aid starting when the port injectors are delivering less fuel because of high temperatures and vapor in the lines.

6.2 Post-Start/Warm-Up

When the key is released to run, enrichment changes from engine-crank needs to engine-run needs. On K-basic and K-lambda, the control-pressure regulator warms up, gradually increasing control pressure on the plunger. This cuts back the enrichment factor in a few minutes.

KE post-start enrichment is determined by the control unit based on input from the engine temperature sensor. Increased actuator current keeps the engine running, right after starting. Then enrichment tapers off during warm-up, depending on engine temperature. See Fig. 6-2.

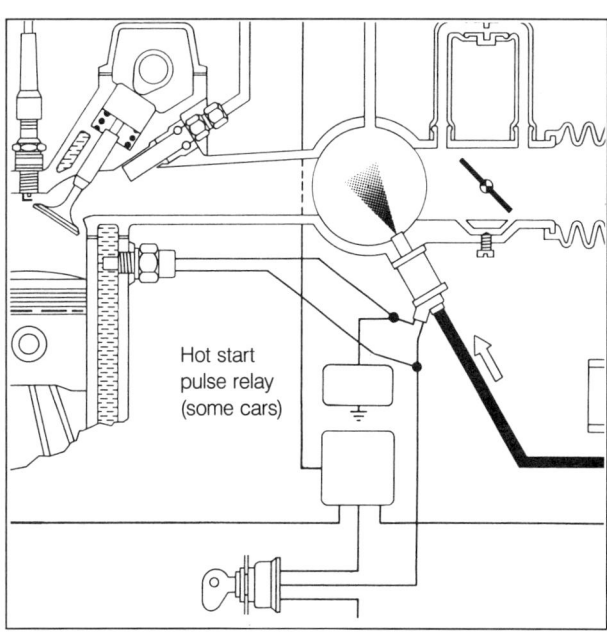

Fig. 6-1. In some VWs and Audis, hot-start pulse relay adds fuel by energizing cold-start injector during hot-engine cranking.

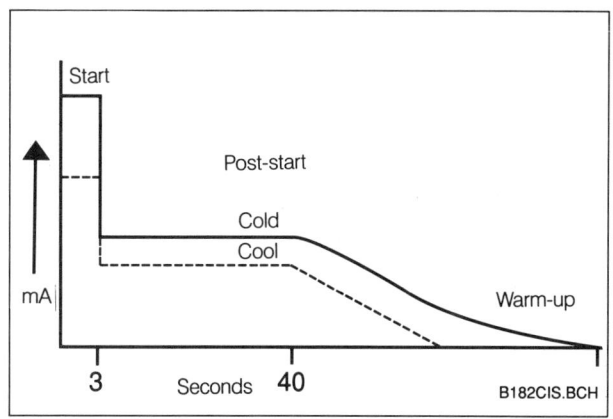

Fig. 6-2. KE enrichment at start and post-start depends on temperature. Warm-up enrichment also depends on time since starting.

6.3 Acceleration

In all systems, rapid opening of the throttle increases air flow and causes the air-flow sensor plate to overreact, in a manner similar to that of pulsed systems vane-type air-flow sensor flaps. Overswing of the sensor plate allows the plunger to momentarily rise higher and enriches the mixture enough to prevent lean-mixture flat-spots and hesitation. See Fig. 6-3.

5

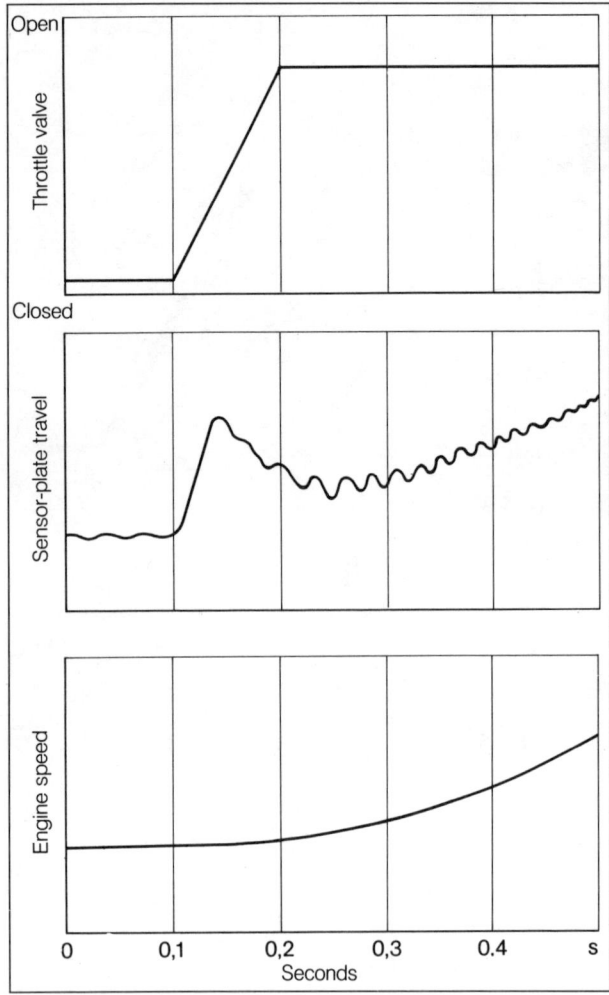

Fig. 6-3. Sudden opening of throttle causes overswing of sensor plate, adding acceleration enrichment for K-basic and K-lambda.

Part-Throttle Acceleration

In K-basic and K-lambda, cold-engine overswing is greater than that of a warm engine because reduced control pressure on the plunger allows more lift. In addition, some control-pressure regulators sense manifold pressure to further reduce control pressure and add acceleration enrichment.

KE systems use an additional sensor, the sensor-plate potentiometer (also called the air-flow sensor position indicator), to signal the need for extra enrichment during cold-engine part-throttle acceleration. The potentiometer signals how fast and how far the sensor plate moves as air flow increases. See Fig. 6-4.

The potentiometer has an arm with sliding contacts that is connected to the sensor-plate pivot. As the arm moves, it changes a voltage signal. See Fig. 6-5. From this input, the control unit sends a compensation output to the pressure actuator. For about one second, pressure-actuator current

increases, depending on sensor plate movement and on engine temperature and rpm. The greatest enrichment is when the engine is cold; middle enrichment when the engine is cool; no enrichment when engine temperature is above 175°F (80°C). Mercedes-Benz has added an Intake Air Temperature Sensor to further compensate acceleration-enrichment for cold air, a refinement for better driveability.

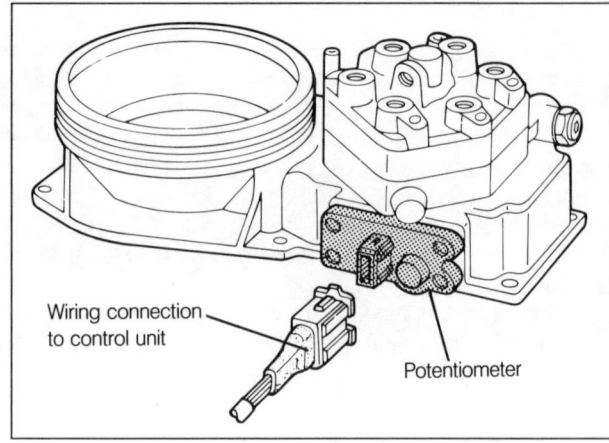

Wiring connection to control unit

Potentiometer

Fig. 6-4. Location of potentiometer on KE mixture-control unit.

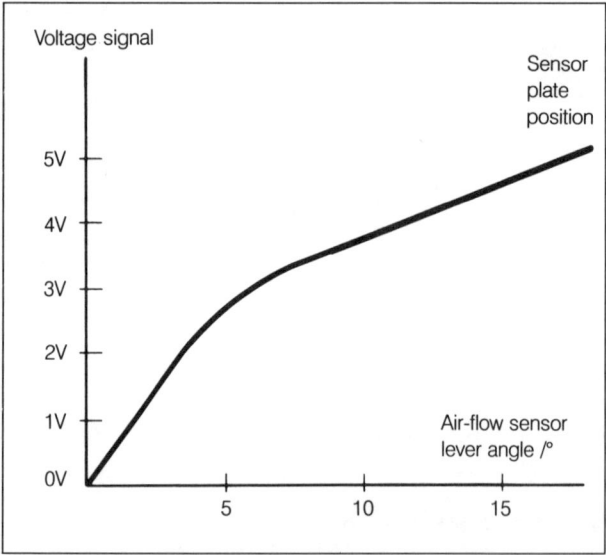

Fig. 6-5. KE adds signal from sensor-plate potentiometer to enrichment from sensor-plate overswing. The more the lever angle changes, the greater the voltage signal to the control unit.

Full-Throttle Acceleration

On K-basic and K-lambda, enrichment is supplied only on systems with control-pressure regulators linked to manifold pressure.

KE enrichment usually continues as long as the throttle switch signals a full-load condition, adding enrichment by control of the pressure actuator. Some engines, particularly VW, need full-throttle enrichment only during limited rpm ranges, so the control unit is programmed accordingly. The control unit adds about 3 mA to the actuator current to compensate the mixture for full-load acceleration, depending on the full-load signal, on rpm and on engine temperature.

6.4 Lambda Control

K-lambda closed-loop operation is described under **5. Control Systems**. Note that closed-loop operation switches to open-loop under these conditions:

- lambda sensor cool, not ready to send valid signals

- engine cool, usually below about 105°F (40°C)

- WOT, full-load

- coasting cut-off

- acceleration enrichment (some cars)

KE systems also operate closed-loop with a lambda sensor. The lambda-sensor signals are processed with the other inputs to determine pressure-actuator current. The plate valve is always moving; either slightly toward the inlet or slightly away from it, the result of constantly changing pressure actuator current. With the engine warm, and the lambda sensor hot, look for actuator current to change back and forth in a narrow range, typically from 8 to 12 mA. This fluctuation during closed-loop operation is normal; the system is adjusting to maintain the ideal air-fuel ratio. During open-loop operation, with the lambda sensor disconnected and the engine warm, actuator current should be about in the middle of the fluctuations, according to engine specifications. When you see no change in actuator current, you know the system is running open-loop.

6.5 Cut-off

(KE only)

In KE, to cut off fuel injection, such as during coasting or for rpm limitation, actuator current is reversed to move the plate valve completely away from the inlet. With no limit to the amount of actuator fuel flowing through the lower chamber, both chambers rise to system pressure. Each spring pushes its diaphragm up, as shown in Fig. 6-6, closing the outlet and cutting off injector fuel. On some cars, such as Mercedes, rpm limitation is also handled by cutting power to the fuel-pump relay. On most cars, coasting cut-off is inhibited during cruise-control operation.

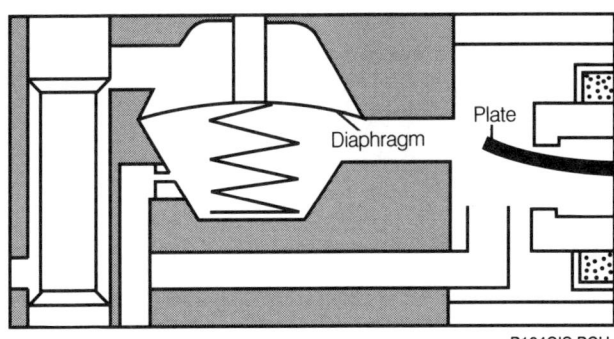

B184CIS.BCH

Fig. 6-6. To cut off fuel in KE systems, pressure actuator current is reversed (to approximately −40 mA), pushing plate away from inlet. Zero differential pressure causes lower-chamber spring to raise diaphragm, shutting off fuel to injectors.

6.6. Altitude

On some K-basic and K-lambda systems, altitude compensation is built into the control-pressure regulator, as described in **5. Control Systems**.

In KE systems on higher-priced cars, an altitude sensor signals the control unit to reduce pressure-actuator current to reduce fuel delivery when the air is thinner. The decrease is about 1 mA actuator current for each thousand feet above sea level. KE-Motronic systems with adaptive air-fuel mixture control "learn" to adjust themselves and therefore need no altitude sensor.

7. KE3-JETRONIC AND KE-MOTRONIC

KE3-Jetronic and KE-Motronic systems combine the control of KE-type fuel injection with the control of ignition timing. The difference between the two systems is that KE3-Jetronic has two separate control units — one for fuel injection, one for ignition timing — and operates with standard KE-Jetronic fuel injection. KE-Motronic is a true "Motronic" system; it has both fuel injection and ignition control combined into one control unit, and has a special data-point map in the control unit that determines actuator current for all operating conditions. Ignition-timing control for both KE3-Jetronic and KE-Motronic also uses maps. For more information on timing control, Motronic systems, and control-unit maps, see the Motronic section of chapter 3.

The first application of KE3-Jetronic is on some 1987 Audi 5000s, and all 1988 5-cylinder Audis. Audi calls this system CIS-E III. The first use of KE-Motronic is on the 1988 4-cylinder Audi 80; Audi calls this system CIS-Motronic.

In the two system diagrams, shown in Fig. 7-1 and Fig. 7-2, you can see the ignition timing input (the Hall-effect sender in the ignition distributor) that makes these systems different from KE-Jetronic. The ignition outputs to the ignition coil with power stage are: 1) close-circuit to begin charging the coil and 2) open circuit to fire the spark plug.

KE3-Jetronic and KE-Motronic

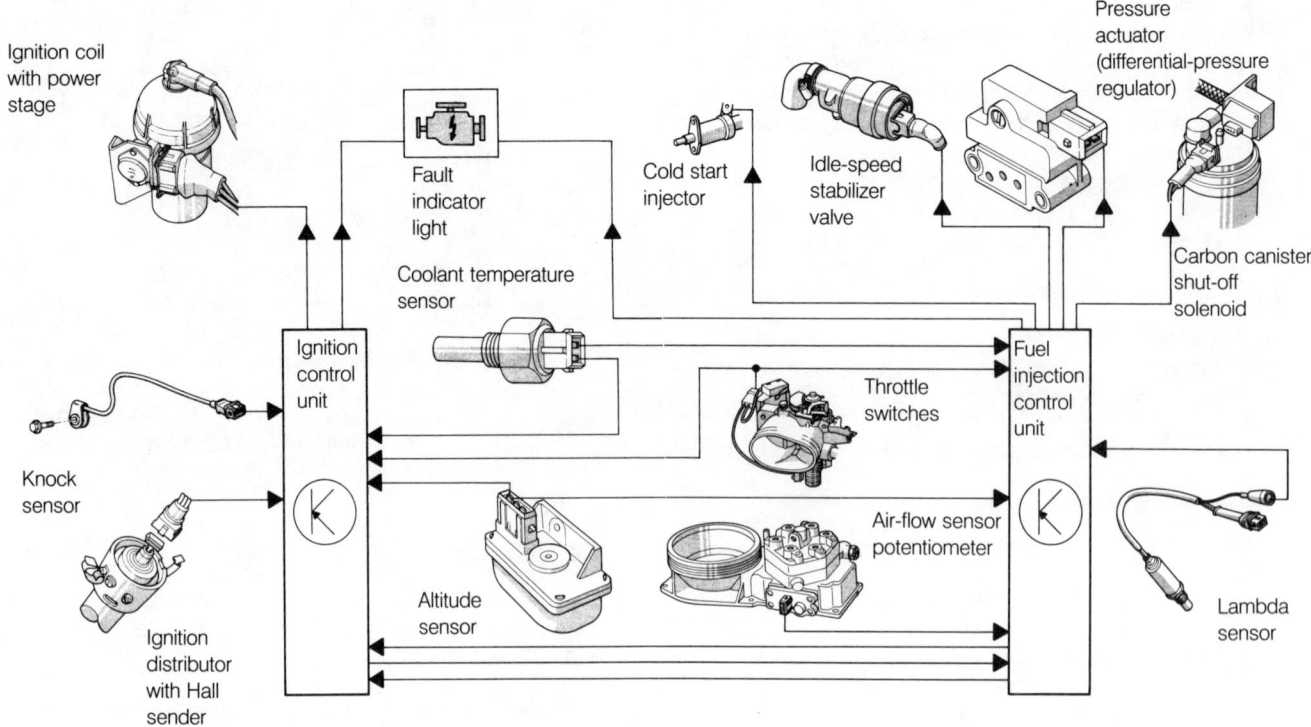

Fig. 7-1. KE3-Jetronic system used in recent Audi and Mercedes includes diagnostics, but controls fuel injection with normal Jetronic control unit. Ignition is controlled by separate control unit.

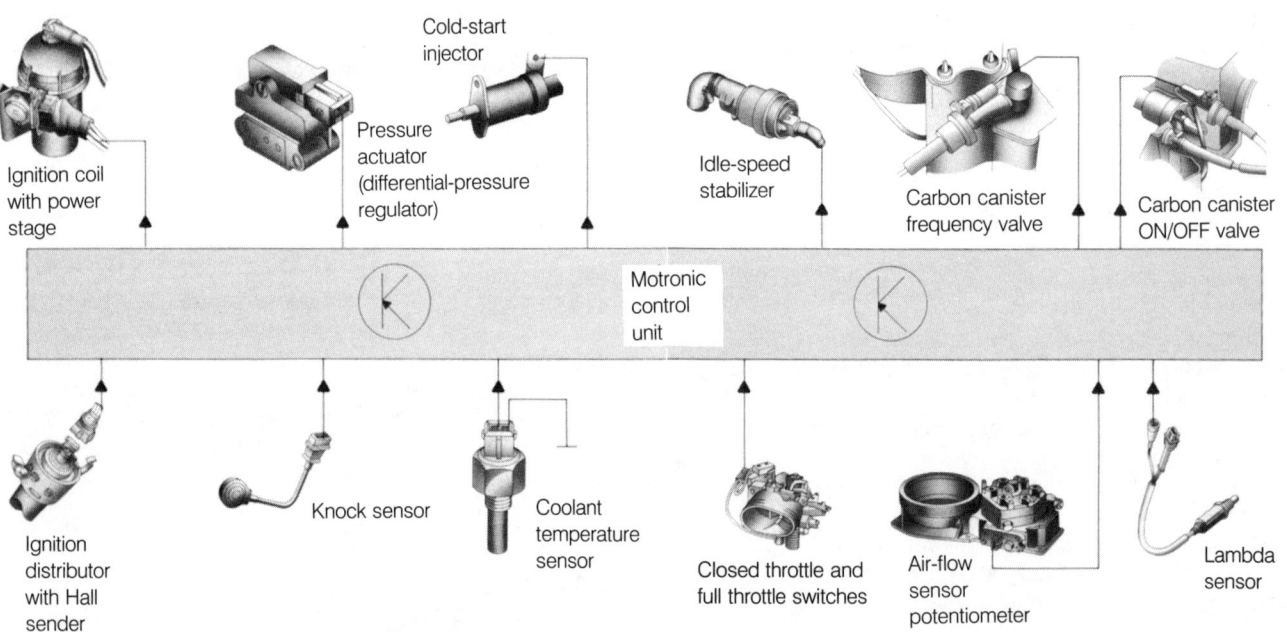

Fig. 7-2. KE-Motronic has special Motronic control unit with special internal data maps to determine operating points for actuator current and other functions. Inputs include Hall sender in distributor, and knock sensor. Outputs include ignition coil for ignition timing control.

KE3-Jetronic and KE-Motronic

The other KE3-Jetronic and KE-Motronic inputs are the same as KE-Jetronic: coolant temperature sensor, closed throttle (idle) and full-throttle switches, air-sensor potentiometer, and lambda sensor. Most of these inputs are used to control ignition timing as well as fuel injection.

The outputs are also the same: pressure actuator (called a differential-pressure regulator by VW/Audi) current, and idle-speed stabilizer control. In addition, the control unit manages the vapor-canister frequency valve and the canister on-off valve; parts of the evaporative emission control system.

Actuator Current

Both systems have a different range of control-unit current for the pressure actuator. For KE3-Jetronic, a typical rich limit is +10 mA, with a lean limit of −10 mA. The nominal adjustment point is 0 mA, to provide for the widest range of adaptation.

For KE-Motronic, a typical rich limit is +23 mA, with a lean limit of −16 mA. This provides for the operation of the adaptive circuitry. The system can compensate for broader changes beyond just lambda control, including all engine operating conditions, altitude, intake leaks and other variations. A typical actuator current for a new engine running open-loop (lambda sensor disconnected) might be 2 mA. As the engine ages, perhaps with minor intake air leaks, typical open-loop actuator current might be 7 mA. See Fig. 7-3.

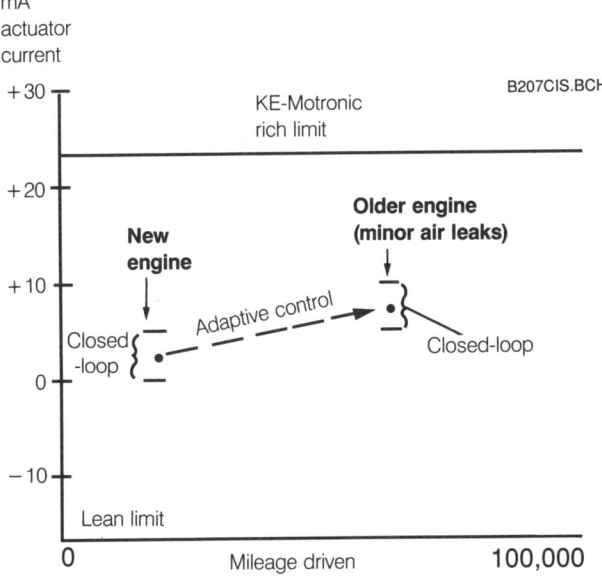

Fig. 7-3. KE-Motronic adaptive control can adjust actuator current to match changing conditions. Older engine may need greater actuator current for enrichment to counter minor air leaks.

KE-Motronic Adaptive Control

Adaptive control means that the control unit "learns" to make automatic mixture adjustment. The control-unit circuitry senses nominal warm-engine cruise values. It stores those values as the normal open-loop operation, and then uses those values as a basis for both closed-loop and open-loop actuator current. The control unit continually "relearns" the compensations necessary for changes in engine condition, or even intake air leaks.

The air-fuel ratio values are stored in the control unit's volatile memory, so they are lost any time the battery is disconnected. If the engine runs differently after service, even if that service was as seemingly unrelated as the installation of a radio or theft alarm, be ready to drive the car normally for about 10 minutes to re-store the nominal air-fuel mixture values in the volatile memory of the KE-Motronic control unit.

Fault Memory
(Diagnostic Codes)

Diagnostic codes (fault memories) are stored in the control unit. The control unit continually checks all systems for incorrect signals. If one is found, the fault is stored in the control-unit fault memory as a series of electrical pulses. On KE3-Jetronic systems you can read these fault-memory pulses at the indicator light on the dash panel as shown in Fig. 7-4. On KE-Motronic systems, you connect a test light between the battery and the test lead next to the fuel distributor on the engine. See Fig. 7-5. The fault codes are read out by inserting a spare fuse into the top of the fuel-pump relay.

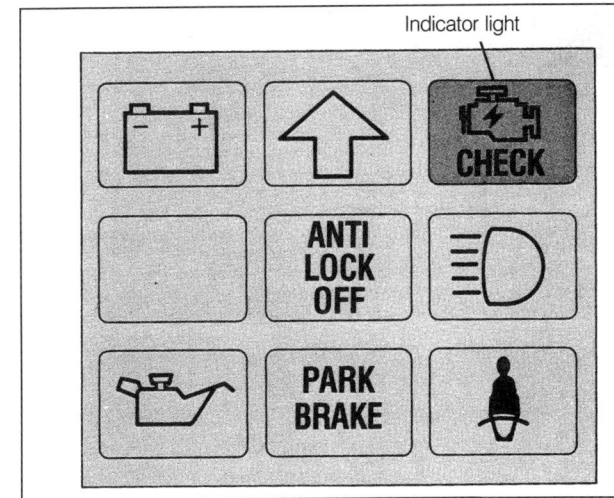

Fig. 7-4. On KE3-Jetronic, diagnostic codes from the control unit fault memory are read out at the dashboard.

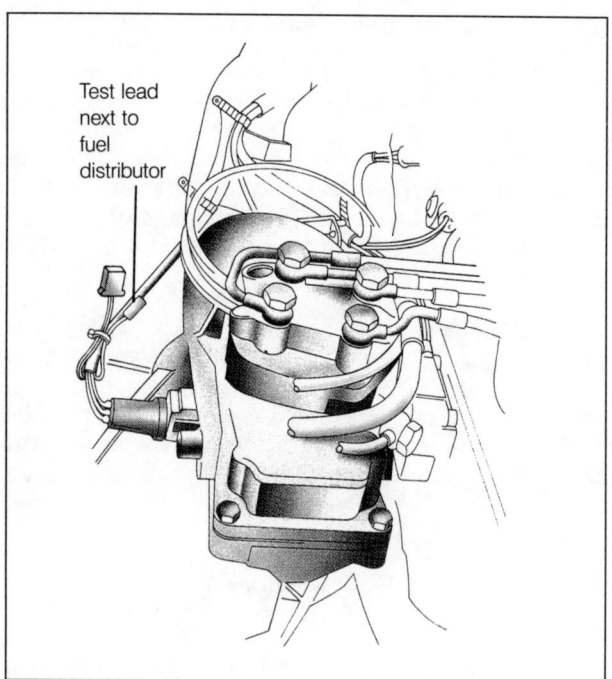

Test lead
next to
fuel
distributor

Fig. 7-5. On KE-Motronic, diagnostic read-out of fault memory is by voltage pulses at test lead near mixture-control unit.

KE3-Jetronic codes in all cars except California are erased when the ignition is turned off. Fault codes are restored by driving for 5 minutes (at at least 3,000 rpm, and once at full-throttle) or by cranking the engine for 6 seconds. If the fault is emission related, the indicator light on California cars remains on as long as the fault is present; in all other cars the code is stored, but will not come on until the fuse is inserted. When you activate the fault memory, injection faults read out first, and then ignition faults.

In KE-Motronic, if the fault is emission related, all cars except California will store the fault in memory until it is erased. California cars flash the fault light as long as a fault is present. This helps in finding intermittent faults. For non-emission related faults, all cars store the fault until the engine is restarted. Troubleshooting can be done before engine is shut off, or with ignition OFF. Fault codes can be restored by driving for 5 minutes (at at least 3,000 rpm, and at least once at full throttle), or by cranking for 6 seconds. For more information on the fault memory and reading trouble codes, see chapter 6.

Chapter 6

Continuous Injection –
Troubleshooting & Service

Contents

6

2   CONTINUOUS INJECTION – TROUBLESHOOTING & SERVICE

1. INTRODUCTION

In this section, you'll see the troubleshooting and general service procedures that apply to Bosch Continuous Injection Systems (CIS). These include:

- original K-Jetronic (I'll call it K-basic)

- K-Jetronic with lambda control (I'll call it K-lambda)

- KE-Jetronic and KE-Motronic (I'll call them KE systems)

For more information on how these systems work, see chapter 5, as well as the general discussion of fuel injection in chapters 1 and 2. Unless otherwise noted, all procedures in this chapter apply to all the systems.

Fig. 1-1. A service technician may use specialized equipment, but with ordinary shop tools and your car shop manual, you can check and adjust many parts of your CIS system.

K-Jetronic systems are basically mechanical, using the hydraulics of fuel pressure to control the air-fuel mixture. From 1974 to about 1980, there are no electronics. Beginning about 1980, to meet the tighter emission limits, the lambda control system was added to K-Jetronic for a limited mixture-control function, but still most compensations are mechanical. Beginning about 1984 in KE systems, for improved driveability and response, all mixture compensations are made electronically using sensors inputting to an electronic control unit similar to that of pulsed systems. Since 1988, KE-Motronic combines the control of ignition-timing with fuel injection in the same control unit.

All K systems are driveable in limp-home mode, even when their mixture compensation functions are inoperative. The engine may be difficult to start – or may not start at all – but if the engine is running and warm, it will get you home, even though KE may run with a bit of lean stumble.

1.1 General Procedures and Precautions

Many of the procedures and precautions for troubleshooting and service of continuous systems are the same as those for pulsed systems, so it's worthwhile to read the introductory information in chapter 4. Continuous systems are even more sensitive to the infiltration of dirt – due to the extremely fine orifices in the fuel distributor – so pay special attention to the section on cleanliness. Other specific procedures from chapter 4 may also apply, and will be referenced when necessary.

For many of the procedures in this chapter, I'll use typical fuel pressures, temperatures and electrical values to give you some basis for comparison and understanding. Don't try to service your car with these numbers; specifications vary considerably from one maker to the next – even within a model year – so use only the specifications from your car shop manual.

You'll also see that in a typical specification, two values are given, a checking value and a setting value, with the checking value usually having a broader tolerance. The idea of this is the system will operate within the broader tolerances, so if you just want to be sure of system operation, the checking value is fine. But if you are setting it, you might as well get as close as possible to the middle of the range.

Operating Fuel Pump For Tests

For many of the tests, you'll need fuel pressure in the system, just as you do for pulsed-injection testing. To run the fuel pump without running the engine, remove the fuel-pump relay at the fuse panel. Connect a bridging adaptor – a pair of long leads with a switch and a 16-amp fuse – between the sockets for the fuel pump and the power supply. With the momentary switch, you can turn on the fuel pump by pushing the button when you need pressure to test the system. Let go of the button to stop the pump. Unless your workplace is noisy, you should be able to hear the pump when it runs.

Relieving Fuel Pressure

If you have to open the fuel lines – for testing or to replace components – you should relieve as much of the fuel system pressure as possible beforehand to prevent fuel from spraying all over, especially if the engine is hot. Because the fuel injectors close when system pressure falls to a certain point, it is not possible – as in L-Jetronic – to relieve pressure in the lines by removing the fuel-pump fuse and then running the engine until it stalls. The best way to relieve fuel pressure is to wrap the fuel-line fitting with a shop rag and then slowly open the line. Be prepared to catch as much of the fuel in a container as possible, and quickly wipe up any spilled fuel.

1.2 Tools

Along with your vehicle shop manual, you'll need ordinary shop tools, as well as a digital volt/ohm meter, a test light, and a fuel-pressure gauge. Be aware that the gauge is slightly different from the one used for testing pulsed systems. The gauge for continuous systems has two inlet hoses to a valve so that the flow through the hose can be stopped for different tests. For more information see **5.1 Fuel Pressure Tests**.

Other special tools you may need when working on all systems include a special wrench (shown later in Fig. 7-1) used to adjust the mixture screw and an injector tester (shown later in Fig. 4-5) used to test the fuel injectors. Additional tools for KE systems include only a special engine temperature resistor, shown in Fig. 1-2, and a special wiring adapter, shown in Fig. 1-3.

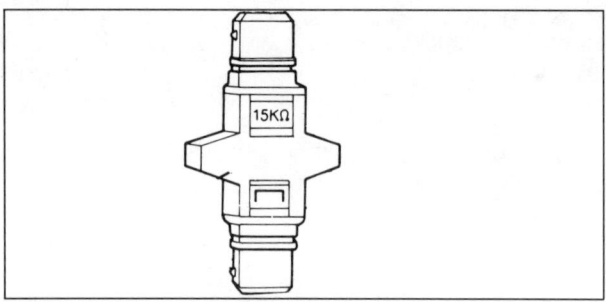

Fig. 1-2. Special resistor used to simulate cold-engine conditions for many tests.

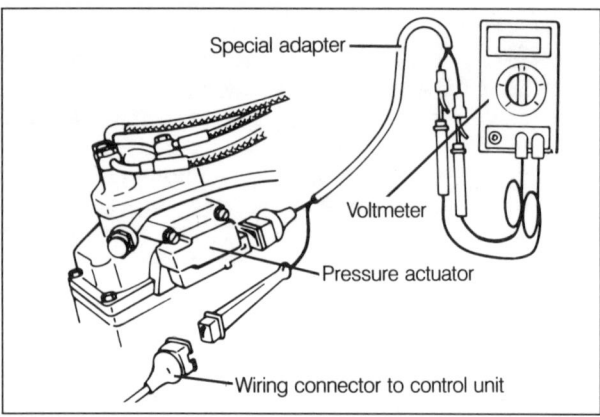

Special adapter

Voltmeter

Pressure actuator

Wiring connector to control unit

Fig. 1-3. Special wiring test adapter used to test KE pressure-actuator current with engine running.

2. TROUBLESHOOTING

This section covers the troubleshooting of continuous fuel injection systems, based on problem symptoms which may be caused by a fault with fuel injection. The main part of this section is a troubleshooting table that will help you to narrow down your tests to specific components or areas of the system.

There are also some general troubleshooting procedures which are easy to perform, and which may save you any further testing of the system. These involve making sure that the engine is in good working order, checking for air leaks in the intake system, and making tests of the electrical system, and

the wiring between components and (where applicable) the control unit. These checks are described in detail in chapter 4, and generally apply to continuous systems.

> **NOTE** ━━━
> The air-flow sensor plate will not hold enough pressure in the air intake to check for leaks using a soap solution. Either block off the air intake or use the drawn-in-solvent method when testing for intake leaks.

On electronic systems, it is not possible to directly test the control unit without highly specialized equipment. If you troubleshoot all other areas of the system and they check out OK, only then should you suspect that there may be a fault in the control unit.

Fuel Condition

Many of the problems you're likely to encounter with the fuel injection system may originally be caused by poor quality fuel. Even small amounts of water can quickly corrode the metal parts, causing restrictions in small passages that reduce fuel flow, or causing the control plunger to stick. Likewise, any dirt that gets into the system through an open fuel line—when changing the fuel filter, for example—can quickly clog the passages and orifices in the fuel distributor. If the fuel system is contaminated, the problem may soon recur. Although many manufacturers now install "lifetime" fuel filters, it's wise to follow a regular schedule of filter changes, especially if you suspect moisture contamination.

2.1 Troubleshooting Basics

The fuel-injection system supplies and meters the correct amount of fuel to the engine in proportion to the amount of air being drawn in to achieve the optimum air-fuel mixture. Any problems with air intake sensing or fuel supply will cause poor running. Any troubleshooting should begin with simple and easy checks of the tightness of the system wiring and the integrity of the air intake system. Proceed from there to more involved troubleshooting.

Generally, fuel injection problems fall into one of four symptom categories: cold start, cold running, warm running, and hot start. Warm running is the most basic condition. Before troubleshooting a condition in any other category, be sure that the system is working well and adjusted properly for warm running.

To simplify troubleshooting, concentrate on those components or sensors that adapt fuel metering for a particular condition. For example, if the engine will not start when cold, the components responsible for cold-start enrichment should be tested first.

Table a lists symptoms of Bosch continuous fuel injection problems, their probable causes, and suggested corrective actions. The boldface numbers in the corrective action column indicate headings in this chapter of the book where the test or repair procedures can be found.

Table a. Troubleshooting Bosch Continuous Fuel Injection

Symptom	Probable cause	Corrective action
1. Cold start – Engine starts hard or fails to start when cold	a. Cold-start valve or thermo-time switch faulty b. Fuel pump not running c. Air-flow sensor plate rest position incorrect d. Fuel pressure incorrect e. Coolant temperature sensor or wiring faulty (KE only)	a. Test cold-start valve and thermo-time switch. Replace faulty parts. 8. b. Check fuel pump fuse and fuel pump relay, as well as fuel pump voltage supply. 4.1 c. Inspect air-flow sensor plate rest position and adjust if necessary. 3.2 d. Test fuel pressure. 5.1, 6.1 e. Check control system. 6.2
2. Hot start – Engine starts hard or fails to start when warm	a. Cold start valve leaking or operating continuously b. Fuel pressure incorrect c. Air-flow sensor plate rest position incorrect d. Insufficient residual fuel pressure e. Fuel leak(s) f. Lambda control faulty g. Fuel injectors faulty or clogged	a. Test cold-start valve and thermo-time switch. 8. b. Test fuel pressure. 5.1, 6.1 c. Inspect air-flow sensor plate rest position and adjust if necessary. 3.2 d. Test residual fuel pressure. Replace fuel pump check valve or fuel accumulator as necessary. 5.1, 6.1 e. Inspect fuel lines and connections. Correct leaks as required f. Check lambda control. 5.2, 6.2 g. Check injectors. 4.3
3. Engine misses and hesitates under load	a. Fuel injector clogged b. Fuel pressure incorrect c. Fuel leak(s) d. Lambda control faulty e. Coolant temperature sensor or wiring faulty (KE only)	a. Test fuel injectors. Check for clogged injector lines. Replace faulty injectors. 4.3 b. Test fuel pressures. 5.1, 6.1 c. Inspect fuel lines and connections. Correct leaks as required d. Check lambda control. 5.2, 6.2 e. Check control system. 6.2
4. Engine starts but stalls at idle	a. Incorrect fuel pressure b. Cold-start valve leaking c. Auxiliary-air regulator/Idle-speed stabilizer faulty d. Vacuum (intake air) leak e. Fuel injectors faulty or clogged f. Coolant temperature sensor or wiring faulty (KE only) g. Control plunger binding or fuel distributor faulty	a. Test fuel pressures. 5.1, 6.1 b. Test and, if necessary, replace cold-start valve. ;bd8. c. Test and, if necessary, replace. 8. d. Inspect intake air components for leaking hoses, hose connections, and cracks or other leaks. Repair as required e. Check injectors. 4.3 f. Check control system. 6.2 g. Check air-flow sensor plate movement. 3.2
5. Engine idles too fast	a. Accelerator pedal, cable, or throttle valve binding b. Auxiliary-air regulator faulty c. Air leaking past throttle valve	a. Inspect for worn or broken parts, kinked cable, or other damage. Replace faulty parts. b. Test and, if necessary, replace. 8. c. Inspect throttle valve and adjust or replace as required. 3.1
6. Hesitation on acceleration	a. Vacuum (intake air) leak b. Fuel injectors clogged c. Cold-start valve leaking d. Control plunger in fuel distributor binding or fuel distributor faulty e. Air-flow sensor plate out of adjustment f. Fuel pressure incorrect g. Idle mixture (%CO) incorrectly adjusted h. Potentiometer faulty or misadjusted (KE only)	a. Inspect intake air components for leaking hoses, hose connections, and cracks or other leaks. Repair as required b. Test injector spray pattern and quantity. Replace faulty injectors. 4.3 c. Test and, if necessary, replace cold-start valve. 8. d. Check air-flow sensor plate movement and, if necessary, replace fuel distributor. 3.2 e. Inspect air-flow sensor plate position and adjust if necessary. 3.2 f. Test fuel pressures. 5.1, 6.1 g. Check and adjust CO. 7. h. Test and adjust or replace as necessary. 6.2

6

continued on next page

Table a. Troubleshooting Bosch Continuous Fuel Injection (cont'd)

Symptom	Probable cause	Corrective action
7. Poor fuel mileage	**a.** Idle speed, ignition timing, and idle mixture (%CO) out of adjustment **b.** Cold-start valve leaking **c.** Fuel pressure incorrect	**a.** Check and adjust. **7.** **b.** Test and, if necessary, replace cold-start valve. **8.** **c.** Test fuel pressures. **5.1, 6.1**
8. Engine continues to run (diesels) after ignition is turned off	**a.** Incorrect ignition timing or faulty ignition system **b.** Engine overheated	**a.** Check ignition system **b.** Check cooling system
9. Low power	**a.** Coolant temperature sensor faulty or wire to sensor broken (KE only) **b.** Fuel pressure incorrect **c.** Throttle plate not opening fully **d.** Full throttle switch faulty or incorrectly adjusted (KE only) **e.** Control-pressure regulator faulty (control-pressure regulators with full-load compensation only)	**a.** Check control system. **6.2** **b.** Test fuel pressures. **5.1, 6.1** **c.** Check throttle cable adjustment to make sure throttle is opening fully. Adjust cable if necessary. **d.** Check throttle switch and adjust if necessary. Replace a faulty switch. **6.2** **e.** Check control-pressure regulator full-load function. **5.1**

3. AIR-FLOW MEASUREMENT

As discussed in chapter 5, the throttle valve regulates the amount of air drawn into the engine; the air-flow sensor measures the air intake and moves the control plunger to meter the fuel. Incorrect adjustment of the throttle valve or air-flow sensor plate, or binding of the control plunger, can cause many problems including rough idle, stalling and hard starting.

Whenever any changes are made to the throttle valve or air-flow sensor, idle speed and mixture (CO) will need to be adjusted also.

3.1 Throttle Valve Basic Adjustment

The throttle valve is adjusted at the factory and does not normally require adjustment. The stop screw should not be used to adjust the idle; its purpose is to prevent the valve from closing too far and damaging the inside of the throttle body. The throttle-valve adjustment procedure is given in case the factory adjustment has been changed.

To correct a faulty throttle valve adjustment, use a screwdriver to back off the throttle-valve adjusting screw until there is clearance between its tip and the throttle valve lever. The screw is shown in Fig. 3-1. It may be necessary to first loosen a locknut. Place a thin piece of paper between the adjusting screw and the throttle valve lever. With the throttle valve closed, turn the screw in until it lightly contacts the paper. From this position, remove the paper and turn the screw in an additional ½ turn.

Fig. 3-1. Typical location (arrow) of throttle-valve adjusting screw.

3.2 Air-Flow Sensor and Fuel Distributor

For some of the tests of the air-flow sensor in this section, you will need fuel pressure on the plunger. You can do this by starting the engine for a minute, then shutting it off.

If you run the pump without running the engine, don't move the sensor plate in the air-funnel. Remember, the pump delivers system pressure, usually about 5 bar (75 psi), and that is high enough to cause continuous injection of raw fuel into the cylinders. After the pump stops, the accumulator provides residual fuel-pressure, about 2 bar (30 psi) for these tests, but that's lower than injector-opening pressure, about 3 bar (45 psi), so the injectors will not deliver fuel when you lift the sensor plate.

The fuel distributor is precisely calibrated when it is manufactured. There are no replacement parts and it cannot be adjusted. If it is faulty in any way, it must be replaced.

Checking Sensor Plate Movement

Using the updraft air-flow sensor as our example, lift the plate slowly to check for smooth movement as shown in Fig. 3-2. There should be even resistance all the way. You can use a magnet on the center steel stud to get the plate started, but the plate itself is aluminum and non-magnetic. When you lift the sensor plate, you are also lifting the control plunger. If the plate lifts freely, then both the sensor plate and plunger are OK. For a downdraft sensor, reverse the step: press down slowly to check movement of the air-flow sensor plate and plunger.

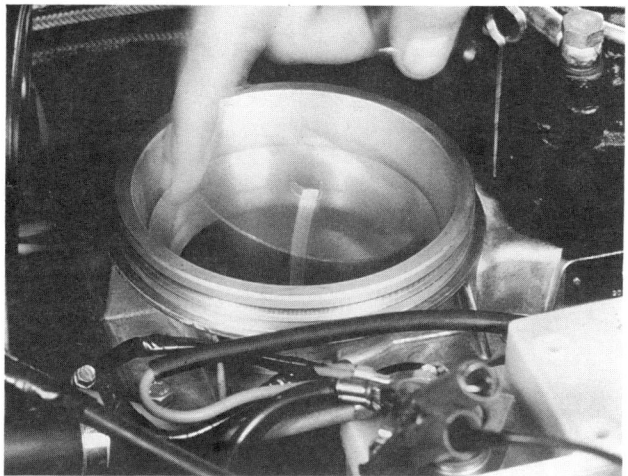

Fig. 3-2. Checking sensor plate movement.

If there is resistance while lifting the plate, either the plate or plunger are sticking. To determine which it is, lift the plate and then let it go. It should fall with a bounce or two. If the plate sticks in falling, it is because there is a problem with the sensor plate or its lever. Check them as described below. If your plate falls freely but feels sticky when lifted, the problem is with the plunger. Clean it as described below and retest it.

Centering Sensor Plate and Lever

If the plate sticks while moving in both directions, it could be off center, touching one side of the cone as shown in Fig. 3-3. To adjust the plate, be sure it is at normal rest position. You can use a special guide ring to center the plate, or you can simply use a 0.1 millimeter feeler gauge: if the gauge slides around the circumference of the plate, then the plate is centered. If you find unequal clearance around the plate, loosen the center screw so the plate can move a little on the arm and recheck for equal clearance all around. When you tighten the screw, be sure the plate is still centered. And be careful not to overtighten.

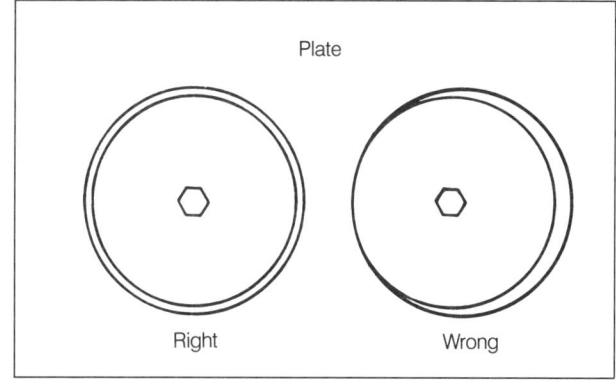

Fig. 3-3. Centering air-flow sensor plate.

Sticking can also be caused by warpage of the air-flow sensor housing which causes the plate lever to bind. You can free a tight sensor-plate lever by loosening the mounting bolts around the housing as shown in Fig. 3-4, and then uniformly retightening them. If that does not solve the sticking, remove the mixture-control unit and check the pivot. If it is dirty, clean it with a solvent and lightly lubricate it. Check for equal clearance on either side of the lever as shown in Fig. 3-5. To adjust the lever, loosen the bolt.

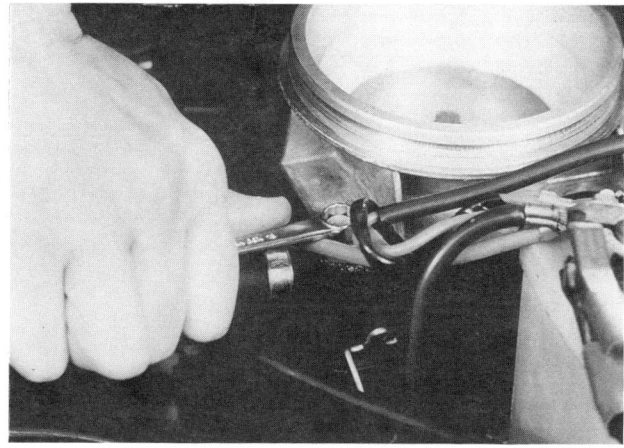

Fig. 3-4. A warped air-flow sensor housing may cause the sensor-plate lever to bind.

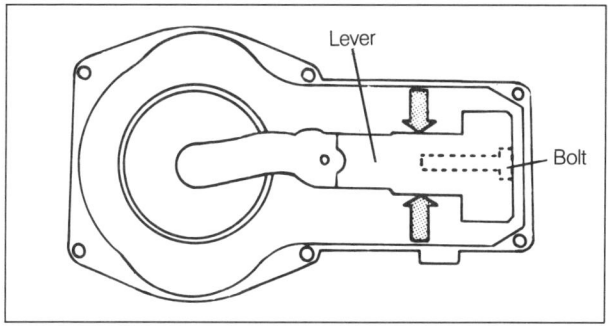

Fig. 3-5. Sensor-plate lever adjustment.

Cleaning Control Plunger

Plungers can stick from fuel gum, or from rust caused by water in the fuel. To free a sticking plunger, relieve the fuel pressure as described in **1.1 General Procedures and Precautions**, then remove the fuel connections, the fuel-distributor bolts, and the fuel distributor.

If the plunger is free, it may fall out, so when you lift the distributor get your finger in there to hold it. If it is sticking, you can blow it out with compressed air as shown in Fig. 3-6, but don't drop it. If the plunger is damaged you have to replace the entire distributor. Also, on 1983 and later systems, don't lose the spring on top of the plunger. With the plunger safely in hand, clean it in solvent. Clean the funnel and sensor plate, too. They can get grimy from crankcase ventilation.

Fig. 3-6. Removing control plunger from fuel distributor. Don't drop it!

Note that on KE systems, a stop screw has to be removed before the plunger can be removed. Measure the recess from the stop screw to the top edge of the gland nut as indicated in Fig. 3-7 so you can reset it, then remove the stop screw and then the control plunger. When you reinstall the plunger, the small shoulder goes in first. Turn in the stop screw until the recess matches your noted measurement.

On all systems, if the plunger still sticks after cleaning, then replace the fuel distributor. Always use a new gasket between the fuel distributor and the air-flow sensor body, and use new gaskets on the fuel lines.

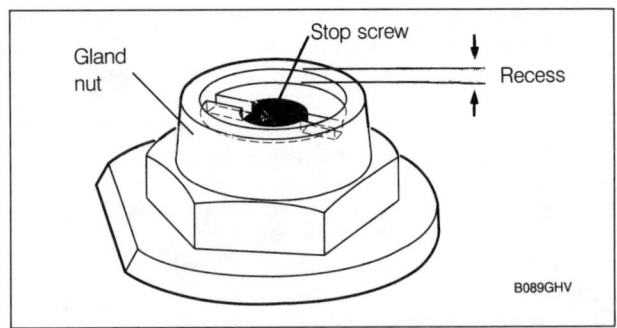

Fig. 3-7. Stop screw and recess on KE fuel distributors.

Sensor Plate Zero (Rest) Position
(K-Basic and K-Lambda only)

With fuel pressure on the control plunger, the sensor plate zero position on updraft sensors should be even with the narrowest part of the air funnel, nearest the fuel-distributor as shown in Fig. 3-8. If the zero-position is wrong, bend the wire clip to adjust, not the arm supporting the clip. You can unbolt the mixture control unit to do this, or reach in past the plate. If you reach in from above the plate with thin-nose pliers, don't nick the plate, or scratch the air-flow sensor.

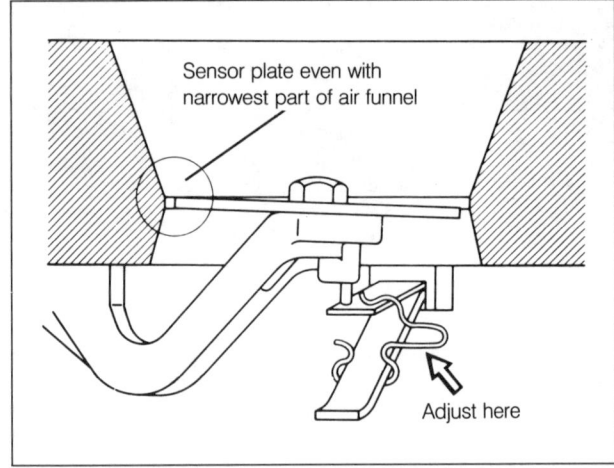

Fig. 3-8. Sensor plate rest position for K-Basic and K-Lambda.

It's OK for the plate to rest as much as 0.5mm (0.02 in.) below the narrowest point, but it should never be above because that would cause hard starting. Air could be pulled into the engine without lifting the plate and the control plunger.

For downdraft sensors, the reverse applies: check zero position at the point farthest away from the fuel distributor.

Sensor Plate Zero (Rest) and Basic Position
(KE Systems)

In KE systems you check two important sensor-plate positions, the zero position and basic position. For more information on these, see chapter 5. There should be fuel pressure on the control plunger for these tests. First, I'll show you how to check updraft air-flow sensors, then downdraft sensors.

In the zero position, the updraft plate must rest below the edge of the funnel so you can see most of the vertical face, as shown in Fig. 3-9. The plate must not rest part way up the vertical face. If you have any doubt, check the specifications in your shop manual and, using a depth gauge, measure the zero position of the updraft plate with your gauge closest to the fuel distributor. A typical depth measurement is 1.9mm (0.075 in.). If zero position is not in specification, bend the wire clip as described previously.

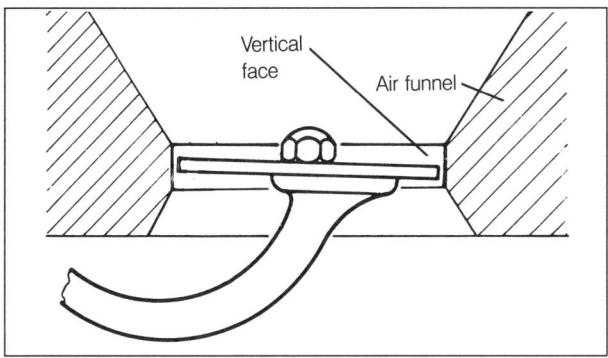

Fig. 3-9. KE sensor plate and vertical face of funnel.

Check the basic position of updraft sensors by lifting the plate slowly until you feel the sensor-plate lever contact the control plunger. This checks the gap necessary to ensure that the plunger sits on its seal. There should be a small amount of freeplay, but no more than 2 mm (0.08 in.) at the edge of the plate. In other words, the freeplay should be enough for the plate to rise and be at the top of the vertical face of the funnel when plunger contact is made. See Fig. 3-10.

If basic position is incorrect, remove the fuel distributor and use a gauge to check the distance between the seating surface and the roller on the sensor-plate arm that lifts the plunger. See Fig. 3-11. Typical distance is 21.2 ± 0.lmm (0.833 ± 0.004 in.). If this is wrong, you may be able to correct the basic position by adjusting the mixture screw on the arm. Check and adjust the mixture (CO) after you have turned this screw. If CO cannot be brought into specification, or if you have problems with starting or driveability, you will have to adjust the plunger seal clearance as described below.

Fig. 3-10. At KE zero position, you should be able to see most of the vertical face of the funnel (double white lines). When you lift sensor plate to basic position, the plate should be at the top of the vertical face (single white line. Note: the lines are for example only.

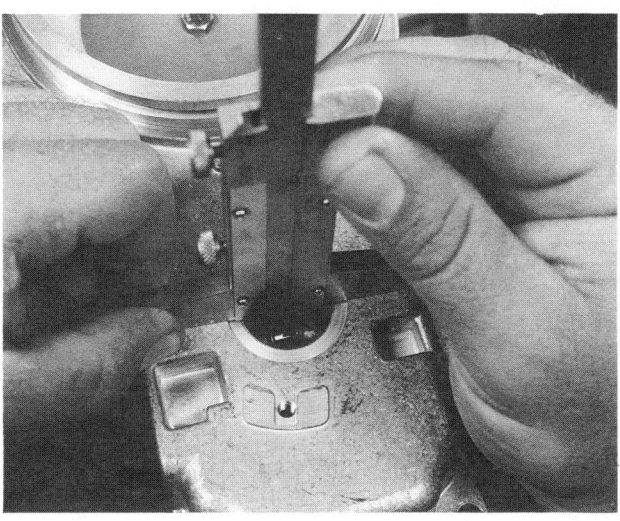

Fig. 3-11. Checking basic position of updraft air-flow sensor.

For downdraft sensors, the zero position is the same in principle. But, with fuel pressure on the plunger, zero position of the plate should be even with the top of the vertical face – or cylinder – of the air-flow sensor that is furthest away from the fuel distributor. See Fig. 3-12. The bolt does not touch the rubber bumper; its upward movement is limited by a spring-stop in the air-flow sensor.

If the zero position is wrong, adjust the guide-pin for the spring-stop. The zero position will usually be too high, so drive the pin down into the air-flow sensor as shown in Fig. 3-13 to lower the zero position. Be careful, since this pin is press-fit. If you drive it too far, or if it becomes loose in the housing, it could fall down the intake manifold and into the engine, causing expensive damage.

6

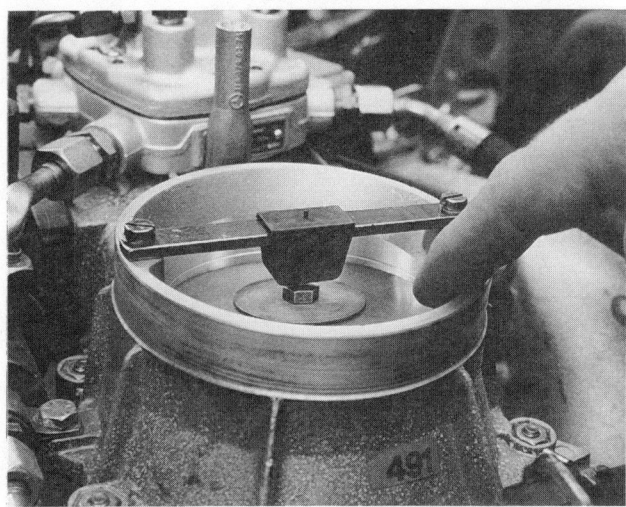

Fig. 3-12. Check zero position of downdraft KE sensor at point furthest from fuel distributor.

Fig. 3-13. Adjusting zero position of downdraft air-flow sensor.

To check downdraft basic position, press down from zero position until you feel the extra resistance that tells you you have touched the plunger. You will not feel it unless you have fuel pressure on the plunger. When you touch basic position, the plate should be at the bottom of the vertical face of the funnel so you can see all of the cylinder, but none of the funnel below. Free travel from zero down to basic must be 1–2 mm (0.04–0.08 in.).

Basic position seems to be more important with the downdraft air-flow sensors, so if in doubt, use a depth gauge to measure from the top edge of the funnel to the center bolt of the sensor plate as shown in Fig. 3-15. Check your manual for the correct specifications. If you can push the plate below the edge of the vertical face before you touch the plunger, clearance is too large and the engine will be hard to start.

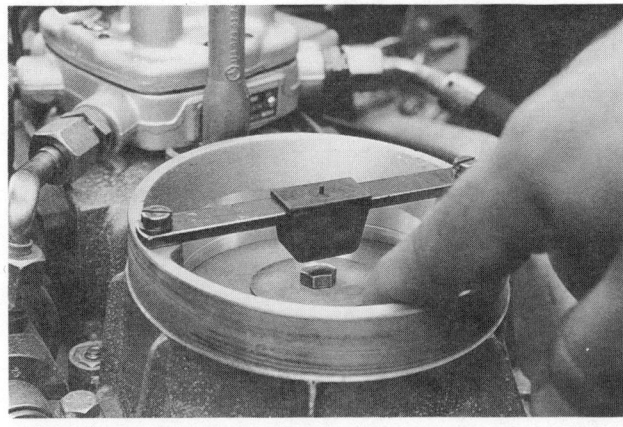

Fig. 3-14. Checking downdraft basic position.

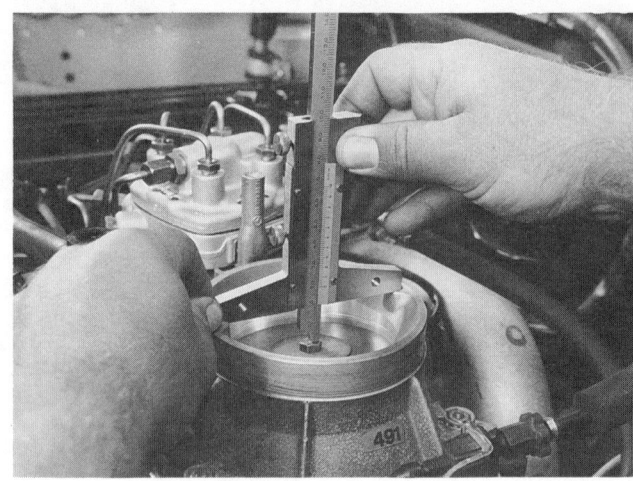

Fig. 3-15. Use depth gauge to check downdraft basic position.

If basic position is incorrect, try adjusting it by turning the mixture screw on the sensor-plate lever. Check and adjust mixture (CO) after this. If CO cannot be brought into specification after this, check the control-plunger seal clearance as described below.

Control Plunger Seal Clearance
(KE systems only)

Whether the air-flow sensor is updraft or downdraft, if zero position is right and basic position is wrong, or you can't adjust CO correctly, the problem is probably the adjustment of the plunger O-ring seal. The plunger may seat too high or low — or not at all. Remove the fuel distributor to check that. Your shop manual should have the specifications for the screw recess. Turn the slotted screw as necessary, as shown in Fig. 3-16. One-quarter turn of the nut changes clearance between the plunger and the sensor-plate lever by 1.3mm. You should not have to do this often, but when you do, it's one of the most important adjustments of the KE-Jetronic system. Of course, you will need to recheck CO so you know the mixture is correct.

Air-Flow Measurement

Fig. 3-16. Adjusting KE fuel-distributor seal clearance at plunger slotted screw.

4. FUEL SUPPLY

Poor fuel delivery can cause starting problems, hot or cold, as well as rough idle, poor mileage, and limited maximum rpm. You can check fuel systems to answer such questions as: Is the fuel pump delivering what the engine needs? If not, why not? Is the fuel flow in the fuel distributor correct? If not, why not? Are the fuel injectors operating correctly?

The procedures for testing these things depend on whether the system has:

- a system-pressure regulator without a push-valve (K-basic, 1974 to 1977);

- a system-pressure regulator with a push-valve (K-basic, 1978 & later; all K-lambda);

- a separate system-pressure regulator (KE systems).

- If you're in doubt about your system, see chapter 5 for more information.

4.1 Fuel Pump Delivery

First, relieve system pressure as described in section **1.1**, and also remove the fuel-tank cap. This prevents tank vacuum from influencing delivery readings when you run the pump; it also relieves tank pressure. Remember, fuel tanks are normally sealed and they can build up internal pressures of 1-2 psi, enough to send fuel gushing out through an open line when you least expect it. Catch gasoline in an unbreakable container, never glass!

Delivery Test Set-Up

For systems with a push-valve, measure fuel delivery at the return line from the pressure regulator. On systems without a push-valve, measure where the return fuel line comes from the T connection. See Fig. 4-1.

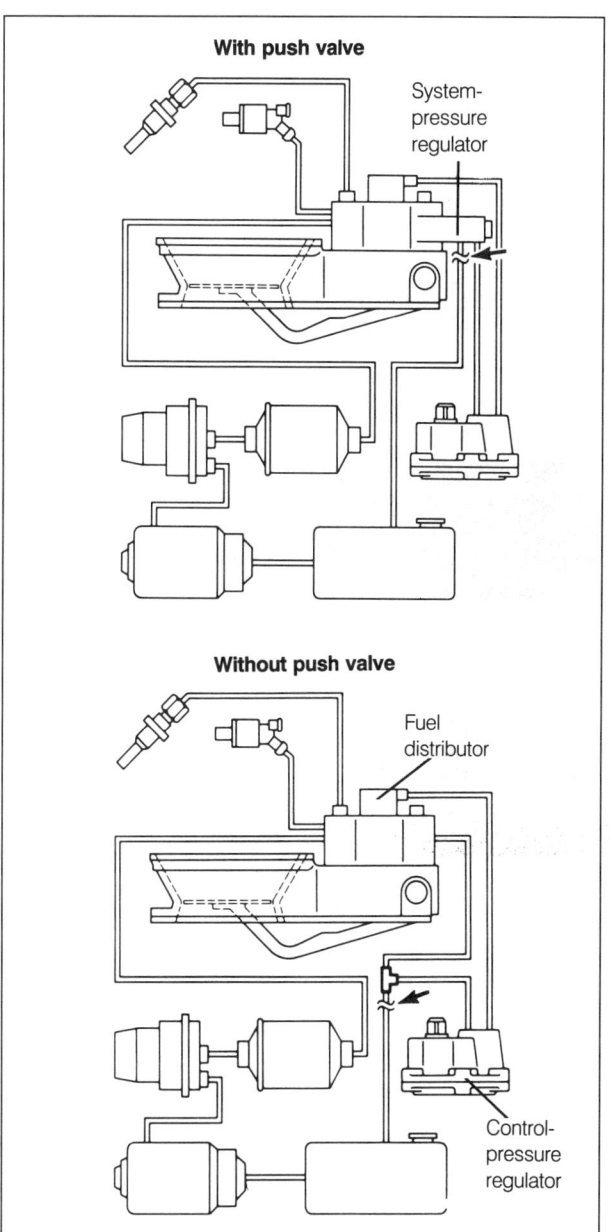

Fig. 4-1. Test point for fuel-pump delivery.

For KE systems, measure delivery at the return line from the system-pressure regulator.

Measure where the flexible line meets the steel tubing. Disconnect the return line and attach a test hose. You're measuring pump delivery at the return line instead of the input to the fuel distributor because fuel delivery is specified at system pressure, against the force of the system-pressure regulator, and would be different if measured at an open supply line.

On K-basic and K-lambda, you'll need to disconnect the electric plugs from the control-pressure regulator as shown in Fig. 4-2. The control-pressure regulator should not be warmed up during this service, so if the engine is warm, you'll have to wait for the block to cool down before you can check fuel delivery.

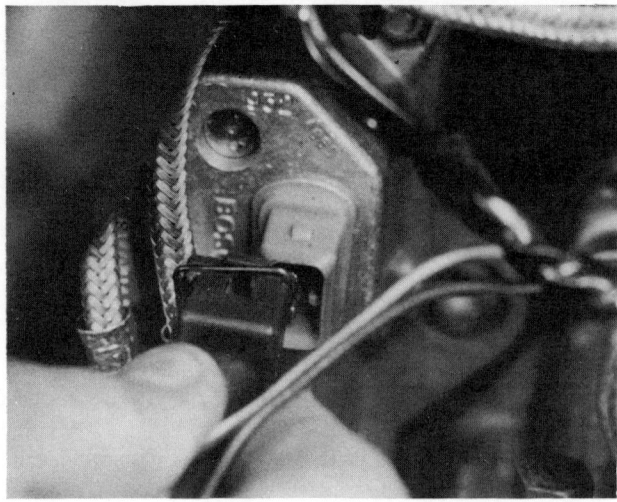

Fig. 4-2. For K-basic and K-lambda delivery test, disconnect control-pressure regulator plug.

Delivery Test

Bridge the safety circuit and have someone turn on the ignition to run the pump while you catch the fuel delivered in a measuring container that's at least 1000 ml (1 qt.). Don't try to do it alone; with all that gasoline in the open on the engine, you don't want to be wishing you had three hands!

You can run the pump for the specified time and measure amount of fuel, or you can run the pump until you see the specified delivery, then check to be sure it took less than the time specified in the shop manual. A typical fuel-delivery specification might be 750cc in 30 seconds or less. If you get that 750cc in 30 seconds or less, the pump is doing fine. If not, what would you check? Either the fuel filter or fuel lines are clogged, or the fuel pump is not operating correctly. These can be checked as described in chapter 4. Whatever you repair or replace, recheck fuel-pump delivery.

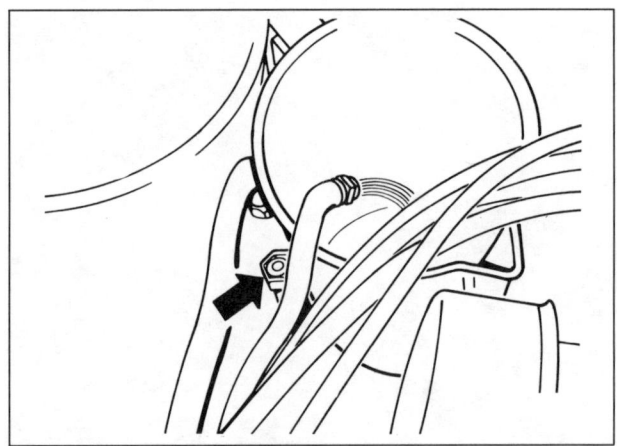

Fig. 4-3. Checking fuel delivery. Arrow indicates fuel return line connection.

4.2 Fuel Pressure Throughflow

If you are having problems with mixtures that are generally too rich or too lean, make a different delivery check before you begin measuring system pressures and control pressures. You want to measure the fuel quantity that passes through the restrictors in the fuel distributor.

In K-basic and K-lambda, if the restrictor has eroded to allow too much fuel into the control-pressure circuit, that extra fuel will increase control pressure and lean the mixture. It is less likely, but possible that the restrictor is clogged; that will decrease control pressure and enrich the mixture.

In KE systems, an eroded or clogged restrictor can affect lower-chamber pressure and the pressure differential. For KE systems, lower-chamber fuel throughflow is tested as part of fuel pressure testing in **6.1 Fuel Pressure Tests**.

Testing Control-Pressure Throughflow
(K-Basic, K-Lambda only)

To measure throughflow passing the control-pressure restrictor, connect a line to the control-pressure fitting on top of the fuel distributor to deliver fuel into a plastic container. See Fig. 4-4. If the control-pressure quantity through this restrictor does not meet delivery specifications (typically 160 to 240cc in 60 seconds), replace the fuel distributor. Too much flow is just as wrong as too little. If you don't check this, you may unnecessarily replace a good control-pressure regulator without even solving your problem.

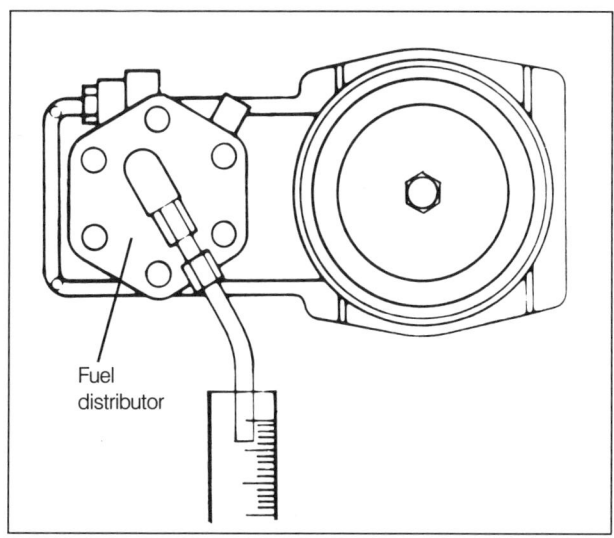

Fig. 4-4. Control-pressure throughflow being checked.

4.3 Fuel Injector Testing

The most common fuel injector problems are low fuel flow or uneven spray pattern, both caused by contaminated fuel or by carbon deposits. Excessive heat at the tip of the injector causes the carbon deposits to form on the atomizing needle. Some signs of faulty injectors are rough idle, hesitation, and knocking at full-throttle. Continuous fuel injectors can be cleaned while installed, as pulsed injectors are, using special additives or cleaners; for more information on fuel additives see chapter 3.

Continuous injectors can be quick-tested for clogging while attached to their fuel lines. But to fully test and clean continuous injectors, they must be removed and installed on a special injector tester, shown below in Fig. 4-5. If you can borrow one, you can learn a good deal about your injectors. You cannot disconnect one injector at a time, looking for rpm drop as an indication of a bad injector, as you can with pulsed injection systems. However, you can short out the spark plugs, one at a time, looking for rpm drop at idle: if the rpm does not drop, the injector may be clogged. Remember, if the car has an idle-speed stabilizer, disconnect it so it does not correct rpm.

When removing the injectors from the car to run these tests on the injector tester, first relieve the fuel pressure. Hold the hex on the injector while loosening the lines, and be careful not to kink the steel lines. Begin the contamination test after filling the injector and injector tester with test-fluid. Use only clean test-fluid in the tester, never gasoline. Use Bosch test fluid VS 14 942-CH (5 973 340 650), or Shell Mineral spirits, or equivalent.

If any injectors are replaced or cleaned of contamination, idle speed and mixture should be checked and adjusted as necessary. Use new O-rings when you install the injectors in the engine, and moisten the O-rings with fuel. You can feel them snap into place. Make sure they're all the way in. If an injector is not seated, the engine will intake false air.

NOTE ▬

The injector pressures listed in this section are for most K-basic and K-lambda cars. Most KE-Jetronic and KE-Motronic injectors operate at higher pressures, typically, 3.3 bar (48 psi), so check your manual.

Contamination Test

Attach the injector loosely to the tester, as shown in Fig. 4-5. With the valve closed to protect the gauge, pump several times to purge air-bubbles from the line; then tighten the injector test line and open the valve.

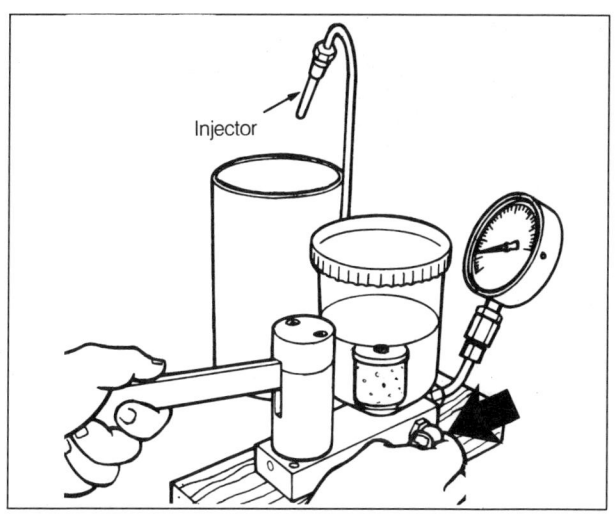

Fig. 4-5. Use continuous fuel-injector pressure tester to test and clean fuel injectors. Position injector to spray into container. Arrow indicates valve.

Slowly build up pressure to 1.5 bar (22 psi). If the injector leaks at this pressure, it could be a sign of contamination. To clean the injector, close the valve and make several sharp pumps of the tester, then open the valve and repeat the contamination test. If the injector does not leak, go on to the next test. If it still leaks, replace the injector.

6

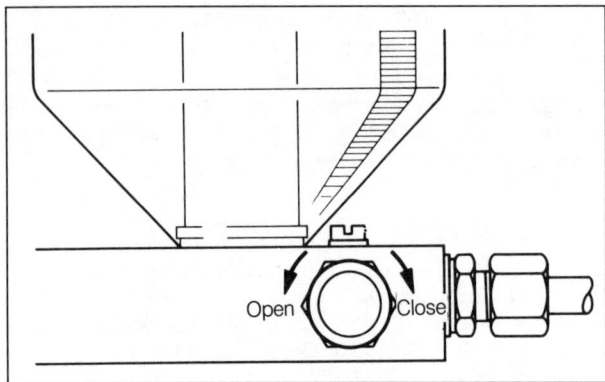

Fig. 4-6. Injector-tester valve. Close valve to protect tester gauge from high pressures.

Opening Pressure Test

Close the tester valve and bleed the injector by pumping the handle rapidly. The injector will squeak at higher and higher frequencies. When you hear the frequency become steady, that's the sound that tells you all the air is out of the system.

Open the valve, and with a firm, slow stroke (about two seconds each stroke), check the injector opening pressure. When it opens, you'll hear it, and you'll see the gauge needle drop, leaving the highest pressure on the resettable needle. A typical opening pressure might be 3–4 bar (44–58 psi); this is an important specification, tied in with the system accumulator and control-pressure regulator, so don't guess; see your shop manual. Write down the actual opening pressure of each injector; you'll need it for the leak test, next.

Leakage Test

Each injector should remain closed without leaking at a pressure 0.5 bar (7 psi) below its measured opening pressure from the preceding test; not from what is specified in the manual, but from each actual injector-opening pressure. Wipe the injector tip dry, and then build up pressure for each injector as you have figured it from your previous test. As you hold that pressure for 15 seconds, watch the injector tip; if you see any drip, replace the injector.

Spray Pattern Test

For your spray pattern test, close the tester valve and stroke about once per second. You should hear a steady chatter sound as the injector sprays. And you should see a well-atomized spray as shown in Fig. 4-7. A one-sided spray pattern is acceptable if within a total spray angle of about 35 degrees, and is well atomized.

Fig. 4-7. Acceptable injector spray patterns. Top pattern is ideal. Bottom pattern is OK if fuel is well atomized.

Fig. 4-8 shows unacceptable spray patterns. If the spray is straggly or thin, if it forms drops at the tip, or if it does not atomize, replace the injector.

Fig. 4-8. Unacceptable injector spray patterns.

5. CONTROL SYSTEMS (K-BASIC, K-LAMBDA)

This section covers the testing of the K-basic and K-lambda control systems that compensate the air-fuel mixture for all operating conditions.

5.1 Fuel Pressure Tests

Fuel pressure influences all engine operating characteristics, such as idle, acceleration response, starting and warm-up, engine power, and emissions.

You will measure three significant K-basic and K-lambda fuel-pressure values: 1) System pressure – the basic fuel pressure created by the fuel pump and regulated by the pressure regulator in the fuel distributor, 2) Control pressure – determined by the control-pressure regulator, and 3) Residual pressure – the amount of pressure the fuel accumulator and fuel pump check valve maintain in the system after the engine (and fuel pump) is shut off. The tests should be done in sequence, after the pressure gauge has been installed.

Control-Pressure Regulators

Your pressure tests will be affected by the type of control-pressure regulator fitted to your car. Some have a vacuum connection to the intake manifold for full-load enrichment. For more information see chapter

5. Testing the full-load enrichment function is part of the test. See Fig. 5-1 through Fig. 5-5 and compare them with your control-pressure regulator to see if the full-load test applies.

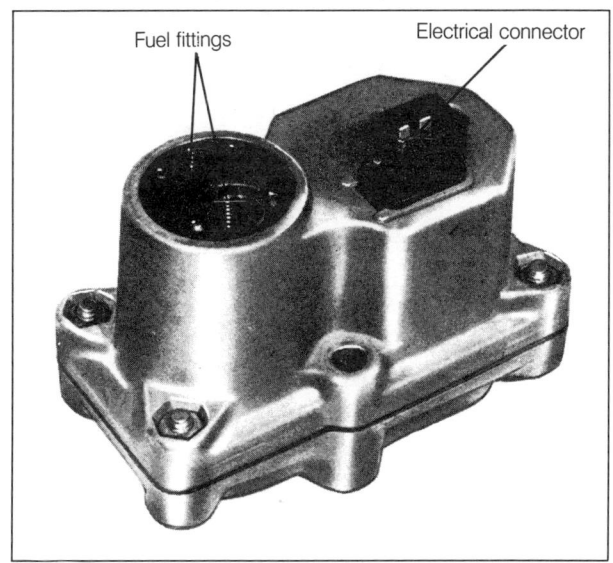

Fig. 5-1. To identify basic control-pressure regulator, look for two fittings for fuel – inlet and return – and electrical connector for heating coil of bimetal strip.

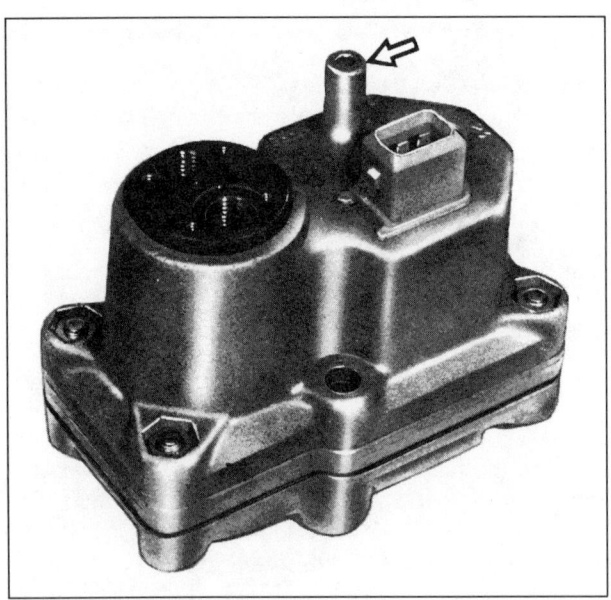

Fig. 5-2. Control-pressure regulator with vacuum-controlled full-load enrichment. On this type, vacuum connection to intake manifold is on top (arrow).

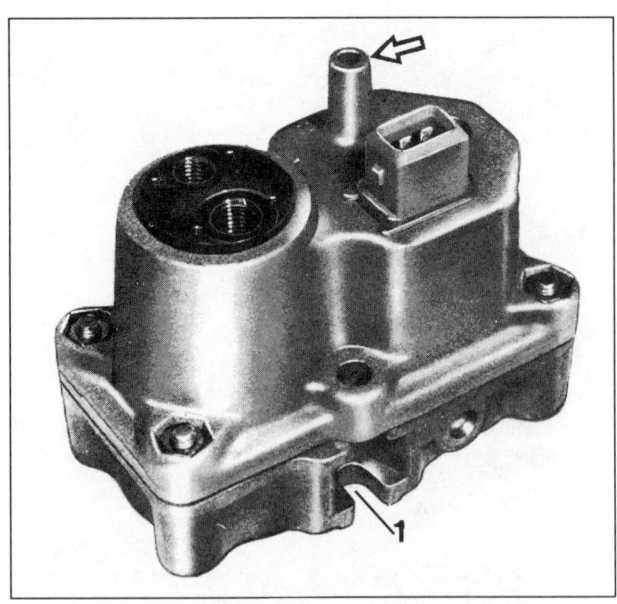

Fig. 5-4. Control-pressure regulator with full-load enrichment and altitude compensation. Vacuum connection is on top (arrow). Also note housing difference (**1**).

Fig. 5-3. A second type of control-pressure regulator with vacuum-controlled full-load enrichment. Vacuum connection to intake manifold is at bottom (arrow).

Fig. 5-5. Second type of control-pressure regulator with full-load enrichment and altitude compensation. Vacuum connection to intake manifold is at bottom (arrow).

Installing Pressure Gauge

Install the pressure gauge between the control-pressure line of the fuel distributor and the control-pressure regulator. See Fig. 5-6. The gauge should read to at least 6 bar (90 psi). If you can't rent the special gauge, you can make one with fuel-line end fittings with M12/1.5 threads. The gauge valve should be on the control-pressure side.

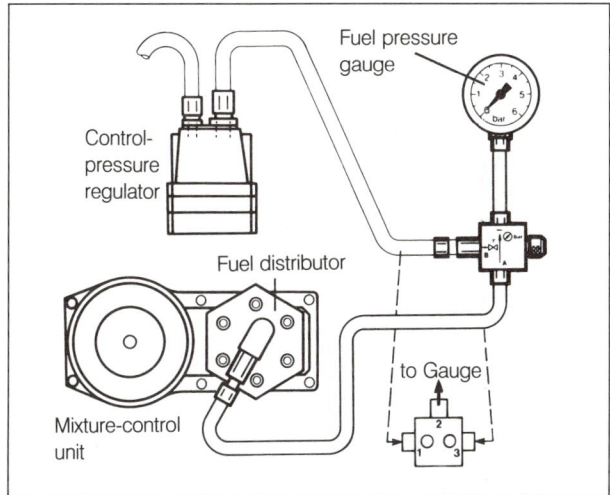

Fig. 5-6. Fuel pressure gauge installed for K-basic and K-lambda fuel pressure tests. Two different types of control valves and their connections are shown.

Even though the gauge is hooked to the control-pressure circuit, the gauge reads system pressure when its valve is closed. This is because the control-pressure regulator is closed off from the fuel distributor; no fuel is returned to the tank. You can measure system pressure at any temperature. When you open the valve, the control-pressure regulator now returns fuel to the tank, so the gauge reads control pressure as it is being regulated, according to temperature. See Fig. 5-7.

Clean off the top of the fuel distributor before undoing the fuel lines and installing the gauge so you don't let any dirt into the system. Disconnect the electrical connectors from the control-pressure regulator and the auxiliary-air device. You want them cold for these and other tests. With the safety-circuit bridged, run the pump to bleed the system of air. With the gauge below the valve as shown in Fig. 5-8, open and close the valve several times, about 10 seconds in each position. You cannot get good readings with air in the lines.

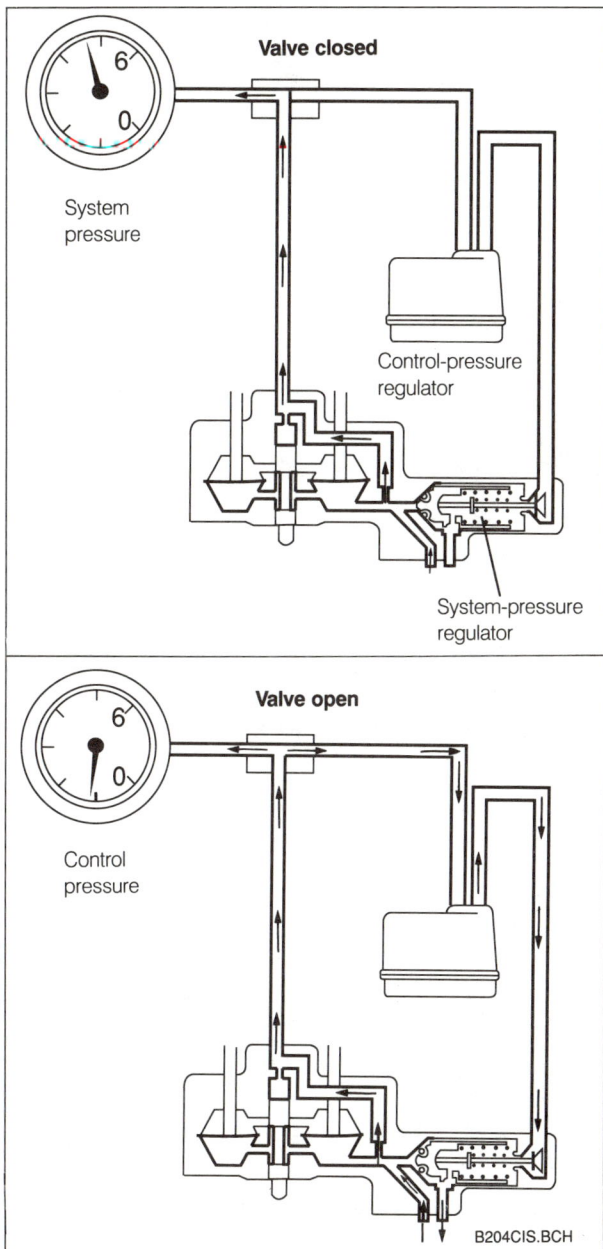

Fig. 5-7. With valve closed (top), control-pressure circuit is closed off; gauge reads system pressure. With valve open (bottom), control-pressure circuit is open; gauge reads control pressure.

6

Control Systems (K-Basic, K-Lambda)

Fig. 5-8. Air in fuel lines can affect pressure readings, so bleed the system at the gauge before making pressure tests.

System Pressure

Measurement. With the gauge valve closed to read system pressure, bridge the fuel pump safety circuit and run the pump. The gauge should read within the limits given in your shop manual, as shown in Fig. 5-9. What if system pressure is not in specification? If you've tested fuel-pump delivery as shown in **4.1 Fuel Pump Delivery** and it's OK, then you should adjust system pressure as described below.

Fig. 5-9. Typical system pressure for K-basic and K-lambda systems is 5.2 bar (75 psi).

Adjustment. If the pressure is not in specification, you can adjust it by adding or subtracting shims in the system-pressure regulator. See Fig. 5-10. Each 0.1 mm shim changes system pressure 0.15 bar (2.2 psi). Use new O-rings and copper gasket when you reinstall the piston.

NOTE ––––

If system pressure cannot be adjusted to specification, or if you drop the piston, you'll have to replace the entire fuel-distributor assembly because these parts are mated for the necessary fine tolerances.

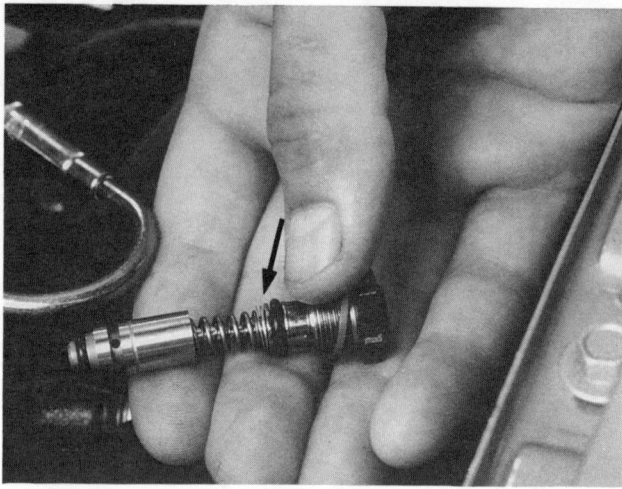

Fig. 5-10. System-pressure regulator removed from fuel distributor. Location of pressure regulator is shown in chapter 5. Add or remove shims (arrow) to change system pressure by 0.15 bar (2.2 psi) increments.

Control Pressure

Cold. Control pressure is related to temperature. If the regulator is cool enough for your hand to touch, it's cool enough to test. You can't fool K-basic and K-lambda by disconnecting the coolant-temperature sensor because they don't have one.

Control pressure goes up when temperature goes up. Higher control pressure reduces slit size and reduces enrichment. If you study a typical control-pressure specification graph as shown in Fig. 5-11, you can see that when the control-pressure regulator is at ambient (or surrounding) temperature in the shop, probably about 70°F (20°C), cold control pressure for the car being checked should be between 1.4 and 1.8 bar (20–26 psi).

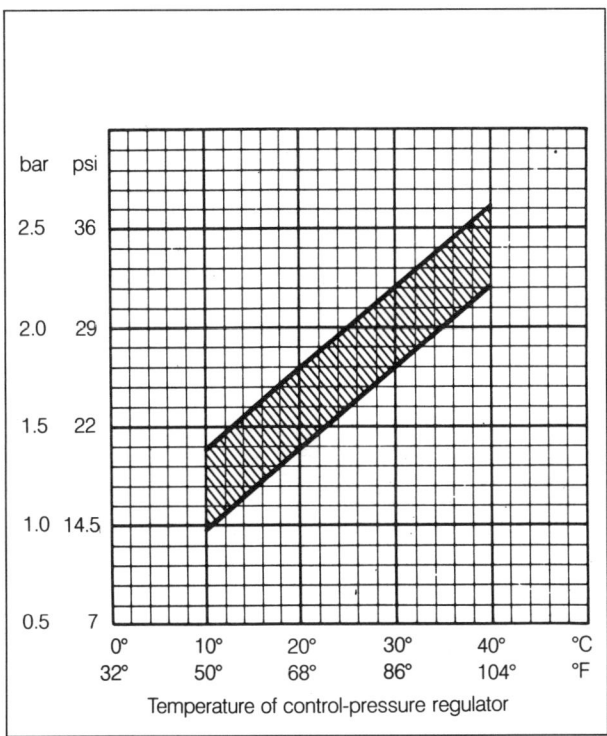

Fig. 5-11. Graph of control pressure based on ambient temperature. Control pressure rises as control-pressure regulator warms up.

With the fuel pump running, open the pressure-gauge valve so that fuel now flows through the control-pressure regulator. The gauge should show a drop from system pressure to control pressure. Typical cold-control pressure is about 1.5 bar (22 psi). The control-pressure regulator is reducing the pressure; that will give the engine the richer mixture needed for cold operation.

Fig. 5-12. When pressure-gauge valve is opened with fuel pump running, system pressure should drop to cold control pressure, approximately 1.5 bar (22 psi).

Warm. For the warm control-pressure test, reconnect the wiring connector to heat up the regulator. With the pump running, the pressure should slowly rise to a typical specification of between 3.4 and 3.8 bar (49–55 psi). Starting from ordinary shop temperatures, it should take one to two minutes for the pressure to rise from cold-control to warm-control, so don't get impatient.

Fig. 5-13. When control-pressure regulator warms up, control pressure should rise slowly, in perhaps one or two minutes.

If control pressure did not rise properly, check for restriction in the return line. Also measure the supply voltage to the heater of the control-pressure regulator as shown in Fig. 5-14. It should be at least 11.5 volts power supply. If return line is clear and the voltage checks out OK, replace the control-pressure regulator.

Full-Load/Altitude. If you have a control-pressure regulator with a vacuum hose to the intake manifold, you also test control pressure by applying vacuum to the outlet leading to the manifold. See Fig. 5-15. Test according to the instructions in your shop manual, with the regulator both cold and warm. Remember that when a vacuum specification reads 485–600 mbar, that's the same as 0.485–0.6 bar (7–9 psi), or about 16 in. Hg vacuum.

Test the regulator diaphragm leakage by pumping down to the specified vacuum, and then checking the pressure drop; if it leaks down more than 0.l bar (3 in. Hg) in 15 seconds, your regulator diaphragm is leaking. Replace the regulator.

6

Control Systems (K-Basic, K-Lambda)

Fig. 5-14. If control pressure is incorrect, check supply voltage to control-pressure regulator, and check fuel-distributor return line for clogging.

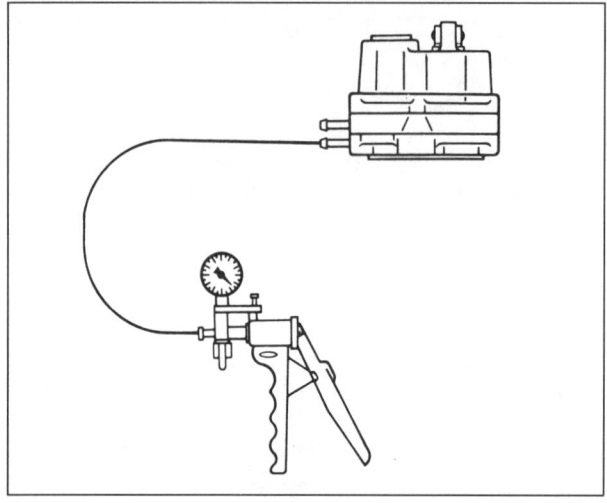

Fig. 5-15. For control-pressure regulators with full-load enrichment or altitude compensation, test using vacuum pump.

Residual Pressure

The residual-pressure test measures the amount of system-pressure leakage. When you turn off the pump after your warm control-pressure test, system leakage should be small enough so that pressure stays at a specified pressure for a specified time: usually 3 bar (43.5 psi) for 10 to 20 minutes.

If residual pressure falls below the minimums with the gauge valve open, the leak is probably in the control-pressure circuit, at the push-valve O-rings. See Fig. 5-16. If it is an older system without a push valve, the leak is probably at the control-pressure regulator.

Rerun the leakage test with the valve closed to shut off any possible leakage through the control-pressure regulator. If residual pressure is now OK, replace the regulator. If you still observe low pressure on the gauge, the regulator is OK, but there is system-pressure leakage.

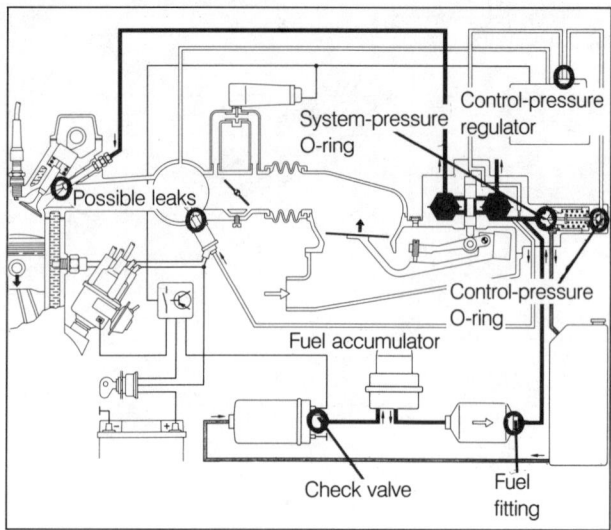

Fig. 5-16. Fuel leaks at many different locations can cause drop in residual pressure. Also check all fuel-line unions and fittings for signs of leakage.

In the system-pressure circuit, the cold start injector or port injectors could be leaking, or it could be the system-pressure regulator O-ring, or it could be the fuel-pump check valve, or it could be other components and connections. See Fig. 5-16. Check the check valve as described below. Run the pump to build pressure in the system and look closely for fuel leaks at fuel-line unions. To quick-check the fuel accumulator, remove the screw from the end of the accumulator. If fuel leaks out, the accumulator diaphragm is faulty and the accumulator should be replaced.

Fuel-Pump Check Valve. The fuel-pump check valve is part of the outlet fitting on the pump, as shown in Fig. 5-17, but early models may vary, so check your shop manual. To check the valve, disconnect the fuel supply line to the fuel distributor and attach the pressure gauge. Make sure that the gauge valve is closed. Run the fuel pump until pressure reaches 5 bar (78 psi) and shut off the pump. If residual pressure falls below specification, replace the pump check valve. Before you remove the fitting, remove the gas tank cap, clamp the pump inlet hose so no fuel will escape, and clean the fittings. Install a new seal and check valve and torque to specification. Check all connections with the pump running and redo the leakage test.

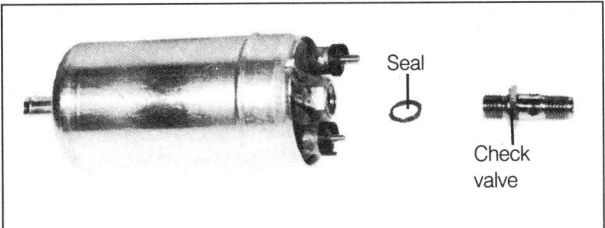

Fig. 5-17. Typical fuel pump with removable check valve.

Remove the gauge and reconnect the fuel lines when finished. With the gauge, you've checked K-basic and K-lambda system pressure, and warm and cold control pressure, as well as residual pressure.

5.2 Lambda Control (K-Lambda only)

In this section, you'll check the K-lambda control system by measuring dwell of the lambda-control valve under various conditions. You are looking to see how the computer controls the duty cycle of the lambda-control valve.

Thermoswitch

Most cars have a thermoswitch for open-loop operation of the lambda system before the lambda sensor is warmed up. A faulty thermoswitch will cause engine stumbling during warm-up. To test the thermoswitch, first check for the switchover temperatures marked on the switch. They are usually about room temperature (60°F – 15°C, or 78°F – 25°C) indicating partial engine warm-up.

You can check the thermoswitch installed or removed, using an ohmmeter as shown in Fig. 5-18. When below the marked temperature, there should be continuity in the switch, indicating that the contacts are closed. When above this temperature, there should be no continuity, indicating that the switch is open. If the switch fails the tests, replace it.

Checking Lambda Control

You can check lambda control-unit function and regulation by simulating thermoswitch and lambda-sensor signals, and then measuring the lambda-control valve duty cycle. You can use a special Bosch tester, or use a dwell meter connected to the test connector in the engine compartment. See your shop manual for its location. If you use a dwell meter, you'll have to convert the specifications from the duty-cycle scale (0–100%) to the equivalent degrees of dwell (0–90°) on the 4-cylinder setting. Do the tests in sequence.

If you get incorrect readings for any of these tests, check the lambda-control system wiring for continuity, and make sure the wiring connections are tight and free from corrosion, then repeat the tests. If the readings are still incorrect, the lambda control unit may be faulty.

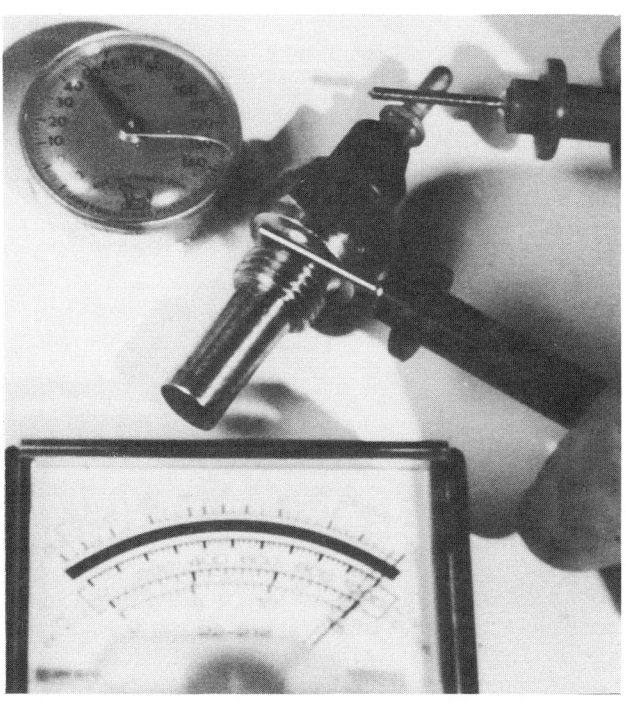

Fig. 5-18. When lambda thermoswitch is warm, it should be open, so there is no continuity as shown. When switch is cold, there should be continuity with switch closed.

Open-Loop, Cold. To check reaction to the thermoswitch signal, start by simulating a cold engine: Disconnect the wiring connector to the thermoswitch, and then short across the two terminals of the connector with a jumper wire.

With the ignition on, you should see the tester needle at about 60% duty cycle. See Fig. 5-19. It's telling you the Lambda control-valve is open more time than normal to enrich the mixture for a cold engine. The needle should not swing. This is cold-engine "open-loop" operation; a fixed pre-programmed signal from the control unit is giving a steady rich mixture.

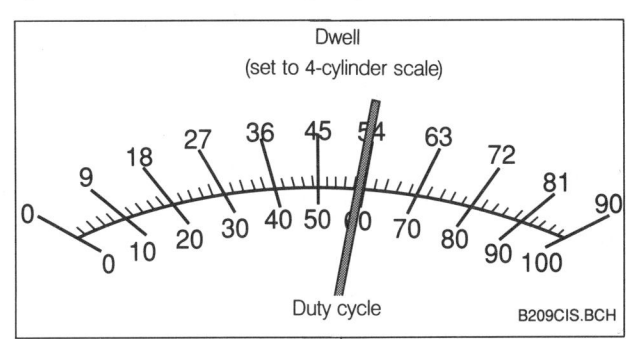

Fig. 5-19. When testing lambda-control system, you can check duty cycle or dwell. Remember that 100% duty cycle equals 90° dwell. Here, for open-loop operation on cold engine, duty cycle is 60%, dwell is 54°.

6

Open-Loop, Warm-Up. Remove the short from the thermoswitch cable so it is open, as if the thermoswitch has opened its contacts. You are signalling the control unit that the engine is partially warmed up so the control unit cuts back enrichment and sends a steady open-loop 50% signal, right in the middle. Again, the needle doesn't swing toward a rich or lean signal because the lambda sensor is too cool to send good signals. When you've completed the warm-up test, reconnect the thermoswitch.

In addition to cold lambda-sensor operation, steady, open-loop signals may come from the control unit as a result of inputs from other sensors, such as a wide-open throttle, or engine speed, depending on the car.

Rich Stop. This test checks control-unit reaction to a zero-voltage output from the lambda sensor. That signals a very lean mixture, which drives the air-fuel ratio towards rich. To test, disconnect the lead to the lambda sensor and ground it, as shown in Fig. 5-20. This is the same as if the lambda sensor were putting out zero voltage. With a no-voltage signal to the control unit, it should pulse the lambda valve to fully enrich the mixture. The duty cycle will read about 95%. The more time the valve is open, the more the mixture will go toward rich. In this case, 95% duty cycle is the rich stop; lambda control cannot go any richer.

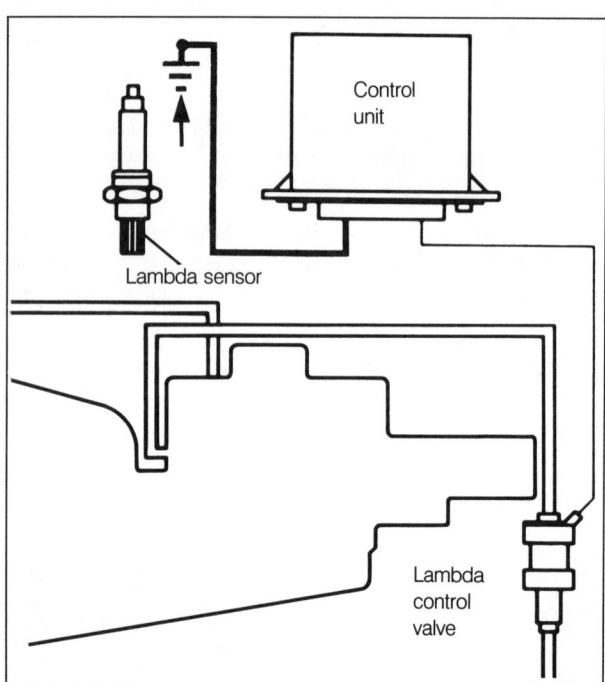

Fig. 5-20. To test if lambda control will enrich mixture in reaction to a no-voltage (lean mixture) lambda-sensor signal, disconnect lead from lambda sensor and ground to engine (arrow).

Lean Stop. This test checks control-unit reaction to a rich mixture. Instead of grounding the lambda-sensor wire, touch it to the positive terminal of a 2 volt dry cell as shown in Fig. 5-21. Ground the negative terminal of the battery to the engine. This

simulates positive voltage generated by the lambda sensor, indicating the mixture is rich. With a rich signal, the control unit should send a duty cycle signal of only about 10%. The less time the valve is open, the less fuel injected. The mixture goes toward lean, the lean stop. You've seen the two lambda control extremes, so reconnect the lambda sensor connector.

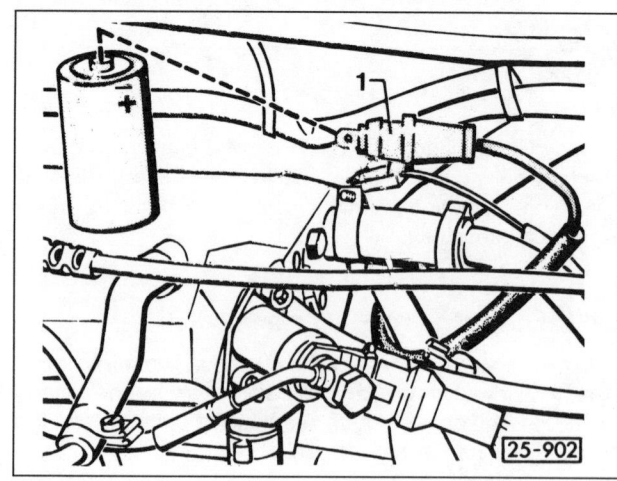

Fig. 5-21. When you apply voltage to the disconnected lambda sensor wire (**1**), control unit should reduce lambda valve duty cycle to fully lean the mixture.

Open-Loop/Closed-Loop Changeover. To test the changeover from open-loop operation to closed-loop, run the engine for a few minutes, until it is above the thermoswitch temperature. Turn it off for a few minutes so the lambda sensor will cool down, then start it again. Right after start-up, you should see a steady needle in the middle, for a minute or so. The sensor is not yet hot, so the electronic control-unit is sending the fixed 50% open-loop signal to the lambda control-valve.

When the sensor is hot enough to send the right kind of "hot" signals, the control unit switches over to operate in a closed-loop mode. You should see the needle fluctuating at the middle, typically between 45–55%. See Fig. 5-22. A swinging needle indicates closed-loop operation — normal idle, warm engine, warm lambda sensor.

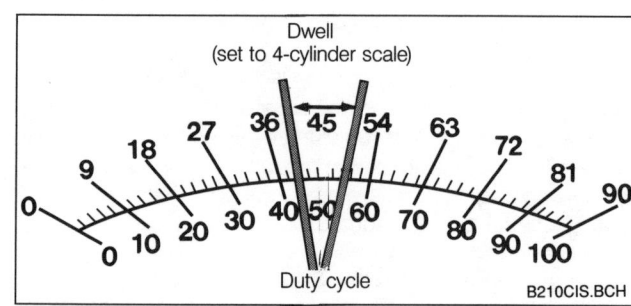

Fig. 5-22. During closed-loop operation, lambda-control valve duty cycle fluctuates as lambda-control system averages mixture around ideal ratio.

6. CONTROL SYSTEMS (KE-SYSTEMS)

This section covers the testing of those KE control systems that compensate the air-fuel mixture for all operating conditions.

6.1 Fuel Pressure Tests

As with K-basic and K-lambda, fuel pressure in KE systems influences all engine operating conditions. You'll measure three significant KE pressure values: 1) System pressure – the pressure created by the main fuel pump and regulated by the diaphragm pressure regulator, 2) Differential pressure – the pressure difference between the lower chambers of the fuel distributor and system pressure, controlled by the pressure actuator, and 3) Residual pressure – the pressure maintained in the system by the fuel accumulator and the fuel-pump check valve after the engine (and fuel pump) is shut off.

Most of your pressure checks on KE systems are quite different from those with K-basic and K-lambda. You will first measure system pressure, as specified in your shop manual, and then subtract from that measurement the actual lower-chamber pressure to see if the differential pressure is as specified.

Before you run your pressure checks, remove the gas tank cap. Remember, the cap seals the tank, so negative pressure (vacuum) could reduce the flow, and positive pressure (from vapor build-up) to about 1–2 psi could force fuel through open lines. Do the pressure tests in sequence.

Measuring Pressure-Actuator Current

For the pressure tests, you'll need to install a wiring test adapter to connect your digital meter to the pressure actuator. The adapter is shown earlier in Fig. 1-6. With it, you'll read the milliamps (mA) of current controlling the mixture by controlling the pressure differential. To install the adapter, remove the wiring connector from the pressure actuator, and install the adapter between the connector and the pressure actuator. Then connect the voltmeter probes to the adapter. See Fig. 6-1.

Installing Pressure Gauge

Before installing the pressure gauge, relieve pressure in the lines. In KE systems, if you do this by opening a fuel line, always relieve pressure first at the upper fitting as shown in Fig. 6-2, and catch any fuel with a cloth. If you open the lower-chamber fitting first, you will not relieve all pressures, and fuel will spray out when you open the upper fitting. You can also open the start-valve fitting; in some cars, notably Mercedes, you'll find a special test connection for attachment of the gauge.

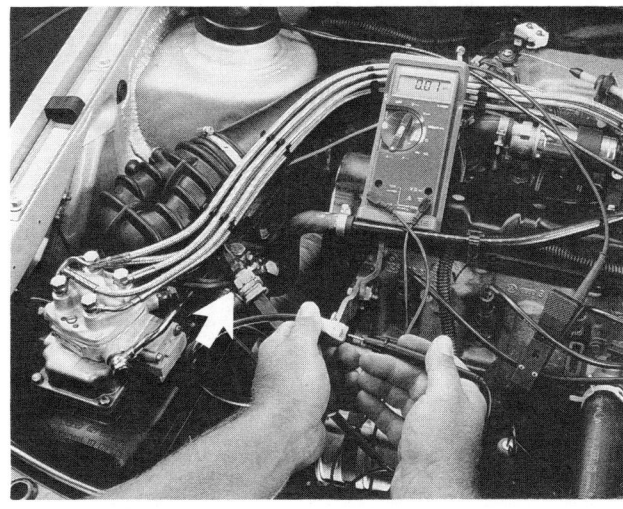

Fig. 6-1. Read pressure actuator current with test adapter connected between pressure actuator and wiring connector (arrow).

Fig. 6-2. On KE systems, if you relieve pressure by opening a fuel fitting, use the upper fitting.

Connect the pressure gauge so the fitting controlled by the shut-off valve is attached to the upper connection as shown in Fig. 6-3. That will measure system pressure. Use new seal rings to contain the pressures. Connect the other fitting to the lower-chamber test connection as shown in Fig. 6-4. Check your shop manual to see if special fittings are necessary.

6

Fig. 6-3. Install gauge line controlled by shut-off valve to upper fitting of fuel distributor.

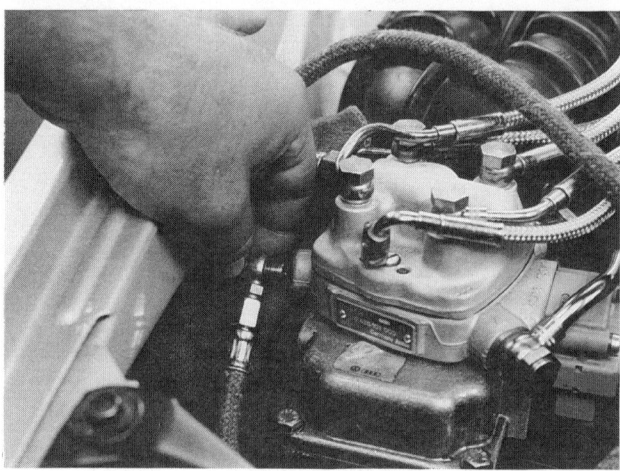

Fig. 6-4. Connect other gauge line to lower-chamber test connection. On some cars, special adapters may be necessary.

System Pressure

With the pressure-gauge valve open and the ignition off, jump the safety circuit as described in **1.1 General Procedures and Precautions** to run the fuel pump. See Fig. 6-5. You should see system pressure on the gauge; 5.5 bar (80 psi) is a typical KE-Jetronic pressure. If system pressure is not up to specification (not between 5.2 and 5.6 bar), check fuel-pump delivery, pump-supply voltage, filter, and supply and return lines for restrictions, just as in K-basic and K-lambda. If everything else checks out, replace the KE-pressure regulator; it is not adjustable.

NOTE ——

All of the gauge readings shown in this section are related to 5.5 bar (80 psi) system pressure. This is for example only. Some KE-Jetronic and KE-Motronic systems since 1988 operate with system pressures of 6.3 bar (91 psi). So check your shop manual.

Fig. 6-5. Test system pressure with gauge valve open and fuel pump running. Ignition should be off so that the pressure actuator does not influence reading.

Lower-Chamber Pressure (Differential Pressure)

Warm. Turn on the ignition and close off system pressure with the valve. Run the pump. The gauge now reads warm lower-chamber pressure. With the voltmeter reading about 10 mA of actuator current, the pressure should read about 5.1 bar (75 psi). Check your manual for the specified pressure difference—say 0.4 bar (5 psi) in this case. Four-tenths bar-pressure differential means subtract the specified drop from your measured system pressure: 5.5 minus 0.4 equals 5.1. If the pressures are correct, go to the cold-differential pressure test. If the pressure is incorrect, first check throughflow as described below. If the throughflow is OK and the actuator current is correct, the pressure actuator is faulty.

Restrictor Throughflow. If the pressure differential is too small, it may be because the actuator-fuel restrictor in the fuel distributor is clogged. To check it, measure the lower-chamber fuel throughflow. Remove the lower-chamber line from the fuel gauge as shown in Fig. 6-7. Make sure the gauge valve is closed. Direct the hose into a container and run the pump. See your shop manual for the specified time and flow; 140 cc (4.7 oz.) per minute is typical. Throughflow is about one-tenth of pump delivery, so you won't see a big stream.

Fig. 6-6. Check differential pressure with the gauge valve closed.

Fig. 6-8. Disconnect temperature sensor wire and bridge terminals with 15K-ohm resistor or special tool (shown). This will fool the control unit into believing engine is cold.

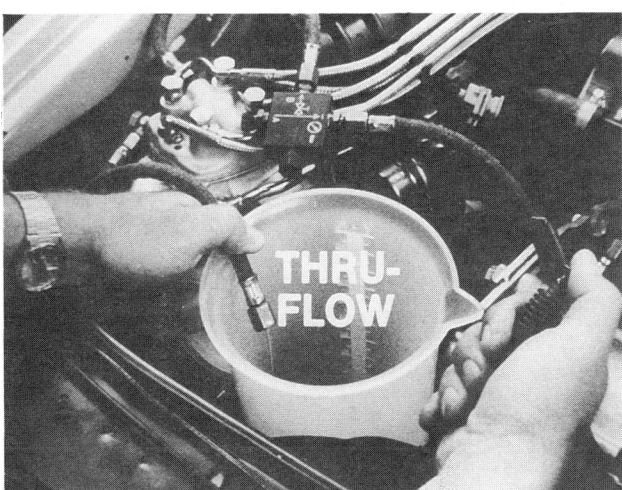

Fig. 6-7. Remove hose from pressure gauge and run pump to check troughflow from the lower chamber. A clogged restrictor in the fuel distributor can affect differential pressure.

Fig. 6-9. Cold-differential pressure is usually 1 bar (14.5 psi) lower than system pressure. Typical corresponding actuator current is 60 mA. Check your manual for correct specifications.

If throughflow is OK, reconnect the hose for the next test. If it's not OK, you can try to clean the restrictor by removing the fuel distributor and blowing carburetor cleaner through the lower-chamber circuit. Retest warm differential pressure. If it's still not OK, replace the fuel distributor.

Cold. To check the cold-differential pressure, plug a 15K-ohm resistor into the engine-temperature connector to send a cold-engine signal to the control unit. See Fig. 6-8. Disconnect the lambda sensor, turn the ignition on, and run the pump. The lower-chamber pressure difference from the system pressure you measured should be about 1 bar (15 psi) with the actuator current at about 60 mA. See Fig. 6-9. If your system pressure is 5.5 bar (80 psi), cold KE lower-chamber pressure will be 4.5 bar (65 psi). If the pressure differential is incorrect and actuator current is OK, replace the pressure actuator. If the differential and actuator current are incorrect, either the wiring or the control unit is faulty.

Residual Pressure

The KE-system residual-pressure test checks system leaks just as the K-basic and K-lambda test does. This test is especially important if you experience hard starting with a warm engine. Open the gauge valve, run the pump briefly, then shut it off. Twenty minutes later, you should still read at least 2.6 bar (38 psi) residual pressure.

Control Systems (KE-Systems)

Fig. 6-10. If residual pressure falls below 2.6 bar (38 psi) after 20 minutes, look for leaks in the fuel system.

If residual pressure is too low, there are many possible sources of leaks:

- external leaks at all fuel-line fittings

- the injectors (including cold-start)

- the sensor-plate basic-position adjustment

- pressure-regulator diaphragm

- the fuel accumulator diaphragm (quick-check the diaphragm by removing the screw from the end of the accumulator if fuel leaks out, the diaphragm is faulty and the accumulator should be replaced.)

- the fuel-pump check valve (see below).

Fuel-Pump Check Valve. The check valve is part of the outlet fitting on the pump. See earlier Fig. 5-17. To check the valve, disconnect the fuel supply line to the fuel distributor and attach the pressure gauge. Make sure that the gauge valve is closed. Briefly run the fuel pump until pressure reaches 5 bar (73 psi) and shut off the pump. If residual pressure falls below specification, replace the pump check valve. Before you remove the fitting, remove the gas tank cap, clamp the pump inlet hose so no fuel will escape, and clean the fittings. Install a new seal and check valve and torque to specification. Check all connections with the pump running and redo the leakage test.

After you remove the fuel-pressure gauge and reconnect the lines with new washers, recheck the system under pressure for leaks. Also remove the wiring test adapter and reconnect the actuator wiring connector.

6.2 Control

This section covers simple tests you can run to troubleshoot faults, such as poor driveability or a failed emissions test, caused by the KE control system and its sensors. You won't find all of these control functions on all cars, so again, check your shop manual to see which ones apply.

Some KE-Jetronic and KE-Motronic systems have on-board diagnostics with fault memory which can indicate trouble in the system. If your system has this, consult section **6.3 Fault Codes** before proceeding with these tests.

To check KE control, connect the wiring adapter and your digital meter to the pressure actuator to read the milliamps (mA) of current controlling the mixture. For more information see **6.1 Fuel Pressure Tests** above. Those are the first steps in this series of checks for control operations. For best results, perform the tests in sequence.

If any of the test results are incorrect, check the wiring between the components responsible for the function and the control-unit connector for continuity before assuming that the control unit is faulty.

Post-Start/Warm-Up

For this check, pull the pump fuse to keep the engine from starting, and remove the engine-temperature sensor connector to simulate that the engine is cold. When you crank the engine briefly, you should see post-start enrichment, typically 140 mA. Leave the ignition on. For about 30 seconds, post-start enrichment continues, then slowly reduces to warm-up enrichment, until the meter reads about 80 mA. Remember to check the specifications for your car.

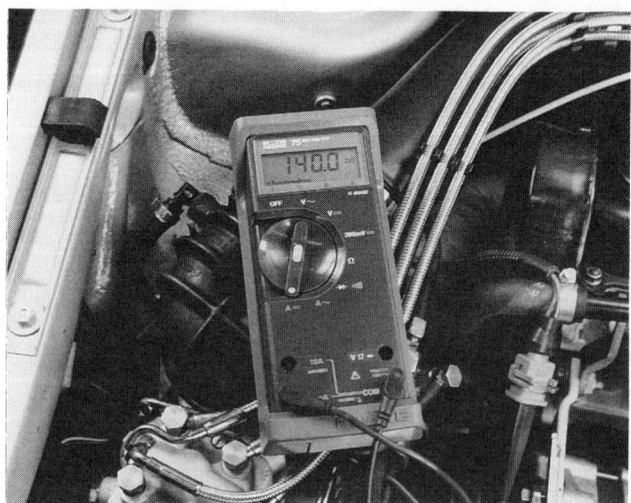

Fig. 6-11. For cold post-start, actuator current should typically read 140 mA.

If the control current is correct but the engine still has poor driveability during warm-up, check the resistance of the engine-temperature sensor as described in your manual.

Cold Acceleration

For cold-acceleration enrichment, quickly lift the sensor-plate as shown in Fig. 6-12 while holding the throttle open to open the idle-switch. You should see over 140 milliamps for about one second, indicating maximum enrichment; then it'll drop off quickly to warm-up enrichment. If you get no reading or the reading is incorrect, check the potentiometer as described below.

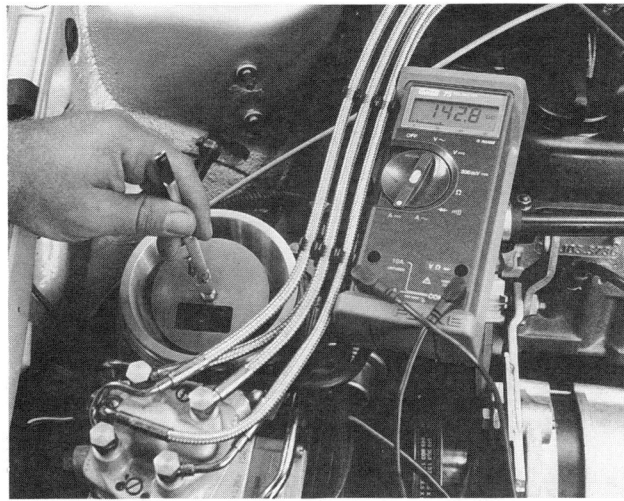

Fig. 6-12. Check cold-acceleration enrichment by quickly lifting sensor plate with throttle open. Incorrect milliamp reading could mean the potentiometer is out of adjustment.

Potentiometer. KE potentiometers measure the position of the sensor plate. The only purpose of this is to provide information for enrichment during cold acceleration. If your engine stumbles during cold acceleration, the potentiometer may be bad or out of adjustment.

To check the potentiometer, remove its wiring connector and, following the wiring diagram in your shop manual, check the resistance between the three potentiometer terminals. Your shop manual should have the resistance specifications, but the main point is that the resistance should change from specification when the plate is raised even slightly above basic position. If the readings are incorrect, loosen the fastening screws enough to rotate the potentiometer in the screw slots. Tighten and retest. The adjustment is very fine, so it might take a few tries to get it right.

Full-Load

For the full-load check, bridge the engine-temperature sensor lead to simulate a warm engine. Reinstall the fuel-pump fuse and start the engine. With the engine running about 4000 rpm, note the actuator current, then close the full-load switch. The milliamp reading should jump 3 mA, say from about 9 mA (switch open) to 12 mA (switch closed). If the reading is incorrect, the full-load switch may be faulty.

Fig. 6-13. With a warm engine above 4000 rpm, close full-load switch and watch voltmeter to note the change in actuator current.

To test the switch, remove its wiring connector and test at its terminals with an ohmmeter. You should see continuity with the switch closed, and no continuity with the switch open.

RPM Limitation

Briefly run wide-open to check rpm limitation. The engine should surge – indicating limitation – and the control current should read a negative value (approximately −20 to −45 mA), allowing actuator fuel to flow at system pressure into the lower chambers. The diaphragms rise and cut off injector fuel.

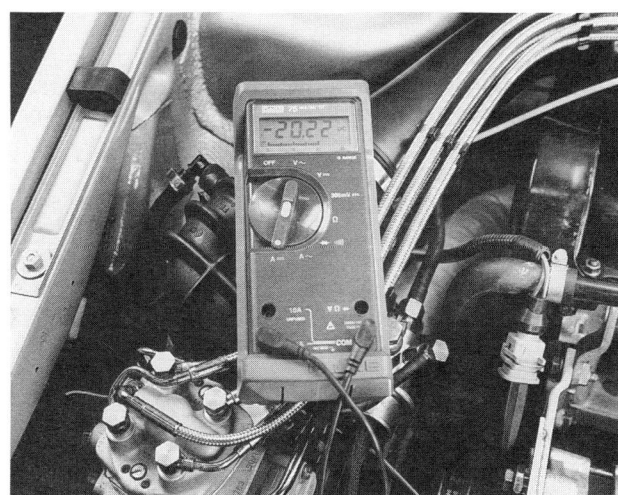

Fig. 6-14. On cars with rpm limitation, pressure-actuator current is negative, in this case −20 mA, so that differential pressure is zero, cutting off injected fuel.

Control Systems (KE-Systems)

Coasting Cut-Off

Rev the engine and then release the throttle so it will snap closed. You should see a quick reading of negative current. It's the same action as rpm limitation, closing off the fuel, but it's momentary, and you should see positive current as soon as the rpm approaches idle.

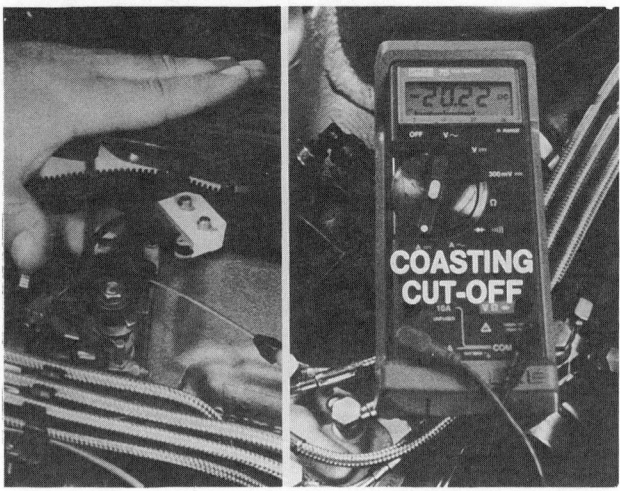

Fig. 6-15. To check coasting cut-off, rev engine and release throttle so it snaps closed. Look for negative actuator current to cut off injected fuel.

Checking Lambda Control

To check control-unit operations for the lambda-sensor circuits, go open-loop by disconnecting the sensor with the engine running. You should see a steady actuator current, typically 10 mA. Note that actuator current will be less if your system has an altitude sensor and if you are above 2000 feet (600 m) altitude; if so, disconnect the altitude sensor for these tests.

Next, ground the lambda-sensor terminal that leads to the control unit as shown in Fig. 6-17. This simulates a lean mixture. The actuator current should increase to 20 mA, a 10 mA increase as the control unit drives the system to the rich stop; you'll see the current increase in about 20 seconds.

Finally, remove the ground and wait about 10 seconds for the system to stabilize. Then, using a dry cell as shown in Fig. 6-18, apply about one volt-plus to the lambda-sensor connector. This will simulate the lambda sensor sending a rich signal to the control unit. The actuator current should fall toward zero as the system is driven to the lean-stop.

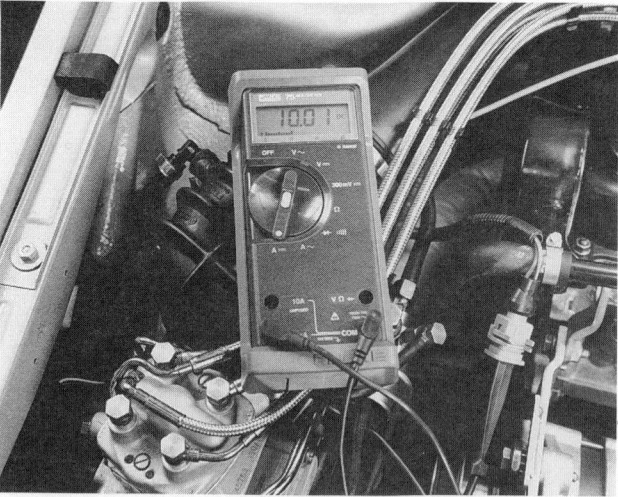

Fig. 6-16. Typical actuator current, engine warm, open-loop (lambda sensor disconnected).

Fig. 6-17. Ground the lambda-sensor terminal to simulate lean mixture. Look for increase in actuator current to rich stop.

Checking Lambda Sensor

You've now seen that the control unit correctly controls the actuator current for different operating conditions.

This test checks that the lambda sensor is sending a voltage signal for the control unit to monitor. With the wiring test adapter installed and the lambda sensor connected, run the engine at idle and look for the actuator current to fluctuate up and down in a narrow range. If the current fluctuates, the sensor is OK; if it doesn't fluctuate, replace the lambda sensor.

If the engine does not have a heated lambda sensor, you may have to idle the engine for a few minutes or hold it at 3000 rpm for about a minute to warm the sensor.

Those are some of the simple ways you can check KE mixture control and the lambda-sensor circuit.

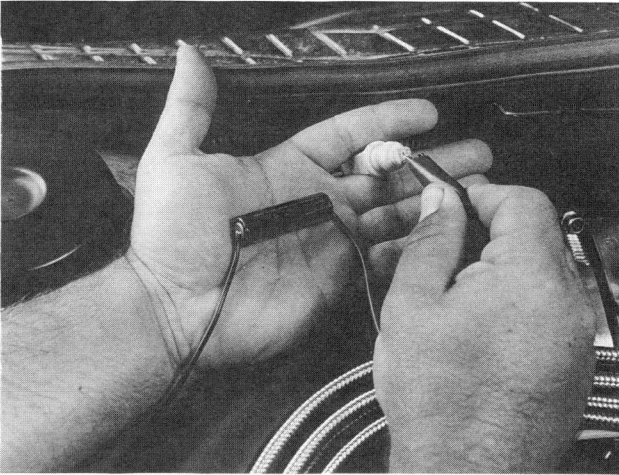

Fig. 6-18. Apply voltage (as from a dry cell) to the lambda-sensor connector that leads to the control unit. This simulates rich mixture. Look for actuator current to fall towards zero, the lean stop.

6.3 Fault Codes

Before troubleshooting, adjusting, or repairing KE3-Jetronic or KE-Motronic systems, read out the fault memory – the diagnostic information stored in the memory of the control unit. Not only will this speed your work, but unless you read these out first, you may experience bogus fault codes, stored as a result of checking or adjustment work that may exceed control limits.

Look over the description of the fault memory in chapter 5, recalling that California cars differ from 49-state cars, and that emission-related faults are stored differently from non-emission-related faults. In these procedures, I'll talk about 49-state cars.

Test Set-Up

Before reading out the fault codes you should:

● check the ground connection to the intake manifold

● make sure the engine is fully warm (upper radiator hose too hot to hold)

● run the engine briefly to 3000 rpm, then

● briefly press the accelerator pedal to the floor, and

● let the engine idle for 2 minutes.

Some cars have a test lead for reading out the fault codes located next to the fuel distributor as shown in Fig. 6-19. Connect a test light between this lead and the positive terminal of the battery. On other cars, notably with KE3-Jetronic, you'll see the fault codes on an indicator light on the instrument panel as shown in Fig. 6-20.

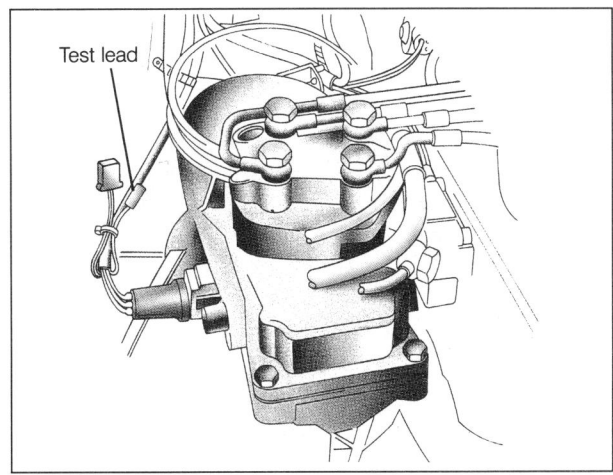

Fig. 6-19. On most systems, read the fault codes at a test connector next to the fuel distributor.

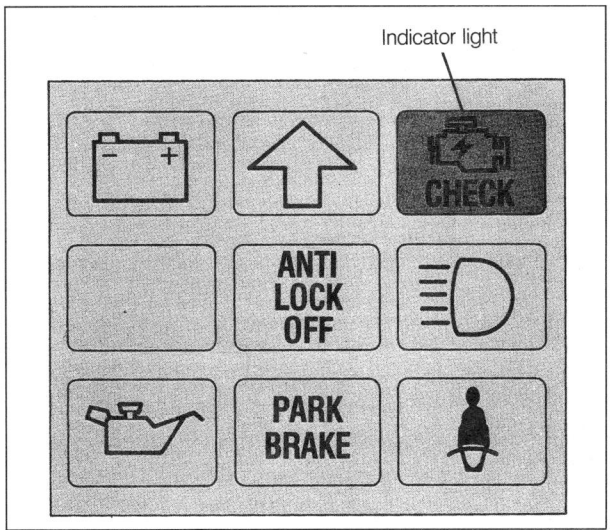

Fig. 6-20. On KE3-Jetronic systems, most notably Audi, read fault codes from flashing dashboard light.

To get the code to read out, insert a spare fuse into the top of the fuel-pump relay for at least 4 seconds. See Fig. 6-21. When you remove the fuse, the fault code will flash as follows:

1. Start signal: ON, then OFF for 2.5 sec.

2. Fault code, 4 digits
  — 1st digit – ON, 2.5 sec. OFF
  — 2nd digit – ON, 2.5 sec. OFF
  — 3rd digit – ON, 2.5 sec. OFF
  — 4th digit – ON, 2.5 sec. OFF

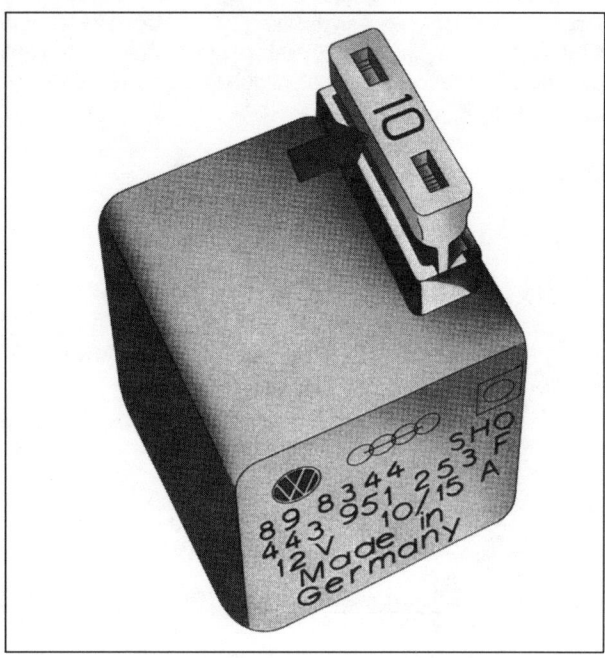

Fig. 6-21. To read out fault codes, insert spare fuse into fuel-pump relay.

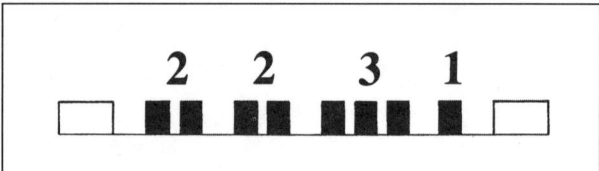

Fig. 6-22. For trouble code, read interval of flashing light; number of flashes indicates digit of fault code.

Each 4-digit code will keep repeating until you insert the fuse again. When you remove the fuse the second time, you'll see the second fault-code flash, continuing until you insert the fuse again. After you've read out all the fault codes, you'll see the end-of-fault display:

  — 2.5 sec. ON, 2.5 sec. OFF

Codes

Codes marked * are stored in permanent fault memory, and will be retained until erased. All others will be stored in temporary memory and will be stored even after turning OFF the engine, but will be erased on restart. If codes are erased, drive the car for about 5 minutes. If you have a no-start, crank the engine for about 6 seconds. California cars are different; check your shop manual. **Table b** lists the four-digit codes and their fault equivalents.

Table b. Fault Codes

Code	Fault equivalent
0 0 0 0	End of fault
1 1 1 1*	Control unit, replace
2 1 1 3*	No speed signal from distributor; check Hall sensor and circuits. OR air-flow sensor plate not moving freely; adjust potentiometer or lever.
2 1 2 1	Idle switch; check switch, circuit.
2 1 2 3	Full throttle switch; check switch, circuit.
2 1 4 1*	Knock control at max. retard; test compression, change fuel octane, adjust timing, check knock-sensor wires.
2 1 4 2*	Knock sensor signal; test sensor. See chapter 4
2 2 3 1	Idle-speed stabilizer adjustment limits exceeded; too-fast idle. Adjust throttle for full closing, check for vacuum leak, check ignition timing
2 2 3 2	Air-flow sensor potentiometer; check wiring
2 3 1 2	Coolant-temperature sensor; check wiring, check resistance
2 3 4 1*	Lambda sensor control limit; check CO, check lambda-sensor wire, check lambda-sensor control, check start valve, check evaporative system, check for vacuum leaks
2 3 4 2	Lambda sensor control; check wire, check sensor
2 3 4 3	Rich limit; the pressure regulator has exceeded +10mA for more than 5 minutes, closed-loop; check idle.
2 3 4 4	Lean limit; the pressure actuator has exceeded −5mA for more than 5 minutes, closed loop; check for vacuum leaks.
4 4 3 1	Idle-speed stabilizer
4 4 4 4	No faults recognized

Permanent fault codes should be erased after any adjustment or repair; temporary codes are erased by restart. To erase KE-Motronic permanent fault codes:

1) turn ignition OFF

2) insert fuse

3) turn ignition ON

4) after about 4 seconds, remove fuse, observe code 0 0 0 0

5) insert fuse; after about 10 seconds, remove fuse

To erase KE3-Jetronic permanent fault codes:

1) turn ignition OFF

2) insert fuse

3) turn ignition ON

4) after about 4 seconds, remove fuse

5) insert, remove fuse three times until indicator flashes
    4 4 4 3

6) insert fuse

7) after about 4 seconds, remove fuse observe code 0 0 0 0

8) insert fuse

9) after about 10 seconds, remove fuse

7. IDLE RPM AND MIXTURE (CO)

After any service, check idle speed and mixture (CO) if they're adjustable. For all of the CO tests you'll need access to an exhaust-gas analyzer. They're expensive, but your local service station or vehicle inspection station may be willing to rent you some time on theirs.

Follow the emission label to set your car to the right specification. Also follow the underhood decal instructions about exhaust-gas recirculation (EGR), the air pump, the vacuum limiter and other equipment, such as idle-speed control.

This label will also advise you of the exhaust-gas sampling point, usually a pipe off the exhaust manifold. That's more accurate than the tailpipe because you'll be sampling exhaust gases as they come from the engine, before they get cleaned up in the catalytic converter. The hose from the exhaust-gas analyzer fits right onto the pipe. It must fit snugly so there is no exhaust leak. If you can only sample CO at the tailpipe, note that the CO specifications will be much lower.

Adjusting Mixture (All Systems)

Before you adjust the mixture of any of the systems, keep this in mind: turning this screw changes the plunger lift for a certain air flow. See Fig. 7-1. Turning the screw clockwise makes the mixture richer (higher CO); turning it counterclockwise makes the mixture leaner (lower CO).

After each adjustment, remove the adjustment wrench and then accelerate the engine to stabilize the mixture before you take a reading on the analyzer; just placing the tool in the screw will change the CO reading. Every time you accelerate, the updraft air-flow sensor and its arm must rise. If you left the tool in place, you'd bend it. Be sure you get that tool out!

On most systems, the mixture-adjustment hole in the air-flow sensor is covered by a tamper-proof plug. To remove the plug, drill a 2.5 mm (³⁄₃₂ in.) hole in the plug. Only drill 3.5 to 4.0 mm (⁹⁄₆₄ to ⁵⁄₃₂ in.) deep. Then use a sheet metal screw to

extract the plug. Do not allow any metal shavings to fall into the air-flow sensor. In most states, emissions regulations require that you install a new plug after making adjustments.

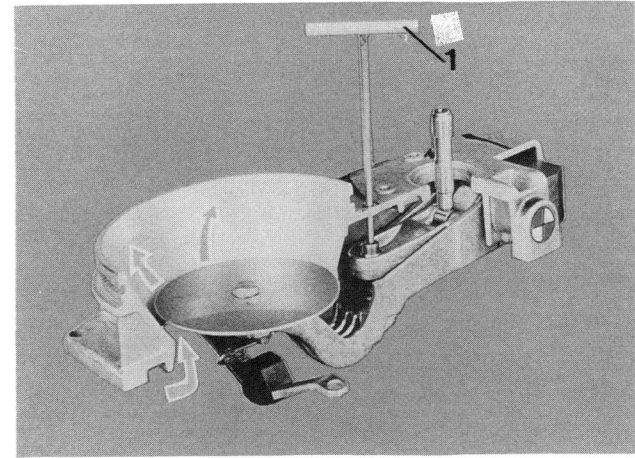

Fig. 7-1. Adjust mixture by using special tool (**1**) to change relationship of control plunger and sensor-plate lever.

NOTE ▬

The U.S. government and some states consider it "illegal tampering" for the car to leave a commercial repair facility without the anti-tampering plug correctly installed.

NOTE ▬

Before adjusting mixture on any system, it is a good idea to change the oil and filter. Fuel in the crankcase oil can alter the CO readings.

7.1 K-Basic

Idle Adjustment. Using a tachometer, check the idle rpm with the engine warm. The idle-speed screw shown in Fig. 7-2 adjusts the amount of air bypassing the closed throttle. It's easy to adjust idle rpm because it doesn't depend on the mixture adjustment.

Mixture Adjustment. With the adjustment wrench inserted as described above, start by screwing it out; always start mixture-adjustment from the lean side.

For downdraft mixture adjustment, remove the access screw to reach the adjustment, but replace the access screw to block false air before reading CO on the analyzer. On some downdraft types, there's no plug to remove; but remove the wrench from the adjustment screw each time you accelerate the engine.

Fig. 7-2. Adjust idle rpm at bypass screw on throttle body.

7.2 K-Lambda

Idle Adjustment K-lambda rpm adjustment is the same as K-basic.

Mixture Adjustment. You can adjust mixture using CO readings, or you can use the lambda tester to read the lambda-control valve duty cycle, and then use the exhaust-gas analyzer to be sure the system is working properly. If you use a dwell meter connected to the test connector, remember that you'll have to convert duty-cycle percent readings to degree-dwell readings.

With the engine warm, operating closed-loop the needle should be swinging in a narrow range. If, instead of swinging in the middle, it's well above 60%, it means that the lambda system is trying to control a mixture that is way too lean. See Fig. 7-3. Before adjusting the mixture, first check for air leaks that could be causing that lean mixture. If you're sure the air system is tight, turn the mixture adjustment screw clockwise to enrich the mixture.

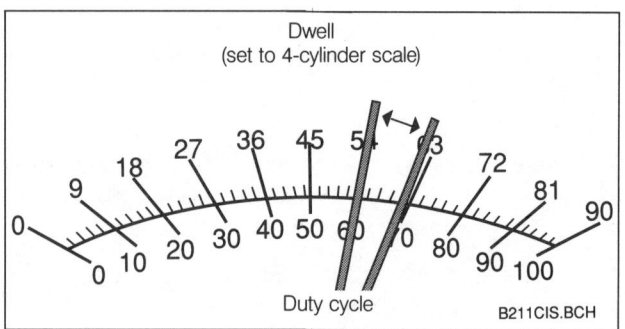

Fig. 7-3. If the needle swings high on the scale when the system is operating closed-loop, the mixture is set too lean; enrich the mixture.

On the other hand, if the needle swings well below 40%, it means the lambda system is trying to control a mixture that's way too rich. See Fig. 7-4. After checking the engine, such as

for oil on the air filter from crankcase verification, check the cold-start injector to be sure it's not leaking. Lambda control can do a great job but only with proper basic settings and good engine condition. If they're OK, turn the mixture screw counterclockwise to lean the mixture.

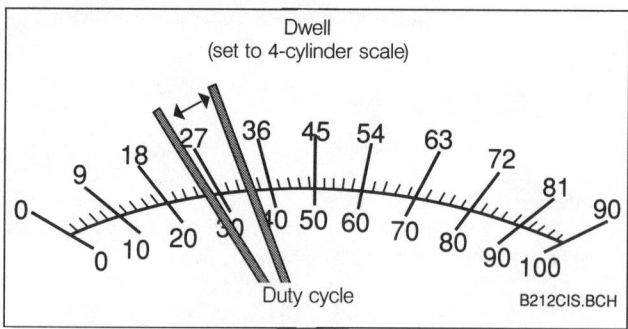

Fig. 7-4. A low swinging needle (duty cycle) means the mixture is set too rich; turn the screw counterclockwise to lean the mixture.

When you see the normal fluctuation of the needle around 45 to 55%, between rich and lean, just right, you've set K-lambda mixture properly. Together, the dwell readings and the CO meter will show you when the mixture is set correctly and both the engine and lambda system are OK.

7.3 KE Systems

For KE systems, begin by checking timing and idle-rpm against the engine specifications. Run up engine rpm for about 30 seconds to be sure the lambda sensor is hot. If there's a heating circuit for the sensor, you don't have to. Make sure the sensor is connected for closed-loop operation. On KE3 and KE-Motronic systems, read the fault codes first, as described in **6.3 Fault Codes**, before making any adjustments.

> **NOTE**
>
> Some KE-Motronic systems are adaptive; they have no adjustments for idle rpm or mixture.

On KE systems you read actuator current as well as CO, so install the wiring test harness at the pressure actuator as shown in Fig. 7-5, so you can read the current. With the engine idling, current should swing in a narrow range, back-and-forth a milliamp or two. That's a sure sign the system is operating closed-loop. The CO reading must also be in spec. A typical KE reading is 9–11 mA; a typical KE-Motronic fluctuation is from 0 mA to +5 mA.

If idle adjustments are necessary, start with mixture. KE is adjusted just as on K-basic and K-lambda systems, using the mixture screw. Adjust until both CO and actuator current are in spec.

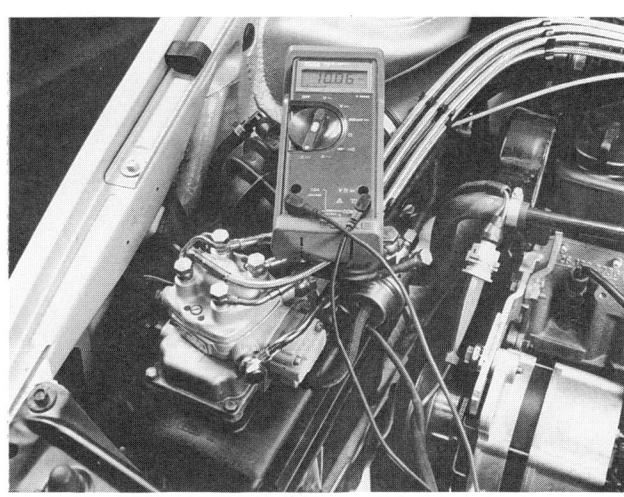

Fig. 7-5. Check KE mixture with exhaust-gas analyzer, and with voltmeter to read actuator current.

Note that in some cars, you adjust mixture closed-loop for the specified actuator current, fluctuating; then you check CO to be sure it's in specification. In other cars, you adjust mixture open-loop for a specified CO; then check closed-loop to see that the actuator current is in specification. Closed-loop, set actuator current, check CO. Or open-loop, set CO, check current. See Fig. 7-6.

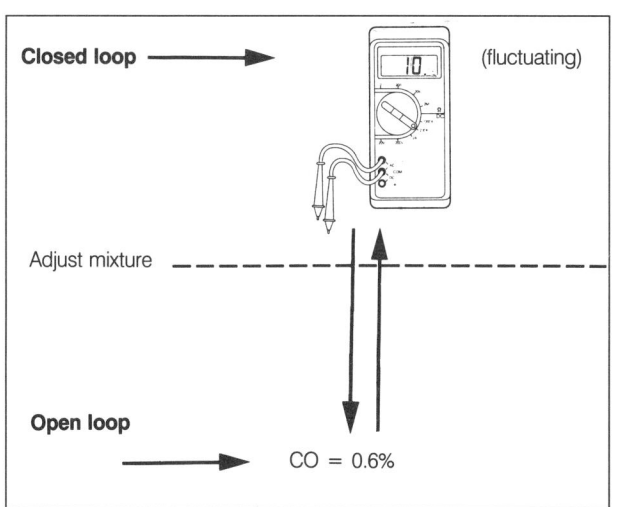

Fig. 7-6. KE mixture can be adjusted open-loop or closed-loop, but the results are still the same.

What if you can't get both actuator current and CO in spec? Check for exhaust leaks that could keep you from matching actuator current and CO. Apply air pressure to the exhaust-gas tap as shown in Fig. 7-7 to help you find them. Exhaust leaks

must be eliminated because leaking air that reaches the lambda sensor when the engine is running can fool the system into delivering a too-rich mixture.

Fig. 7-7. Check for exhaust leaks by pressurizing system at exhaust CO tap and listening for leaks. Exhaust leaks can fool the lambda sensor and cause rich mixtures.

You can adjust KE-Jetronic idle-rpm with the idle screw if the engine has an auxiliary-air valve. Adjust it the same as you do on K-basic and K-lambda.

If the KE system has an idle-speed stabilizer, you do not change rpm by turning the adjusting screw. The idle rpm is programmed in the control unit to control the idle-stabilizer dwell. Remember, when you adjust the screw bypass, the control-unit changes the dwell signal which alters the stabilizer bypass to hold the rpm.

To adjust idle on systems with an idle-speed stabilizer, you check stabilizer dwell as shown in Fig. 7-8. There should be a test connector in the engine compartment; check your shop manual for location. Turning the idle screw, in effect, adjusts dwell, not rpm. You may adjust it open-loop, with the stabilizer grounded, or as specified. If you can't set dwell to spec., check the stabilizer according to the manual. If it's OK, replace the control unit.

6

Fig. 7-8. To set idle rpm on KE systems with an idle-speed stabilzer, you check stabilizer dwell to set rpm.

8. START AUXILIARIES

The start auxiliaries include those components which increase air flow or add fuel enrichment independent of the control unit. These include the cold-start injector and the auxiliary-air valve.

Test the cold-start injector the same way as described in chapter 4. Just remember that some K-lambda systems have a hot-start pulse relay that grounds the start injector on-and-off during cranking, even when the thermo-time switch circuit is open.

Also test the auxiliary-air valve and idle-speed stabilizer as described in chapter 4.

Chapter 7

Tuning for Performance and Economy

Contents

7

1. INTRODUCTION

In the previous chapters I explained the detailed workings of Bosch fuel-injection systems for passenger cars, and hopefully cleared away much of the mystery surrounding them. Now I will take a look at the car enthusiast's obvious next step—modifying fuel-injection systems for high performance.

After you've read the other chapters you should know enough about the various ways that Bosch systems meter fuel that you can comprehend what the aftermarket tuners and parts suppliers are offering you, and what features of the stock system you might have to give up. You'll know enough to ask questions and to understand the trade-offs that usually accompany fuel-system modifications.

At this point, it is only natural to begin thinking about ways of fine-tuning or modifying your fuel-injection system. First a strong word of caution. Robert Bosch is one of the most technically sophisticated suppliers on the automotive scene today. Bosch pioneered the use of fuel-injection systems in automobiles decades ago, and they have been leaders in the field ever since. It takes a lot more than casual effort or a turn of a screwdriver to improve on the Bosch expertise. Many attempts result only in reducing overall performance and creating expensive headaches.

There are exceptions, of course, particularly when seeking substantial power gains from a highly modified engine. In some cases, modification of the fuel-injection system is necessary to realize the full benefits of engine modifications such as more radical camshafts, cylinder head porting, or forced induction.

Fig. 1-1. High-performance tuning is not hit-or-miss. To maintain driveability along with performance, as on this BMW engine, requires much testing.

This chapter looks at high-performance and racing applications of Bosch fuel-injection systems. I'll discuss the different applications and many current approaches to modifying Bosch fuel injection. You'll see when and why some modifications may be necessary, and why some others are likely to be a waste of time.

Fig. 1-2. Barber-Saab racing fleet uses stock LH-Jetronic fuel injection on turbocharged Saab engines.

At first glance, it may seem that I'm out to discourage fuel-injection system modifications, but that is hardly the case. Time spent under the hood, investigating and experimenting with fuel injection modifications is fascinating and educational. My intent is simply to help you to avoid wasting time and money, to avoid some of the more common and costly mistakes, and to give you the basic knowledge necessary to build the best running fuel-injection system for your needs. For many applications, the stock fuel system is really the best system so the best recommendation that this book can make is to leave it as is.

Of course, in the presence of a waving green flag the whole scene changes. Some exciting examples of Bosch fuel injection can be found on practically any racetrack. A quick glance into the engine compartment of almost any world-class racing car—from wild World Championship rally cars to GT prototypes to Formula 1 cars—will show that some very serious manufacturers and engine builders are prime users of the incredible Bosch fuel-management expertise. The turbocharged engines used in Formula 1 have produced over 1,000 horsepower from a mere 1.5 liters displacement with Bosch fuel injection, so don't let anyone convince you that bolting on a set of monster carburetors is the only route to maximum horsepower.

2. THE LEGAL ISSUES

OK, this subject is not the most fun, but it is a realistic problem. All fuel-delivery systems, carburetors and fuel-injection systems alike, are now considered to be part of an engine's emission-control equipment. Modifying the fuel-injection system is, therefore, modifying the emission-control system, and that raises some questions.

Practically speaking, if the car is to be registered and driven on public roads, the laws in your state and the legality of your modifications may have to become more important than ultimate performance.

7

Street Cars and Emissions

As this is written, at least 37 states require some sort of inspection program which could declare a car ineligible for use on public roads if it is found to be modified, or if it fails to pass an exhaust emissions test.

California regulations are especially tough, defining illegal tampering as "missing, modified, or disconnected smog control systems or parts." No matter how clean your exhaust is, passing a Smog Check in California includes passing a visual inspection. Any missing or modified parts must be restored to their original, functioning condition. The cost incurred by the owner to bring a non-tampered engine into compliance is limited by law, but if the inspection reveals evidence of tampering, there is no limit; the owner must bring the engine into compliance no matter what the cost.

Fig. 2-1. Depending upon the laws in your state, any modifications to the fuel system (such as the KE-Jetronic fuel-injection system shown) might be considered illegal tampering!

Not all states are as closely regulated, and from a practical standpoint the legality issue may mean little in a state with no enforcement program. The timeworn cliche, however, is true: Clean air is everyone's responsibility.

If the moral argument isn't convincing, then consider this; before automotive fuel injection ever found its way into a passenger car, it was used on racing cars by manufacturers seeking a more controllable and more efficient alternative to carburetors. Later, the first passenger car applications were on high-performance models. Remember the Mercedes-Benz 300SL "Gullwing" or the "fuelie" V8 Chevrolets of the 1950s? The point is this: Precise control of fuel delivery results in more complete combustion and a more efficient engine. Efficiency is the one basic building block for optimum horsepower, clean exhaust, and maximum fuel economy.

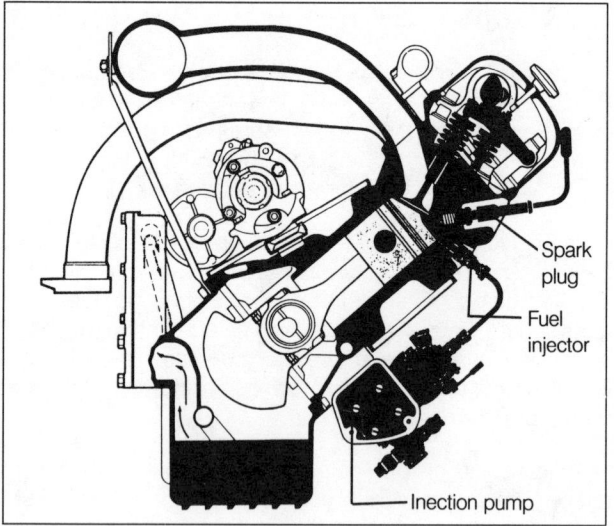

Fig. 2-2. The first uses of gasoline fuel-injection systems in passenger cars were in high-performance applications, like this Mercedes 300SL engine with direct mechanical injection.

A final note: The fuel and emission control system on your car may carry a 50,000 mile warranty or even more. The manufacturer may void that warranty if he decides that your modifications interfere with the performance or reliability of the original equipment. In one extreme case a manufacturer voided warranty coverage after installation of an aftermarket car-phone. It seems the installer had inadvertently drilled through a circuit in the Anti-lock Brake System (ABS). What can I say? Read the fine print and know what you might be giving up.

Altering Air-Fuel Ratios

In practice, the best set-up for clean exhaust and the best set-up for maximum power are slightly different. I'll discuss the differences in detail later in this chapter. For now, just remember that maximum power output demands a slightly richer air-fuel mixture—more fuel for a given amount of air.

Adjustments to fuel mixture would, at first glance, seem to be one aspect of fuel injection that is ready-made for fine-tuning to increase power output. To some extent, Bosch is already ahead of you. Most Bosch fuel-injection and engine-management systems recognize full-throttle acceleration as a special condition with special requirements. Under normal, part-throttle running conditions these systems precisely adjust the air-fuel mixture for good performance with minimum exhaust emissions. Then, at wide-open throttle, they provide a richer mixture—more fuel—to meet the brief demand for maximum power. Emissions are increased at wide-open throttle, but the trade-off is acceptable because of the short periods of time actually spent at full throttle.

Fig. 2-3. Many systems employ a throttle switch that signals the ECU when full-throttle enrichment is required.

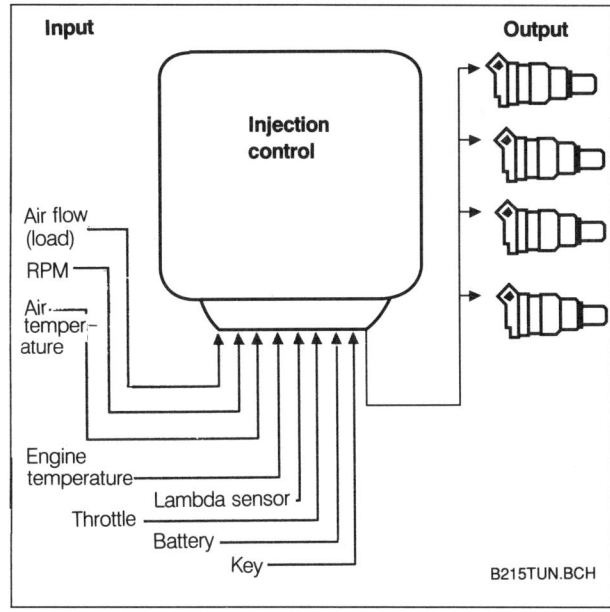

Fig. 3-1. Fuel-injection systems respond to varying operating conditions by monitoring the factors which affect the engine's fuel requirements.

The message of this section is this: For registered cars, driven on public road, emission control laws are a necessary consideration. Many of the modifications discussed in this chapter will not be legal for street use. Before beginning any modifications, know how the car will be used and know what the laws are in your state. Also remember that, when properly adjusted, the original Bosch system may be difficult to improve on, especially in street applications where exhaust emission control is a concern.

Racing Applications

Racing is another matter. Legality is only an issue in terms of the applicable rule book, and emissions are not usually a factor at all in racing. Especially in production-based racing classes, the rules governing fuel-injected cars will usually include some distinctions between modifications which are allowed and those which are not. The best advice is to study the rule book carefully before modifying the fuel-injection system. If possible, consult someone who is knowledgable about race-preparation of cars in your particular class.

3. DRIVEABILITY

One of the most significant advances that fuel-injection systems offer over carburetors is improved driveability—the ability to deliver smooth-running performance under a wide variety of operating conditions. A fuel-injection system measures many of the factors which affect driveability—engine temperature for example—and compensates for different conditions to deliver the appropriate air-fuel mixture.

The stock fuel-injection system is a compromise design which balances power output against concerns of driveability, fuel economy, and exhaust emission control. When it comes to modifying fuel injection to deliver more performance, or to match engine modifications, you will almost always be making trade-offs. Many people overlook just what the trade-offs are. The important point is this: Not all modifications will retain the driveability or fuel economy of the original, unmodified system.

Fooling The Computer

Modifications which attempt to trick the electronic control unit (ECU) by manipulating its input signals may produce some performance gains under full-throttle acceleration, but these same modifications might also result in less-than-perfect air-fuel mixtures under other operating conditions.

Wolfgang Hustedt is Bosch Motorsports Manager. He says, "Don't try to fool the computer. Can you enrich by playing games with the sensors? That's kid stuff which seldom gets results. The effect of the sensors on the enrichment is usually linear, so if you want rich at speed, you'll probably get too much rich at lower speeds." In other words, any marginal improvement may well be accompanied by problems elsewhere.

Other modifications might bring the cold-start injector into play, enriching the mixture by adding additional fuel under certain conditions. But the cold-start injector was never designed to deliver fuel efficiently to a warm-running engine. Its atomization characteristics are poor. In some applications its location results in poor fuel distribution. At anything less than full throttle, expect uneven fuel distribution, manifold wetting, flat spots, and stumbles.

7

Bigger Is Better (Or Is It?)

Installing a larger throttle valve is a good example of a modification which can cost driveability. Accurate fuel metering depends in part on smooth and consistent intake air flow. A larger throttle valve may cause changes in air velocity or uneven flow distribution, resulting in lean and rich variations in fuel delivery and stumbling or jerky throttle response.

Steve Dinan, President of Dinan Engineering, an aftermarket tuning firm in California, says, " . . . power gain is elusive. Somebody can show you tests with more power at one point on the curve, but you don't know that you are losing somewhere else. For overall power improvement, everything before turbos is much pain for small gain: bigger airbox, larger throttle body, larger injectors, removing the converter, porting usually result in no overall gain . . . You might gain peak power, but driveability glitches can develop, such as poor idle, loss of low-end torque, and high HC [hydrocarbon exhaust emissions]. Some of the most involved work we do is trying to restore driveability to a (car) that has been fooled with for a little gain at one point of the curve."

You Have To Decide

I'll discuss the details of some of these modifications later in this chapter. For now, just remember that any set-up is a compromise. Part-throttle responsiveness, fuel economy, and low-speed torque are all balanced against power at wide open throttle. Most high-performance enthusiasts are willing to put up with some sort of driveability problem in exchange for increased power, but it is something you have to decide for yourself. If the car has to start and run in cold weather, idle, and run smoothly in stop-and-go traffic, then you need to carefully consider the driveability trade-offs.

4. HIGH PERFORMANCE BASICS

In a practical sense, the fuel injection system is just one more engine component, and all of the components must work together in a balanced way to achieve peak performance and efficiency. So, before getting into the nuts and bolts of fuel injection modification, a quick review of internal combustion engine basics will help keep our goals in perspective.

Air Flow and Volumetric Efficiency

Most importantly, keep in mind the concept of an internal combustion engine as an air pump. A piston, traveling downward on its intake stroke, creates low pressure or vacuum in the cylinder. Air rushes in, is mixed with fuel and burned to produce power, and is then expelled to make room for more air. If you can increase the air flow through the engine, you can burn more fuel and produce more power.

In theory, the amount of air which is taken in by an engine is determined by displacement and rpm. In practice, two factors reduce the theoretical maximum: (1) Valve timing limits the amount of air which can be taken in on the intake stroke or pumped out on the exhaust stroke. The term used to describe how well the engine pumps air — the true value as compared to the theoretical 100% — is "volumetric efficiency". (2) In the real world, automotive engines are not very efficient air pumps. Volumetric efficiency is reduced on the intake side by the air filter, the air-flow sensor, the throttle valve, and the intake manifold and ports. They impede the free flow of air into the combustion chamber. Volumetric efficiency is further reduced by the restrictions of the exhaust system — exhaust manifolds, catalytic converters, mufflers and tailpipes.

With these things in mind, it is easy to see that nearly all the hot-rodder's or racer's horsepower tricks have one common goal: to increase air flow through the engine by increasing volumetric efficiency at one part of the power curve (at the expense of other parts). The gains may be tailored to the middle rpm range to improve torque, or to the high rpm range to maximize peak horsepower, but the idea is the same. Higher lift and longer duration camshafts, larger valves, ported cylinder heads, larger carburetors, aftermarket intake manifolds, low-restriction exhaust headers, and even dual exhaust systems all have the same job: to reduce air-flow restriction and allow more air to pass through the engine. Superchargers and turbochargers have the same purpose, except that their job is to force more air through.

The Fuel System

Naturally, when engine improvements allow increased air flow through the engine, the fuel system must compensate. It must deliver a proportionally greater amount of fuel to maintain the proper air-fuel ratio, or the engine will run lean.

The air flow and fuel delivery capabilities of the stock fuel-injection system have been chosen to correspond to the performance demands of the engine — its volumetric efficiency. Modifying the engine changes the engine's volumetric efficiency and, therefore, the demands being placed on the original fuel-injection system.

In practice, this may or may not be a problem. Some of the stock Bosch systems are quite flexible, able to compensate for some pretty impressive flow increases. More on that later in this chapter.

For now, just keep in mind this fundamental question: When you modify the engines does the increased air flow and increased demand for fuel exceed the limits of the stock fuel-injection system? If so, some fuel injection modifications may be necessary and worthwhile.

As for stock or only slightly modified engines, modifying the injection system to get more power is a different story. This is where significant power gains are elusive, and where it is easy to do more harm than good. Even the earliest, most basic Bosch fuel-injection systems are precise and highly optimized, especially when compared to carburetors. Decades of Robert Bosch development expertise are not easy to improve on.

The systems are already very good at measuring intake air flow and precisely metering fuel. They adjust to changing engine characteristics and compensate for different operating conditions. The only real opportunity for gain—and small gain at that—lies in fine-tuning the air-fuel mixture.

Air-Fuel Ratio and Performance

As discussed in Chapter 2, all engines need a proper mix of air and fuel to achieve complete combustion. In gasoline engines, this ideal (stoichiometric) air-fuel ratio is about 14.7:1 — approximately 14 kg of air are required for complete combustion of 1 kg of gasoline. This ratio is also described as λ (lambda) = 1.

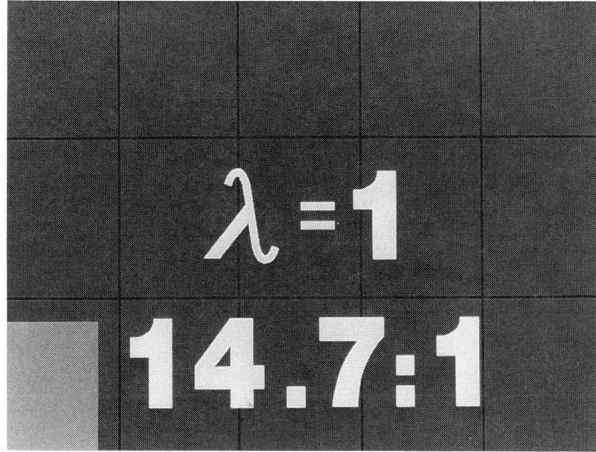

Fig. 4-1. For gasoline-fueled engines, the most complete combustion occurs at the ideal (stoichiometric) air-fuel ratio of about 14.7:1 (approximately 14 kg air for every 1 kg fuel).

The stoichiometric ratio, however, is not necessarily the optimum ratio for peak power or for minimum fuel consumption. The graph in Fig. 4-2 shows the relationships between power, fuel consumption, and air-fuel mixture. Peak power is achieved with a slightly richer air-fuel mixture (when λ is approximately 0.9). Minimum specific fuel consumption is achieved with a slightly leaner mixture (when λ is approximately 1.05).

The fuel-injection system is set up to operate in a narrow range around what is approximately the stoichiometric air-fuel ratio ($\lambda = 1$). As the graph in Fig. 4-2 shows, this provides the best compromise between maximum power output and minimum fuel consumption. Most importantly, operating in this narrow range is essential for minimizing exhaust emissions.

Fig. 4-2. Power and specific fuel consumption both vary as a function of air-fuel ratio.

In theory, it is possible to fine-tune the air-fuel mixture, either to maximize power or to minimize fuel consumption. This is tempting, of course, but take a closer look.

First, most systems already provide some kind of mixture enrichment at full throttle, so some of what you could hope to gain by optimizing the mixture for maximum power is already there.

Second, reconsider the curves on the graph. In the areas of interest—near the maximum power point and the minimum fuel consumption point—those curves are relatively flat. Even if the system can be adjusted to deliver the perfect mixture (just at the point of maximum power), the gain promises to be pretty small. There are no huge amounts of horsepower to be unlocked here! And if you miss and go too rich, its easy to end up de-tuning instead of improving! Of all the methods which try to optimize the air-fuel mixture for peak power, the ones that actually provide more gain than pain are likely to have extensive dynamometer testing and road testing behind them.

A third important consideration for street cars is exhaust emissions. On those cars equipped with a lambda sensor and catalytic converter, the relatively clean exhaust resulting from combustion at the stoichiometric ratio is necessary for proper operation of the catalytic converter. For all cars, any significant deviation from $\lambda = 1$ increases engine exhaust emissions dramatically. As the mixture becomes rich, hydrocarbons (HC) and carbon monoxide (CO) go up. As the mixture becomes lean, oxides of nitrogen (NO_x) increase very rapidly. See Fig. 4-3.

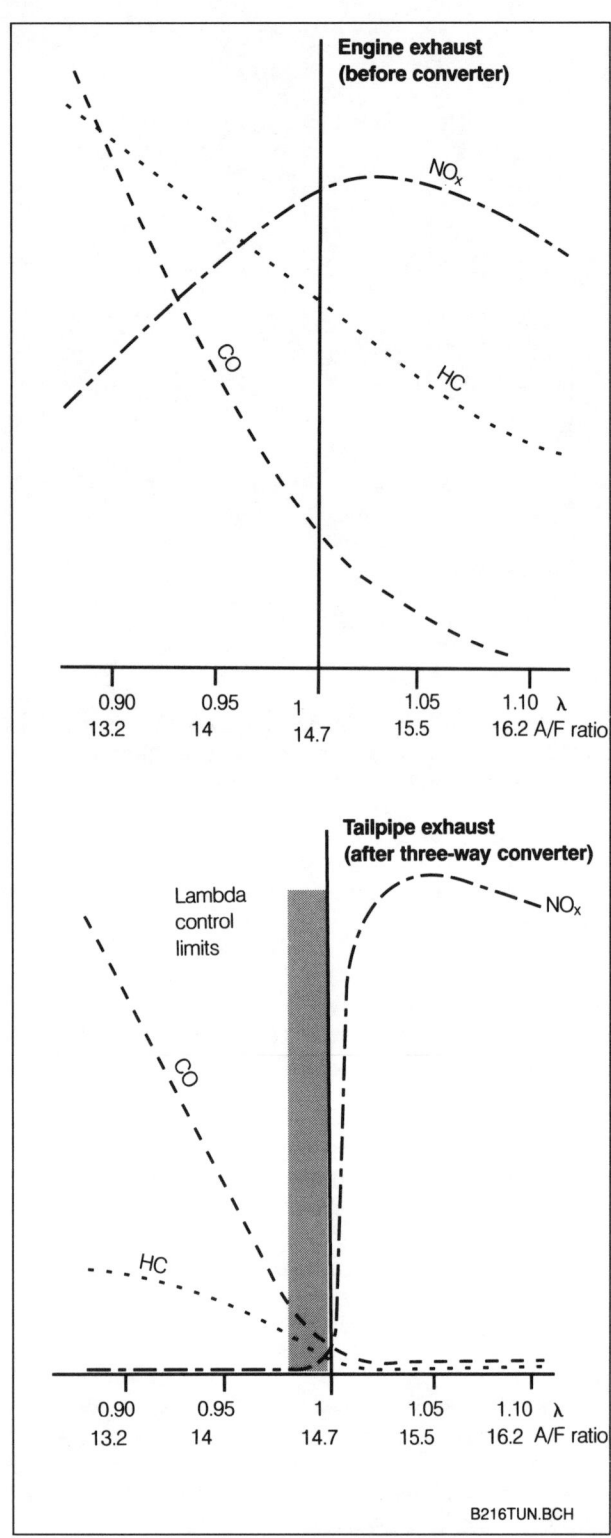

Fig. 4-3. The stoichiometric air-fuel ratio (λ = 1), results in most complete combustion and minimizes harmful exhaust emissions when the engine has a catalytic converter. Notice the inevitable increase in exhaust emissions as the mixture deviates from λ = 1.

Ignition

Another performance factor that goes hand-in-hand with fuel delivery is ignition timing. Conservative fuel mixture control in the name of low exhaust emissions may also be accompanied by conservative ignition timing curves, primarily because of manufacturer's concerns over the quality of available gasoline.

Many tuners of fuel-injected cars will advocate revised ignition timing specifications to further unlock the performance potential of revised fuel control. This is particularly true for Motronic systems which control fuel delivery and ignition timing in one electronic control unit.

Too much spark advance can bring on problems of its own. The biggest danger is detonation. Modifications to the manufacturer's ignition advance curve may make it necessary to use only gasoline of the highest available octane. In extreme cases, you may need to use an octane-boosting fuel additive. Be sure to weigh the added expense and nuisance against the gains in peak power.

Summary

The bottom line is this: From the factory, Bosch fuel-injection and engine-management systems are optimized for the best compromise among power, driveability, fuel economy, and exhaust emissions. Just "twiddling the knobs" on the injection system stands a good chance of just reducing power—or fuel economy—and definitely threatens exhaust emissions. Only when the engine's volumetric efficiency and breathing characteristics are significantly changed will the fuel-injection system really require major modifications, or be able to benefit from them.

5. SYSTEM BASICS

Each of the Bosch fuel-injection systems covered in this manual has its own unique features and characteristics, and the strategies for modifying them are therefore different. I'll describe the kinds of modifications which apply to each general type of fuel-injection system. I'll begin by discussing the general implications of lambda control and the modification of systems which retain a lambda sensor.

While I've included many possible modifications to Bosch systems, this is by no means a complete list. I've tried to describe the most popular modifications, and to give a cross-section of the methods employed by experienced tuners. Their efforts are not only interesting and exciting. They also provide valuable insight into how the experts think about the problems, and how they approach solving them. These insights, combined with your knowledge of the system basics from the first six chapters, should help you to decide which modifications hold promise for your application.

Some of these modifications are well-known and well-tested. I can give you a fairly accurate idea of the results, if any, that you might expect. For many others, the results are much less certain. I've described these modifications and the theories by which they are supposed to work. I make no recommendations.

Modifications must always be approached with caution. There is always a risk of making the car run worse. There may also be a risk of causing permanent damage to the fuel system, the engine, or other components. Carefully consider the basics that you've learned from this book, what each modification offers, and whether it is right for your particular car. If you have doubts, it is always wise to seek the advice of knowledgable tuners about your particular application.

5.1 Systems with Lambda Sensors

The assumption is that, in most cases, street-driven cars which were originally equipped with lambda sensors (and catalytic converters) are going to retain them, both for legal (emissions and inspection) reasons, and because of the driveability and fuel economy advantages—their ability to continuously fine-tune the air-fuel mixture to match different conditions. The functional details of these systems are well covered in other parts of this book. In the context of high-performance modifications, however, there are some important things to remember.

The basic fuel-injection system meters fuel to air in the best proportions possible, based on its various inputs. In closed-loop operation, the lambda system monitors the exhaust and continuously makes additional fine adjustments to the air-fuel mixture. The exceptions (open-loop operation) occur during warm-up when the lambda sensor is not yet up to operating temperature and, in most systems, at full throttle when lambda sensor control is bypassed in favor of a slightly enriched mixture.

Remember, even if the system is modified to make it capable of providing more fuel, the lambda-sensor system in closed-loop operation will still do just what it was designed to do—continously adjust the air-fuel mixture to approach the stoichiometric ratio, the narrow range around $\lambda = 1$. In short, no gain, except (maybe) in open-loop operation at full throttle.

Normally, this self-correcting capability—automatically keeping the mixture near the perfect stoichiometric ratio—is very desirable. If minor system modifications make the mixture a little too rich at low and mid-range rpm, the lambda-sensor system in closed-loop operation will tend to correct back to the stoichiometric ratio and preserve driveability, exhaust emissions control, and fuel economy.

The problems come when fuel system modifications force the system to the limits of its normal range of adjustment—when the system which is constantly sensing an over-rich mixture and constantly trying to adjust more lean reaches the limits of its adjustment range. In such a case, the modifications and the resulting rich mixture will override the lambda sensor's ability to correct, and will wreak havoc on fuel economy, driveability, and exhaust emissions.

This situation, which can even affect the performance and driveability of unmodified engines with high mileage and wear, is addressed in the very latest Bosch fuel-injection and engine-management systems which feature adaptive control. Instead of working within a fixed range of lambda control parameters, these systems are able to accumulate data during operation and adjust the center of their operating range according to the needs of the individual car and conditions. Within the realm of stock or mildly modified engines, this capability should eliminate the problem of overriding the range of the lambda control. For more specific information on those systems which feature adaptive control, see the information on system operation in Chapter 3 (pulsed systems) or Chapter 5 (continuous systems).

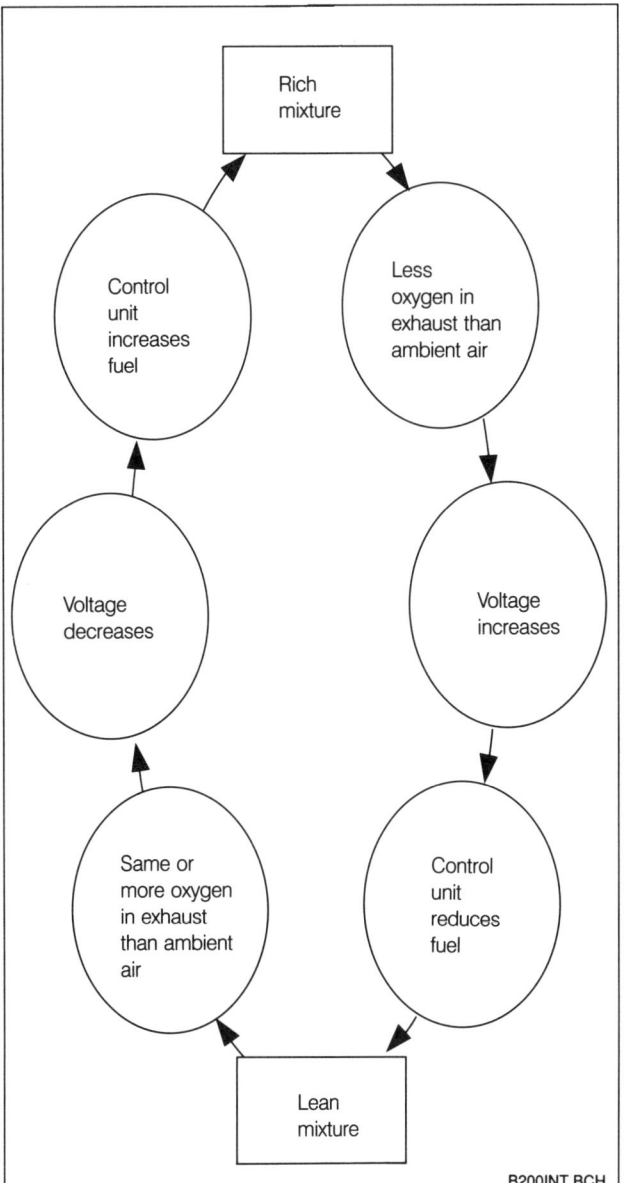

B200INT.BCH

Fig. 5-1. In closed-loop operation, the lambda sensor system will continuously try to correct the air-fuel mixture to a pre-programmed value—no matter what modifications have been made to increase fuel delivery.

7

5.2 Continuous Injection Systems

There is tremendous interest in high-performance modification of the K-Jetronic family of systems, mainly because many of the most popular and affordable sporty cars being "tweaked" by enthusiasts today were originally equipped with K-Jetronic. In fact, much of its popularity with enthusiasts rests squarely with Volkswagen and the water-cooled four-cylinder engine used in Rabbits, Sciroccos, Golfs, Jettas, and other models. Other K-Jetronic applications, on cars such as Porsches and BMWs, have helped earn a good performance reputation for the continuous injection systems.

Darrell Vittone at Techtonics is one of many Volkswagen specialists with experience in the successful use of K-Jetronic and K-lambda systems on modified engines. What that experience has shown is that these systems have air flow and fuel metering capability that far exceeds what these engines require in stock form. They will support considerable increases in air flow resulting from engine modifications such as hotter cams, low-restriction exhaust headers, cylinder head porting, and other common hot-rodding practices. While the exact limits are hard to define, Darrell's own estimates are that an unmodified K-Jetronic system will adequately support the air flow and fuel delivery requirements of a modified Volkswagen four-cylinder engine producing as much as 170 horsepower.

This is a very impressive figure. It means that K-Jetronic systems, in these Volkswagen engine applications at least, can handle perhaps 70% horsepower increases without modifications.

For all practical purposes then, virtually any streetable hot rodding modifications are within the capabilities of the stock system. With modest boost levels, even forced induction is within the limits of stock K-Jetronic.

It is also interesting to note that because K-Jetronic injectors are delivering fuel continously (when system pressure is sufficient to open them), the injector design is simpler than other types. This simplicity lends itself to a more optimized design. Bosch engineers have recognized this, and the injectors developed for the continuous-flow systems have very good fuel atomizing characteristic.

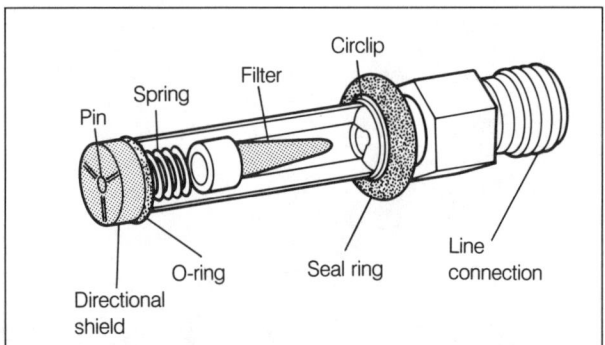

Fig. 5-2. The constant fuel flow of continuous systems allows less complex fuel injectors which are optimized for best possible fuel atomization.

Limitations

Of course, there are limits to the modifications that are compatible with continuous injection systems.

Radical camshaft timing is a problem, specifically long duration with radical overlap. The K-Jetronic air-flow sensor is a sensitive and very responsive device, perfect for precise air flow measurement within the relatively calm and consistant intake-flow dynamics of an unmodified engine. But, a long-duration camshaft at low rpm will reduce manifold vacuum and cause all sorts of lazy, confused air pulsations in the intake-flow path. These pulsations will upset the air flow meter sensor plate and interfere with accurate fuel metering. Idle quality suffers, perhaps dramatically, and not until such an engine is revved past perhaps 2000 rpm will it receive accurate fuel metering and begin to run smoothly.

Even if you can make your modified engine run somewhat smoothly at low rpm, it may still be necessary to make some adjustment to idle mixture to compensate for the altered intake flow dynamics.

Mixture Adjustments

The only real adjustable elements of the stock systems are idle mixture and, on K-basic and K-lambda, the system fuel pressure. Some people have also experimented with altering control pressure to make the fuel distributor respond more quickly to air flow changes.

None of these adjustments is particularly well suited to fine-tuning the fuel mixture at mid to high rpm, since any change which is effective there brings with it a more dramatic change at low rpm and idle. Thus, there are the usual side effects of hard starting, poor driveability, higher fuel consumption, and high exhaust emissions.

Competitors in the now defunct SCCA Rabbit/Bilstein Cup racing series (for near-stock, Volkswagen Rabbit sedans equipped with K-basic) used to carefully adjust the idle mixture to a very rich setting—the series rules prohibited any real modifications. The result was that the large enrichment in fuel mixture at idle brought with it a smaller but meaningful enrichment at high rpm, where it would give a little more power. At idle and at moderate engine speeds the cars ran very rich, but for racing, the slight power increase made this trade-off acceptable.

The same strategy might be useful in any race-only application, as long as such out-of-factory-specification adjustments are not prohibited by the rules. Remember though how easy it is to overdo mixture adjustments and actually reduce peak horsepower. Also consider that an overrich mixture will probably be harmful to mid-range torque, which can be as important as peak horsepower in some racing applications.

Mixture Enrichment

Autotech Sport Tuning offers a unique mixture enrichment device for KE-Jetronic systems. With inputs from the coolant-temperature sensor, the ignition coil, and intake-manifold pressure, this "black box" manipulates the signals to the fuel distributor's pressure actuator to modify fuel flow at high load and high rpm. They claim a 5 to 7 horsepower gain at peak rpm on stock or mildly modified engines.

The unit is not adjustable. The fuel enrichment characteristics have to be calibrated prior to shipment for the specified application. Several "standard" variations are available to match different engine configurations.

Load-Sensitive Enrichment

Bosch makes several different types of control pressure regulators. Some feature a hose connection to the intake manifold and provide mixture enrichment (lower control pressure) as a function of increased manifold pressure. I've described them in Chapter 5 and Chapter 6. They are widely installed as original equipment on more expensive cars with continuous injection systems. If your car lacks this refinement, you may look into replacing your control pressure regulator. You'll need the specifications for control pressure vs. manifold pressure for the unit you're installing, and you'll probably need to experiment.

One interesting variation on the idea of boosting system fuel pressure is the rising-rate fuel pressure regulator created by Micro Dynamics. Instead of a fixed increase in system fuel pressure which will yield enrichment under all operating conditions — probably causing the system to be overrich at part-throttle — this device boosts system fuel pressure only under load, in response to changes in intake-manifold pressure.

Instead of using intake-manifold pressure to alter control pressure — as in a few K-Jetronic applications — this regulator replaces the standard system pressure regulating valve in the fuel distributor. The effect — load-sensitive mixture enrichment — is the same, but with the Micro Dynamics regulator the "base" system pressure is externally adjustable. Under load, increased manifold pressure acts on a diaphragm assembly in the regulator to yield a proportional increase in system fuel pressure. Since it works on Manifold Absolute Pressure (MAP), the system also works under boost from forced induction systems. In addition to providing tuneable load-sensitive enrichment, the rising-rate regulator reportedly also improves throttle response by reacting faster to the changes in load and the demand for more fuel.

The smallest rising-rate fuel pressure regulator increases fuel pressure ½ psi for every 1 psi increase in manifold pressure. This ratio is not adjustable, but the unit is "dialed in" by adjusting the base system pressure. The smallest size regulator is probably sufficient for most street applications, and incrementally larger sizes are available for competition and other more radical applications.

Fig. 5-3. The Micro Dymanics rising-rate fuel pressure regulator replaces the system pressure regulator on continuous injection systems to provide adjustable load-sensitive enrichment and quicker throttle response.

Tim Smith of Veloce Distributing, Micro Dynamics' U.S. distributor, says that the idea is to first select the correct size regulator for the application (with Micro Dynamics' help), then fine-tune by adjusting the regulator pressure to deliver the best mixture in full load, high rpm situations. By selecting the correct regulator and setting it up properly, the full-throttle, full-load fuel mixture should benefit, and system pressure (and fuel mixture) should be at near-stock levels the rest of the time.

On K-lambda systems, of course, the lambda sensor system will compensate for any small mixture problems which might be present in normal part-throttle driving.

Higher system fuel pressure may also be a safety concern. With fuel lines and connections subjected to higher pressure, there naturally is an increased risk of leaks or outright failure. To ensure reliability, the standard Bosch parts are rated for pressures well above the normal operating range but, in theory at least, significantly higher fuel pressures may be a source of problems. Higher fuel pressure also places an increased demand on the fuel pump. This may be a problem if the pump is expected to operate above its designed delivery pressure. Tim suggests that some OEM fuel pumps may not be quite up to the task of supporting higher fuel pressures, but that they have had success using the pumps originally supplied on turbocharged cars such as the Saab Turbos.

7

Another interesting note: Tim also points out that, at least as of this book's publication date, the rising-rate regulator is a legal modification in the Sports Car Club of America's (SCCA's) Improved Touring road racing classes for slightly modified production cars.

5.3 Pulsed (EFI) Injection Systems

In pulsed injection systems, increased demand for air and fuel in modified engines is complicated by the fact that there is no direct mechanical relationship between air flow and fuel metering. There is only the indirect, electronic relationship. Any modification to such a system then, must either (1) alter the inputs to the electronic control unit (ECU), (2) reprogram the ECU to alter its outputs, or (3) make mechanical changes to get different results from the ECU's output.

D-Jetronic

The D-Jetronic system is unique among Bosch systems in that its ECU controls fuel metering based primarily on intake manifold pressure (as well as throttle position and engine rpm). This is an early system, a real improvement over carburetors in its day, but not nearly as precise and sophisticated as the newer systems. Thus, few tuners have been willing to commit time and effort to unraveling the system's performance secrets, if any exist. Fewer still have had any real success. Most just give up and bolt on carburetors for performance applications.

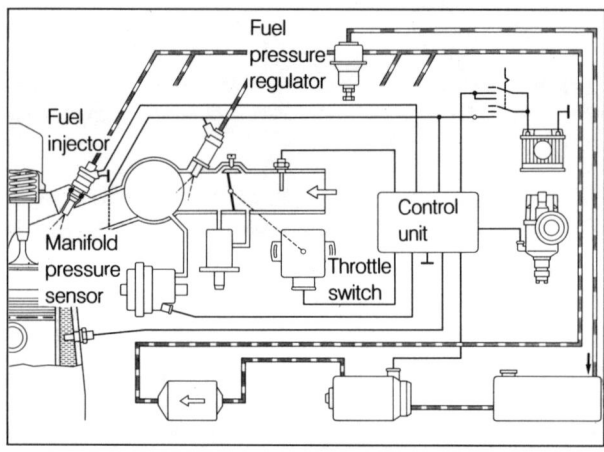

Fig. 5-4. Bosch D-Jetronic fuel-injection system.

Aside from the questionable practices of boosting fuel pressure or simply adjusting for a richer mixture—both of which will have a mainly linear effect over the entire rpm range—the only "opportunity" for tuning D-Jetronic to a modified engine lies in reprogramming the ECU to alter the way it meters fuel for a given set of inputs. This is a complicated undertaking and, once again, in this application very few tuners have ever bothered.

L-Jetronic

The L-Jetronic fuel-injection system is more sophisticated than the older D-Jetronic system. It offers fuel economy, emission control, and even performance advantages that make carburetors a questionable choice.

On the other hand, the stock L-Jetronic system is not particularly adaptable to engine modifications. Engine modifications which improve air flow, such as a ported cylinder head, a hotter cam, or more open exhaust, may simply cause the stock L-Jetronic system to run lean. In fact, the earliest L-Jetronic cars, like most of their mid-1970s contemporaries, were already on the edge of being programmed to run too lean as they came from the factory (this changed with the later addition of lambda control to L-Jetronic systems).

At first glance, it would seem that L-Jetronic—a system which actually measures air flow and proportionally meters fuel—would be able to adapt. On a modified engine, it should be able to measure the increased air flow and meter proportionally greater amounts of fuel. Unfortunately, it's just not that simple.

On stock engines equipped with L-Jetronic as original equipment, the air-flow sensor is most effective at low to moderate engine speeds. Above 3500 to 4000 rpm, the sensor flap has reaches the end of its travel. It is fully open at about the point where maximum volumetric efficiency is reached. Above this rpm point, air flow into the engine per-stroke actually goes down, so you don't want any more fuel for the mixture. In this higher rpm range, fuel delivery becomes a function of engine rpm rather than air flow.

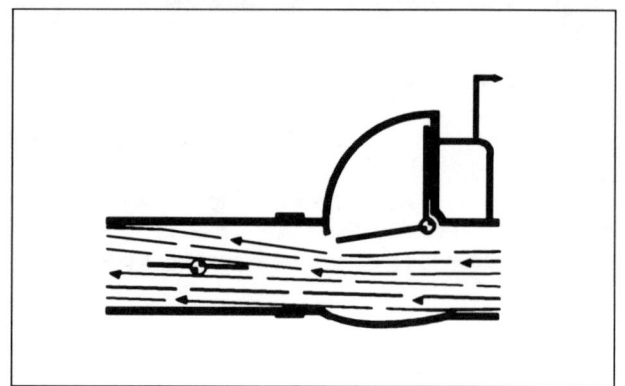

Fig. 5-5. Even on stock systems, L-Jetronic air-flow sensor flap is fully open at 3500 to 4000 rpm. At higher rpm, it is no longer signalling increased air flow to the ECU.

On a modified engine flowing more air through increased volumetric efficiency, the system should be able to measure more air and deliver more fuel by lengthening pulse time accordingly—up to its 3500 to 4000 rpm limit. But, above that point, fuel delivery is controlled by the ECU's original programming, and based primarily on engine rpm. The modified engine may be flowing significantly more air, but the ECU has no way of knowing. The result is leaner mixtures in the higher rpm range.

Even if you could get the ECU to increase pulse time to deliver more fuel to your modified engine, time is another factor which will limit high-rpm fuel delivery. In Chapter 3, I showed you how pulse period—the amount of time available in which to open and close the injector with each pulse—gets smaller as rpm increases. Pulse time becomes a larger and larger proportion of the smaller and smaller pulse period until, at peak rpm, the injectors are open nearly all the time. Even if the ECU could be reprogrammed for increased pulse time and more fuel delivery, there may simply be no more time available.

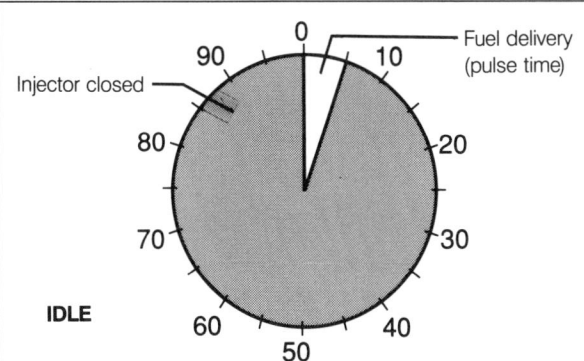

IDLE

At idle, the crankshaft rotates once in 1/10th of a second, or 100 ms. That's the pulse period. The injectors open once per revolution, for a pulse time of 5 ms.

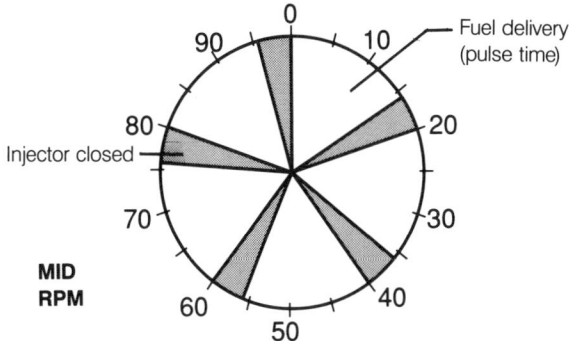

MID RPM

At mid-rpm, say 3000, the crankshaft rotates in a 20 ms pulse period. Full-load delivery for maximum torque at wide-open throttle requires longer injection times than at maximum rpm since at mid-rpm the engine takes in more air per stroke. But a pulse time of 15 ms easily fits into the 20 ms pulse period.

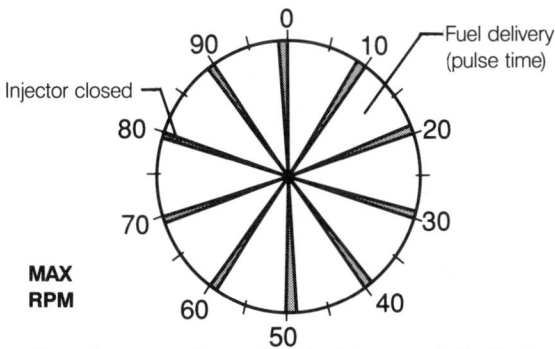

MAX RPM

At maximum rpm, the crankshaft rotates once in 1/100th of a second. Pulse period is now 10 ms; in 100 ms the crankshaft rotates 10 times, and the injectors open ten times. If the pulse time is 8 ms, the injectors barely have time to close before they open again.

Fig. 5-6. Relation between pulse period and pulse time. Each circle represents 100 milliseconds (1/10 sec.).

Increasing Fuel Delivery

For fuel delivery increases at mid-range to high rpm, there remains one kind of solution: make mechanical changes to get different results (more fuel) from the ECU's output. The most popular methods are to increase system fuel pressure and to install larger injectors.

One sure way to increase fuel delivery is to use higher system pressure to pump more fuel through the injectors while they are open. This is normally done by replacing the stock fuel pressure regulator with an adjustable unit. A positive side-effect of increased fuel pressure, almost a "free" benefit, is that forcing the fuel through the same injector at higher pressure also tends to improve fuel atomization. This will tend to improve fuel distribution and combustion efficiency, and may contribute to improved fuel economy.

The benefits of higher pressure are accompanied by some additional concerns, the main one being safety. With fuel lines and connections subjected to higher pressure, there naturally is an increased risk of leaks or outright failure. To ensure reliability, the standard Bosch parts are rated for pressures well above the normal operating range but, in theory at least, significantly higher fuel pressures may be a source of problems. Higher fuel pressure also places an increased demand on the fuel pump. This may be a problem if the pump is expected to operate above its designed delivery pressure. In some instances, fuel pump failures have occurred following increases in fuel pressure on L-Jetronic systems to more than about 4 bar (60 psi).

To achieve significant fuel delivery increases without extremely high fuel pressures, you may also want to install larger injectors. At near-stock fuel pressures, poorer fuel atomization from larger injectors is a drawback, but a combination of larger injectors and boosted fuel pressure can produce big fuel delivery gains with less of the potential for problems associated with excessive fuel pressure.

Once more, however, this is no simple fix. Increasing fuel pressure or installing larger injectors will increase fuel delivery throughout the engine's entire operating range, even though it is really only needed at high rpm. So, with modifications of this type, a corresponding change must be made to maintain correct air-fuel mixtures at the low and middle rpm ranges, either by (1) altering the inputs to the ECU—recalibrating the air-flow sensor—or, (2) reprogramming the ECU to alter its outputs (pulse time).

Recalibrating the Air-Flow sensor

The air-flow sensor translates the mechanical movement of the sensor flap into a proportional electrical signal to the ECU. The ECU uses this signal input as an indication of air flow to help calculate the correct fuel-injection pulse time.

7

From the factory, the relationship between the air-flow sensor mechanism and the output signal (ECU input) is precisely calibrated. By recalibrating the mechanism—tightening or loosening the return-spring tension on the air-flow sensor flap—this relationship can be altered. Adjusting the spring tension will result in a small adjustment of sensor flap position for any given intake air flow, thus producing a greater or lesser signal to the ECU and, therefore, a leaner or richer mixture—in the range up to about 4000 rpm where the air-flow sensor is fully open.

Some very knowledgable independent tuners have had success with this method, but remember that the air-flow sensor is a sensitive, delicate device. Without knowing what you are doing, it is easy to do more harm than good. You must check the fine adjustments and their effects on mixture with an engine exhaust analyzer. It is worth noting that Bosch does not recommend or endorse any kind of adjustment to the internal workings of the air-flow sensor.

Reprogramming the Electronics

As an alternative to recalibrating the air-flow sensor, many tuners choose to customize the ECU—alter the electronics so that the same inputs produce different output and different injector pulse times. As you might guess, this kind of modification requires a great deal of expertise and experience, and probably long hours on the dynamometer as well.

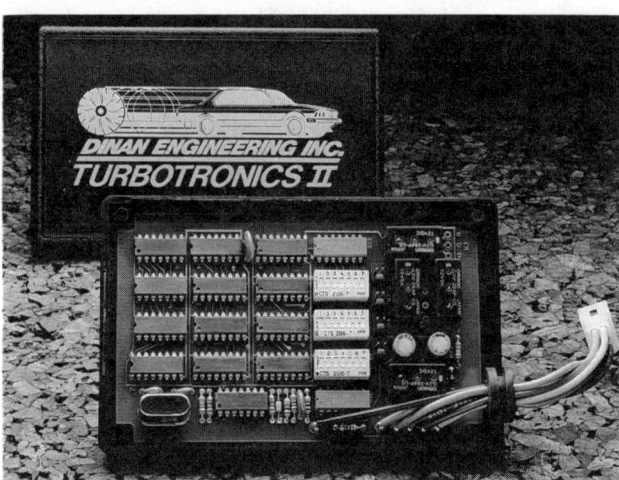

Fig. 5-7. Dinan Engineering offers this aftermarket ECU to manage fuel delivery on highly modified, turbocharged BMWs using larger fuel injectors and increased fuel pressure.

For most of us, a discussion of electronics is primarily limited to commercially available, aftermarket parts—a replacement PROM (Programmable Read-Only Memory) computer chip, or perhaps an entire replacement ECU. The hardest part is knowing what you're getting.

Unfortunately, it is very difficult to hold a computer chip or "black box" in your hand and be able to tell what it can and cannot do, or what the trade-offs or undesirable effects might be. There is plenty of opportunity for a small gain in one area—full-throttle acceleration, for example—at the expense of driveability, fuel economy, and exhaust emissions. Be skeptical. Try to find out as much as possible about the product. Claims may be nothing more than claims. It's real results that count.

Load-Sensitive Enrichment

One interesting answer to the problem of a linear increase in fuel pressure may be the Micro Dynamics rising-rate fuel pressure regulator. Instead of raising the fuel pressure and increasing injector flow regardless of operating conditions, this device increases fuel pressure only when the engine is under load. The regulator responds to changes in intake manifold pressure.

The unit's "base" fuel pressure—the pressure which is maintained under normal light-load or no-load conditions—is externally adjustable. Under load, increased manifold pressure acts on a diaphragm assembly in the regulator to yield a proportional increase in fuel pressure. Since it responds to manifold absolute pressure (MAP), this regulator also works well with forced induction systems. In addition to providing tuneable load-sensitive enrichment, the rising-rate regulator reportedly also improves throttle response by reacting faster to the changes in load and the demand for more fuel.

Fig. 5-8. Micro Dynamics rising-rate fuel pressure regulator installed on a stock pulsed-injection system.

The smallest rising-rate fuel pressure regulator increases fuel pressure ½ psi for every 1 psi increase in manifold pressure. Although this is similar to the load-based increase provided by the original regulator, the point at which the pressure increase begins on the Micro Dynamics unit is adjustable. For more radical applications, units with higher ratios of fuel pressure to manifold pressure are also available. Tim Smith of Veloce Distributing, Micro Dynamics' US distributor, says that the idea is to first select the correct size regulator for the application (with Micro Dynamics' help), then fine-tune by adjusting the regulator pressure so that the injection system delivers the best mixture in full-load, high-rpm situations. With the correct size regulator properly set up, full-throttle, full-load fuel mixture should benefit, and the rest of the time fuel pressure should be at near-stock levels.

In most street applications, of course, the lambda sensor system in closed-loop operation should compensate for any small pressure-related fuel mixture problems which may occur during normal part-throttle driving.

LH-Jetronic

As discussed in Chapter 3, the sophisticated LH-Jetronic system measures not simply the volume of the incoming air, but the mass. This means that the system is much more adaptable to changes in altitude and air density, and therefore more precise over a wider range of operating conditions.

Unfortunately, the system's design works against its successful application to modified engines. The air-mass sensor is not adjustable, so the stock system can properly respond only to modest increases in air flow which are within the flow and measurement capacity of the stock system.

Fig. 5-9. The sophisticated LH-Jetronic air mass sensor has no moving parts which can be recalibrated for applications with increased air and fuel flow.

For application to a modified engine, increased fuel delivery is theoretically possible using the methods described above for L-Jetronic—boosting fuel pressure and/or installing larger injectors. With LH-Jetronic, however, there is no mechanical way to compensate with revised control of fuel metering for low and mid-range rpm. Any significant modifications to engine and fuel delivery components would also require changes to the ECU electronics.

Dennis Watkins of Wray Electronics in Seattle uses Bosch LH-Jetronic hot-wire air-mass sensors in his custom-designed, high-performance fuel-injection systems, in spite of the air-mass sensor's air flow limitations.

Instead of using the air-mass sensor to measure all of the intake air, he fabricates new hardware which uses the original air-mass sensor to measure only part of the air. His custom electronic controls are calibrated to make the necessary conversions and meter the correct amount of fuel for the actual amount of air entering the engine.

Fig. 5-10. Wray Electronics custom air-mass sensor (left) compared to standard Bosch sensor. Wray sensor flows more air and measures mass of only a portion of that air.

Bosch has researched the use of a hot-wire air-mass sensor measuring only a sample of the incoming air and concluded that it is quite difficult to ensure the proportionality of the measured sample to the total air mass. They have yet to enter production with such a system. Wray Electronics, however, has had considerable success in competition applications, on engines up to 1200 horsepower. This is a very innovative way to apply the hot-wire air-mass sensing technology of LH-Jetronic, but making it work is probably way beyond the capabilities of the average hot rodder.

Motronic

As described in more detail in Chapter 3, "Motronic" is a term that actually applies to several distinct types of systems. The original Motronic systems were evolved from L-Jetronic fuel injection. You'll also now find Motronic systems based on the LH-Jetronic and KE-Jetronic fuel-injection systems.

7

In general, the Motronic label refers to an engine-management system—a fuel-injection system which also includes electronic control of ignition timing. In terms of modifying fuel injection for increased fuel delivery to a modified engine, the principles which apply to modification of the basic fuel-injection system also generally apply to the Motronic variations.

The major difference is in the electronics and in the control of ignition timing. Many aftermarket suppliers and tuners modify ignition timing and timing advance curves along with the fuel injection. In the case of the basic fuel-injection systems, these are separate functions and modifications have to be made separately. In a Motronic system, fuel delivery and ignition are controlled together. For Motronic applications, most efforts at reprogramming the electronics (substituting PROMs) make changes to fuel delivery and ignition timing together.

6. AIR FLOW IMPROVEMENTS

Generally speaking, there is little to be gained in the way of intake air flow improvements on Bosch systems. But, for highly modified engines demanding significant increases in inlet air flow, there are some possibilities.

6.1 Continuous Injection Systems

The beauty of the continuous injection system is that the fuel metering unit—the combined air-flow sensor and fuel distributor—is a self-contained, hydro-mechanical system. Unless the system has been worn out, contaminated, or tampered with, accurate air flow measurement and fuel metering is pretty well guaranteed.

I have already discussed how continuous systems adapt to increases in air flow. If even more air flow capacity is required, one possible solution is to install the larger capacity OEM metering unit from a larger engine application. You should keep in mind that using a big metering unit on an engine that is smaller than its intended application may cause some low-speed driveability problems, much the same way that installing big carburetors can. At low rpm, the larger available flow path causes air flow velocity to be reduced. In extreme cases, flow velocity at low rpm may be so low as to impair the accuracy of the fuel metering.

For less modified engines whose air flow requirements are within the capabilities of the stock system, the following modifications are popular.

Air-Flow Sensors

There is no real need to go to a larger air-flow sensor to increase flow capacity, but it may be that similar cars of different model years were fitted with different hardware. The classic example is 1980 to 1983 Volkswagen Rabbits with 1.6 and 1.7 liter engines. They were originally equipped with 60 mm air-flow sensors, while both earlier and later models came with 80 mm

parts. In this case, changing to the larger air-flow sensor seems like a sensible, low-risk move, and the decreased restriction may be worth a slight high-rpm power gain.

Fig. 6-1. Different K-Jetronic metering units may feature different size air-flow sensors. 60 mm sensor (left) was standard equipment on 1980–1983 1.6 and 1.7 liter Volkswagen Rabbits, while 80 mm sensor (right) was standard on 1977–1979 Rabbits and 1983 and 1984 Rabbit GTI.

Throttle Valves

There are a number of larger-than-stock throttle valves on the market. Believable dynamometer test results indicate that the increases are less than spectacular. In any case, the engine is only going to take in as much air as it can. The gains can only come from the slight reductions in intake restriction.

Simply bolting a larger valve onto an unmodified intake manifold yields very small gains, perhaps one or two horsepower at peak rpm. A more involved installation which includes machining the manifold to match the shape of the new valve might be worth as much as three horsepower. Cutting out the booster venturi on the primary side of early K-Jetronic throttle valves has proven to be worth less than one horsepower. Remember, Bosch is already pretty good at this stuff.

Replacing a stock, two-stage progressive throttle valve with a large, single-throat design has similar limitations, and also has the biggest drawbacks in terms of reduced flow velocity and potential driveability effects at idle and part-throttle.

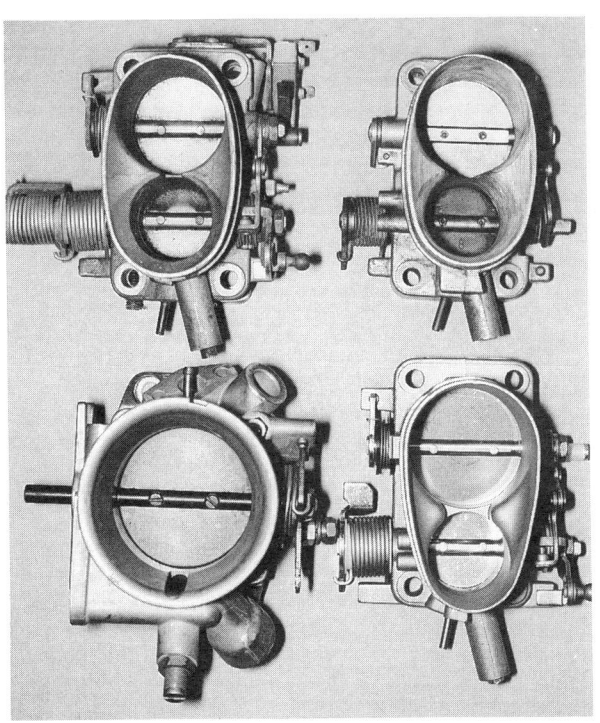

Fig. 6-2. Throttle valves. Rabbit throttle valve modified by Techtonics (top right) shows negligible power gains compared to stock Rabbit throttle valve (top left). Aftermarket throttle valve (lower right) makes claims which are not supported by dynamometer test results. Large single bore throttle valve (lower left) offers some gain at high rpm due to larger area, but will probably cause driveability problems, especially on modified engines.

6.2 L-Jetronic Systems

As I discussed earlier, the L-Jetronic systems at high rpm are limited more by available injection pulse time than by air flow restriction. Only a highly modified engine would stand any chance of benefiting from larger intake air-flow components.

Fitting a larger air-flow sensor which was designed for the greater air flow of a larger engine might increase the system's air flow capacity and marginally reduce intake restriction, but there are other important considerations.

Depending on the application, the throttle valve may actually be the more restrictive component. Compare the cross-sectional area of the fully open air-flow sensor with that of the fully open throttle valve. If the throttle valve has substantially less cross-sectional area, adding a big air-flow sensor alone won't have much effect.

If a bigger air-flow sensor still seems like a good idea, remember the relationship of the air-flow sensor to the ECU. The air-flow sensor influences fuel metering only by its electrical signal to the ECU. A different air-flow sensor will send different signals. Such a swap will probably require recalibration of the sensor mechanism, reprogramming of the ECU, or both. Even then, it may be difficult to restore proper low-and mid-range performance and driveability. This is a big, complicated step.

Fig. 6-3. L-Jetronic air-flow sensors have different sizes, flow capacities, and calibrations for different applications. They are not interchangeable without other system changes.

7. KNOCK SENSOR

Although a knock sensor is not strictly a fuel-injection component, its advantages and benefits bear mention. A knock sensor system uses a simple sensor mounted to the side of the engine block to detect harmful detonation in the combustion chambers. When it does, the system retards ignition timing in small increments until the detonation ceases. It constantly adjusts to allow the engine to run with as much ignition advance as possible without detonation.

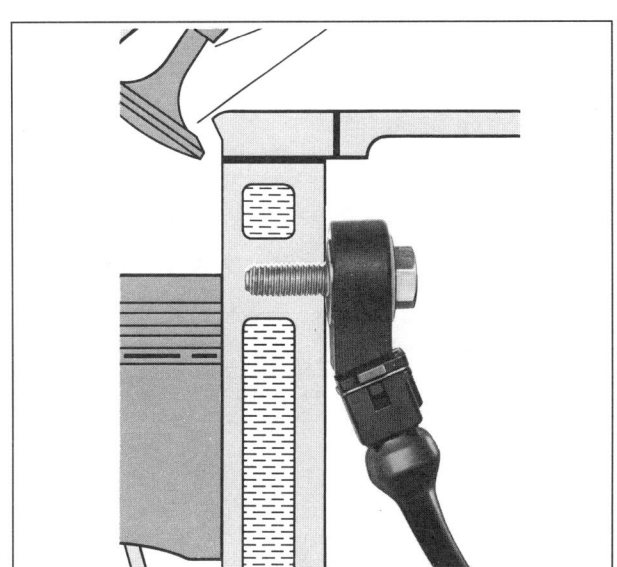

Fig. 7-1. A knock sensor mounted to the engine block detects harmful detonation in the combustion chambers, and signals the knock-sensor control unit to retard ignition timing in small steps until detonation stops.

7

Knock Sensor

Many of the newer cars equipped with Bosch fuel injection also have knock-sensor systems. If not, installation of an after-market system to a car with manually adjustable ignition timing can be a big improvement. Free of the need to guard against detonation, the add-on closed-loop control of spark advance allows you to optimize basic ignition timing for maximum power. The ignition system can be set up so that maximum spark advance is close to the point of light detonation — the most efficient power point — without the need for fudge-factors and retarded timing that may be required by open-loop ignition timing.

With a knock-sensor system, there is actually some advantage to buying higher octane gasoline. The better fuel is more resistant to detonation, and the knock-sensor system will allow more spark advance for more power.

8. TRICKS

So much for the sensible, holistic approach to fuel injection modifications. In this section, I'll discuss some of the tricks which are most often tried in an attempt to get more power from fuel-injected engines.

8.1 Cold-Start Injector

Considering that the cold-start injector always has fuel pressure when the engine (or more correctly, the fuel pump) is running, it is a readily available source of extra fuel. In normal operation, the circuit is powered continously. The cold-start injector admits extra fuel into the intake manifold for starting when the ground side of the circuit is complete to ground. Some manufacturers use a pulse system to inject extra fuel for hot starts as well. One popular idea is to hook up an auxiliary-pulse system to the cold-start injector so that it would supply additional fuel on demand, under full-throttle acceleration for example.

Fig. 8-1. The cold-start injector is similar to an additional (solenoid-type) fuel injector. In theory, you can use it to provide additional fuel; how successful this tactic might be is debatable.

As I mentioned earlier in this chapter, the problem with using the cold-start injector for extra fuel is that it was never designed for this kind of continuous use in a warm engine. It adds fuel from a single location in the intake manifold. Atomization is poor. In some applications, the location of the cold-start injector causes fuel distribution problems. Thus, the result can be widely varying mixtures in different cylinders — causing real problems under load at full throttle.

8.2 Fooling the Computer

In some cases, the fuel-injection system has plenty of extra fuel capacity. The trick is in getting it to deliver a slightly richer mixture. There are a number of "tricks" which, in theory, manipulate inputs to fool the ECU into calling for a richer mixture.

Coolant Temperature Sensor

Some fuel-injection systems use the variable resistance of a coolant-temperature sensor as an input to the ECU. At lower temperatures, the sensor's higher resistance signals the ECU to call for more fuel. The theory is that adding resistance to this circuit will trick the ECU into thinking the engine is cold and that enrichment is needed. The neat way to do this is with a variable, perhaps driver-controlled, resistance.

There is a potential problem. Some ECUs are programmed to ignore sudden changes in the resistance of the coolant temperature sensor circuit. Instead of enriching the mixture, a radical change in resistance might cause the ECU to revert to memory and simply operate based on its last "normal" temperature signal. When using a system with variable resistance, changes to the resistance should always be made with the ignition off.

A variation of coolant-temperature sensor enrichment is offered in kit form by Light Performance Works in Michigan. The Full Throttle Fuel Control (FTFC) kit operates only at full throttle, so that there is no enrichment at idle or part throttle. In theory, this should keep emissions from being affected except at full throttle (remember, it may still be technically considered illegal tampering). The driver-adjustable system gives a claimed two-horsepower gain on an eight-valve (KE-Jetronic) Volkswagen GTI.

Last but not least, the big caution in a set-up like this is to watch for too much enrichment on cars equipped with catalytic converters. Too much fuel will send converter temperatures soaring — in some circumstances high enough to light the car on fire!

Lower-Temperature Thermostat

This is not a terrific idea. Most engines are designed to operate with a coolant temperature of around 200°F (95°C). Unfortunately, the lower temperature is also going to reduce power, increase cylinder wall wear and, in cold climates, affect driveability. The gains are small. It's better to pass on this one.

8.3 More K-Jetronic Tricks

A couple of additional items apply to K-systems in particular.

Eliminating Air Intake Heating

The idea here is to make K-Jetronic systems run better by reducing the temperature of the intake air, thus filling the cylinders with cooler, denser air for more power.

Unfortunately, the benefits to cold weather driveability from leaving the air-intake system intact tend to outweigh any small gain to be had by removing it.

Fuel Shut-Off

Volkswagen engines with KE-Jetronic have a fuel shut-off feature in the ECU which cuts off the fuel supply at 6200 to 6300 rpm. Installing the ECU from a 16-valve model will move the fuel cut-off point to about 7300 rpm.

8.4 Add-On Injectors

One unique approach to providing needed fuel enrichment to modified engines is to forget about modifying the existing fuel system altogether, and add a separate system with the capability to meter additional fuel. This is the approach taken by Micro Dynamics with their Petrol Injection Control (PIC) systems.

The PIC systems employ an additional solenoid-type, pulsed fuel injector, usually mounted in the intake manifold near the throttle valve. In some applications, this extra injector can use the mounting point for the existing cold-start injector.

The basic system, PIC5, meters fuel based on a signal from a pre-set manifold-pressure switch or a full-throttle position switch, and engine rpm. The system is adjustable using a calibration screw on the front panel so that the amount of the extra fuel delivered (the pulse time of the extra injector) can be fine-tuned. The more sophisticated PIC7 system adds a throttle position sensor and manifold absolute pressure (MAP) sensor. The result is a separate fuel delivery system which can be calibrated to deliver the right amount of extra fuel for a modified engine, and provide significantly improved throttle response.

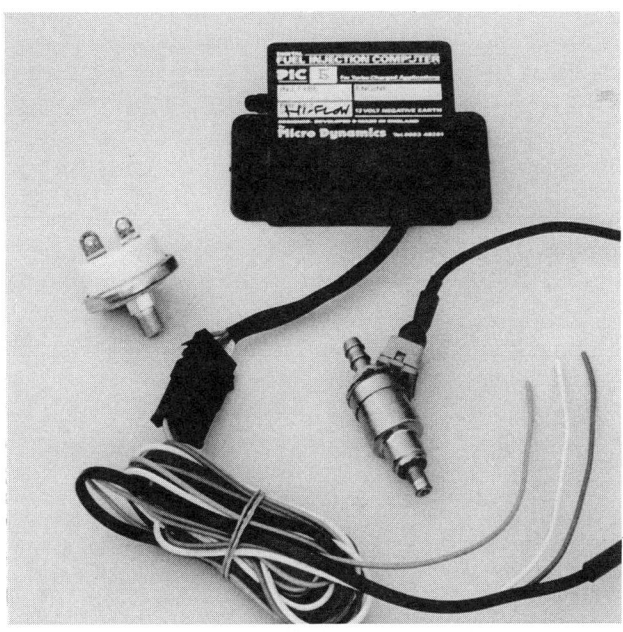

Fig. 8-2. Add-on fuel enrichment system from Micro Dynamics uses an extra injector to provide fuel enrichment based on engine rpm and load.

9. Racing Applications

At almost any major racing event these days, the starting grid is filled with race cars prominently displaying the Bosch name. What fuel systems are they using, and can they be used in other applications? Of course, the really exotic GT prototypes use highly specialized, custom-designed systems that most of us could never afford. But, interestingly, a surprising number of cars using Bosch fuel injection in competition are using the same, mass-produced, street-style systems covered in this book. Here is a brief look at a few racing applications of Bosch fuel injection.

Production-Based Racing Classes

The Sports Car Club of America (SCCA) and the International Motor Sports Association (IMSA) organize a variety of amateur and professional events for what are essentially production cars. The car preparation rules are very restrictive, sometimes limited only to the addition of required safety equipment. Even under the aggressive race conditions that these cars endure, some for as much as twenty-four hours at a time, they are racing and winning with standard production-type Bosch fuel injection.

One example is the Volkswagen Golfs racing in the Pro Stock class of the IMSA International Sedan series, equipped with stock KE-Jetronic fuel injection. Mufflers and catalytic converters are removed, and the teams tune with CO meters — not to monitor emissions per se, but simply as a way to monitor the air-fuel mixture.

They measure the KE-Jetronic pressure actuator current and set the idle mixture at the air-flow sensor, as in the factory procedure, except that they like to change the recommended 10 mA open-loop setting at idle to a more lean 6 mA. The idea seems to be that starting with the lean adjustment causes the lambda sensor system to call for more fuel than if it were starting from 10mA. Of course, if you tried this tactic in a street car the lean operation at the low end of the rpm range would probably result in serious driveability problems.

Fig. 9-1. This Volkswagen Golf competitor in the IMSA International Sedan series uses a carbon monoxide (CO) meter to fine-tune the idle mixture of the stock KE-Jetronic fuel injection system.

Fig. 9-2. Fine-tuning this race car's KE-Jetronic system follows the factory procedure of reading pressure actuator current. The paper-clip connectors (arrow) are not factory-recommended, but they seem to do the job! (Beware of connector and terminal damage).

Formula Cars

A popular movement in many forms of racing in recent years has been the creation of "spec" classes, where all competitors are required to use the same engines, and often the same chassis, as a way of equalizing competition and reducing costs. One such "spec" class car is the open-wheel single-seater used in the Skip Barber Saab Pro Series.

Fig. 9-3. The Barber Saab Pro Series cars are an example of a "spec" class where all competitors use the same production-based powerplant.

These cars all use stock production Saab 16-valve Turbo engines equipped with LH-Jetronic. The only change from the stock set-up is that the systems operate open-loop without Lambda control, and allowable turbocharger boost is increased slightly to about 1 bar (14 psi). The LH-Jetronic fuel injection system is stock.

Fig. 9-4. The engine cover of this exotic looking race car hides a stock, turbocharged Saab engine equipped with Bosch LH-Jetronic fuel injection.

The biggest differences are in the electronics. An ECU might cost $20,000, and it might have five different Erasable Programmable Read Only Memory (EPROM) chips. In addition, most top teams use different EPROMS to specifically tailor the engine's characteristics for unique race conditions or track features—fuel economy for endurance races, top-end power for extra long straights, etc.

Fig. 9-5. The Barber Saabs' few engine modifications include this free-flow air filter. Other than running open-loop—no oxygen sensor—and slightly higher turbocharger boost, the cars use stock production LH-Jetronic fuel-injection systems.

GT Prototypes and Formula 1

For the mega-dollar teams competing with the world's most exotic cars, Bosch makes available sophisticated, race-only fuel injection and engine-management systems. The costs are high—astronomical, in fact—but the fundamental operating principles would still be very familiar to anyone with a basic knowledge of Bosch production-car injection systems.

Fig. 9-7. This Ford Probe GTP car uses Bosch fuel injection. Notice the use of what appear to be two stock injectors per cylinder.

Fig. 9-8. Twin control units in this GTP car operate twin sets of injectors.

Fig. 9-6. Porsche is a prime user of Bosch's most advanced fuel management expertise. Shown here is one example of the very successful model 962 GT Prototype.

Even in the win-at-any-price world of international motorsports, many of the most successful teams turn the engine-management chores over to Bosch. They know that Bosch has the experience and the technical expertise to deliver results.

7

10. RADICAL INJECTION SYSTEMS

As Mercedes race engineers found out in the 1950s with their mechanical direct-injection systems, the precise management of fuel delivery made possible by fuel injection offers the potential for squeezing all the power possible out of a particular engine design. For this reason, new and more creative methods are being developed all the time. I've described a few of the more interesting ones are below.

Dual K-Jetronic

As this is written, Darrell Vittone at Techtonics talks of a turbocharged, 400+ horsepower Volkswagen Scirocco being built in his Riverside, California shop for top-speed runs at Bonneville. The car is designed to use two K-Jetronic systems, working together to satisfy the air flow demands of the big breathing engine. The elegance of such a set-up is that all of the smooth running, driveability, and precise fuel metering of the Bosch-built system are retained, while the air flow and fuel delivery capacity for high-speed, high-boost conditions is approximately doubled.

This idea is a natural for such high-output applications. In fact, some European manufacturers have equipped their high horsepower V-8 and V-12 supercars with dual K-Jetronic systems as an off-the-shelf way to satisfy a big engine's requirements for performance, driveability, and exhaust emission control.

User-Programmable Fuel Injection

In nearly all facets of life, the micro-electronics revolution is firmly established and marching along at a fantastic rate. More and more every day, the vast capabilities of computers and software are becoming less mysterious and more accessible to the end user. Nowhere is this more true than in the field of electronic engine management.

The Haltech system which is available from Cartech, a turbocharging specialist firm in Dallas, Texas, is an ECU that can be completely custom-programed using a personal computer (PC). It includes the programmable ECU, a complete wiring harness, air and coolant temperature sensors, and a throttle-position sensor. It can be used with the stock throttle body and fuel injectors. An air-flow sensor is not required, allowing some reduction of intake-air flow restriction.

The system, which allows precise tailoring of the fuel delivery "map" as a function of engine rpm and manifold pressure, is almost infinitely tuneable. Thanks to easily understood, user-friendly software, programming the system is relatively simple.

Corky Bell of Cartech describes running the car on a chassis dynamometer to create the necessary rpm and load conditions, and then programming the system for each set of conditions using a PC connected to the ECU in the shop. Reports are that a very good, workable set-up can be programmed this way in about an hour. Corky even suggests the possibility of hooking up to a lap-top PC and fine-tuning the system in the car!

Of course, with a vast number of programming choices, there are also a nearly infinite number of opportunities for mistakes. This is a simple system to use, but some tuning expertise is necessary in order to know what the engine's requirements are and to use the system well. Although it is perhaps a little out of reach of the average backyard do-it-yourselfer, this system would seem to offer almost unlimited potential for a knowledgable and talented tuner to program a fuel injection system to suit any application.

Real-Time Digital Fuel Injection

Micro Dynamics' Extended Petrol Injection Control (EPIC) is a digital fuel-injection controller that, rather than metering fuel according to a pre-programmed map, uses sensor inputs and adjustable calibration factors to continuously calculate the amount of fuel needed. Inputs include manifold absolute pressure (MAP), engine rpm, and throttle position. The EPIC system recognizes not only the input signal values, but also the rate of change of those input signals.

The system depends on manifold pressure for its load indication, rather than an air flow sensor. The theory is that by analyzing the input signals' rate of change, and operating without an air flow sensor, the system can respond with fuel delivery adjustments more quickly. But remember that the Bosch ECU responds very quickly too, plenty quickly for any street application.

Fig. 10-1. Micro Dynamics' Extended Petrol Injection Control (EPIC) is a fully adjustable digital fuel-injection ECU that calculates fuel metering requirements instead of referring to a programmed "map" of fuel delivery parameters.

Adjusting potentiometers in the ECU allows the system to be calibrated for the unique requirements of any application or operating condition. The system has been used with great success in Europe on, among others, the wild Group B MG Metro 6R4 rally car, the Lancia Delta Group A factory rally cars, the Rothman's Ford Sierra Cosworths in both rallying and the World Touring Car Championship, and a number of Formula 3 cars.

11. IN CONCLUSION

This chapter has presented a broad spectrum of ideas on the high-performance application of Bosch fuel-injection systems and on modification of those systems. The subjects have ranged from inexpensive modifications to street cars to extremely expensive custom electronic controls made for racing.

The fundamental messages of the chapter are these:

1. In the quest for more power, increased efficiency, or better fuel economy, the stock fuel-injection system is not necessarily the weak link. Sometimes the stock system is the best system, considering the limitations of the engine and the needs of the driver.

2. There are no "demon tweaks" which can miraculously unlock vast amounts of horsepower. The need for fuel-injection system modifications depends directly on the needs of the engine. It all comes down to whether the needs of the engine exceed the capabilities of the injection system.

3. There is no substitute for knowledge — knowledge of the engine's requirements, knowledge of the fuel-injection system's limitations, and an understanding of the real effects of modifications. I've devoted the first six chapters to describing the detailed functions of each Bosch fuel-injection system. Thorough understanding of these functions will be the first and most important step in figuring out what adjustments and modifications will pay off for your car and your individual needs.

12. SUPPLIERS

I provide the following list of suppliers to give you a starting point for investigation of possible modifications to your system. Again, I list these without making any particular judgements about their merits.

Autotech (Hor Technologie)
1800 N. Glassell Street
Orange, CA 92665
(714) 974-4600
(800) 553-1055 (orders only)
Load-and rpm-sensitive enrichment module for KE-Jetronic.

AutoThority, Inc.
3763 Pickett Road
Fairfax, VA 22031
(703) 323-7830
EPROMs for Motronic systems (Porsche)

Cartech
11212 Goodnight Ln.
#200
Dallas, TX 75229
(214) 620-0389
Haltech user-programmable fuel-injection computer

Dinan Engineering, Inc.
81 Pioneer Way
Mountain View, CA 94041
(415) 962-9417
L-Jetronic based Turbotronics II fuel injection

Hypertech, Inc.
2104 Hillshire Cr.
Memphis, TN 38133
(901) 382-8888
Replacement PROMs for Bosch Motronic

Light Performance Works
3905 North Jefferson Road
Midland, MI 48640
(517) 835-2389
Full-Throttle Fuel Control for KE-Jetronic

Motec Systems USA
5692 Buckingham Drive
Huntington Beach CA 92647
(714) 897-5821
(714) 897-8782 FAX
User-programmable engine management systems fuel system components and intake manifolds

Techtonics Tuning
1253 West La Cadena Drive
Riverside, CA 92501
(714) 788-4116

Veloce Distributing (Micro Dynamics)
5003 South Genesee Street
Seattle, WA 98118
(206) 725-6561
Rising-rate fuel-pressure regulators Petrol Injection Control (PIC) add-on injector systems Extended Petrol Injection Control (EPIC) digital controller for fuel injection systems

Wray Electronics
Seattle, WA
(206) 781-1823

7

Table 1. SAE terms for fuel injection and other systems, and Bosch equivalents (where applicable)

SAE abbreviation	Component or system	Bosch term
ACS	Air Conditioning Sensor	
AFS	Air Flow Sensor	
AIR	Secondary Air Injection	
BARO	Barometric Absolute Pressure Sensor	
CAC	Charge Air Cooler	Intercooler
CFC	Coolant Fan Control	
CIS	Continuous Injection System	K-Jetronic, KE-Jetronic
CL	Closed Loop	
CPS	Crankshaft Position Sensor	TDC (reference mark) Sensor
DLC	Data Link Connector	
DLI	DistributorLess Ignition	
DTC	Diagnostic Trouble Codes	
ECTS	Engine Coolant Temperature Sensor	
EVAP	Fuel Evaporative System	
FPR	Fuel Pump Relay	
HO2S	Heated Oxygen Sensor (bypass)	Lambda Sensor (heated)
IACV	Intake Air Control Valve	Idle-Speed Stabilizer
IATS	Intake Air Temperature Sensor	Air Temp Sensor, Temp I
KS	Knock Sensor	
MAP	Manifold Absolute Pressure Sensor	Pressure Sensor
MPI	MultiPoint Electronic Fuel Injection	
MVS	Manifold Vacuum Sensor	
OBD	On-Board Diagnostics	
OL	Open Loop	
O2S	Oxygen Sensor	Lambda Sensor
PCME	Powertrain Control Module	ECU
PNS	Park Neutral Switch	
SMPI	Sequential MultiPoint EFI	
TCC	Transaxle Converter Clutch Switch	
TGSS	Transaxle Gear Selection Switch	
TPS	Throttle Position Sensor	
TWC	Three-Way Catalyst	
VSS	Vehicle Speed Sensor	
WOTS	Wide-Open Throttle Switch	

8

2 APPENDIX

Table 2. Pressure Conversion Table

Bar	kPa	psi	in.Hg.
6.0	600	87.0	
5.9	590	85.5	
5.8	580	84.0	
5.7	570	82.5	
5.6	560	81.0	
5.5	550	79.0	
5.4	540	78.5	
5.3	530	77.0	
5.2	520	75.5	
5.1	510	73.5	
5.0	500	72.5	
4.9	490	71.0	
4.8	480	69.5	
4.7	470	68.0	
4.6	460	66.5	
4.5	450	65.5	
4.4	440	64.0	
4.3	430	62.5	
4.2	420	61.0	
4.1	410	59.5	
4.0	400	58.0	
3.9	390	56.5	
3.8	380	55.0	
3.7	370	53.5	
3.6	360	52.0	
3.5	350	51.0	
3.4	340	49.5	
3.3	330	48.0	
3.2	320	46.5	
3.1	310	45.0	
3.0	300	43.5	
2.9	290	42.0	
2.8	280	40.5	
2.7	270	39.0	
2.6	260	37.5	
2.5	250	36.5	
2.4	240	35.0	
2.3	230	33.5	
2.2	220	32.0	
2.1	210	30.5	
2.0	200	29.0	
1.9	190	27.5	
1.8	180	26.0	
1.7	170	24.5	
1.6	160	23.0	
1.5	150	22.0	
1.4	140	20.5	
1.3	130	19.0	
1.2	120	17.5	35.90
1.1	110	16.0	32.91
1.0[1]	100	14.5	29.92
0.9	90	13.0	26.93
0.8	80	11.5	23.94
0.7	70	10.0	20.94
0.6	60	9.0	17.95
0.5	50	7.5	14.96
0.4	40	6.0	11.97
0.3	30	4.5	8.98
0.2	20	3.0	5.98
0.1	10	1.5	2.99
0.0	0	0.0	0.0

[1] Atmospheric pressure at sea level. Atmospheric pressure at Denver, CO, altitude 5000 feet: 0.83 bar.

Glossary

Absolute Pressure. Pressure measured from the point of total vacuum. For example, absolute atmospheric pressure at sea level is 14.5 psi (1 bar).

Actuator. (See Pressure Actuator)

Actuator Fuel. Fuel flow through the lower chambers of the KE-Jetronic fuel distributor, regulated by the Pressure Actuator.

Adaptive Control. The ability of the control unit to adapt closed-loop control to changing operating conditions such as engine wear, fuel quality, or altitude to improve control of the air-fuel ratio, ignition timing, or idle rpm. Sometimes referred to as self-learning.

Advance. (See Ignition Advance/Retard)

AFC. Air-Flow Controlled, an early term for Bosch pulsed injection systems, particularly L-Jetronic to distinguish it from D-Jetronic.

Air-Flow Meter. In pulsed injection, a device that measures the amount of air the engine is using, to indicate load. The L-Jetronic air-flow sensor and the LH-Jetronic air-mass sensor are examples of air-flow meters.

Air-Flow Sensor. Device used to measure the volume of air drawn into the engine. Volume measurements must be corrected for density (temperature and altitude are factors) to determine the necessary fuel quantity to be injected. See also Air-Mass Sensor.

Air-Fuel Ratio. The amount of air compared to the amount of fuel in the air-fuel mixture, almost always expressed in terms of mass. See also Stoichiometric Ratio.

Air-Mass Sensor. An air-flow sensor that uses changing resistance of a heated wire in the intake air stream to measure the mass of the air drawn into the engine. Also called a Hot-Wire Sensor. See also Air-Flow Sensor.

Air Vane. The internal part of an L-Jetronic or Motronic air-flow sensor that moves in proportion to the amount of air flowing in to the engine.

Ambient Temperature. The temperature of the surrounding air.

Ampere (Amps). A measure of current flow. See also Milliampere (mA).

Atmospheric Pressure. Normal pressure in the surrounding atmosphere, generated by the weight of the air above us pressing down. At sea level, in average weather conditions, atmospheric pressure is approximately 1 bar (about 14.5 psi) above vacuum or zero absolute pressure. See also Barometric Pressure.

Auxiliary Air Valve. A device to provide air to bypass a closed throttle during start and warm-up.

Bar. Metric unit of pressure measurement, used in the measurement of both air and fuel. One bar is approximately 14.5 psi, or 100 kPa.

Barometric Pressure. Another term for atmospheric pressure. Expressed in inches of Mercury (in.Hg.): how high atmospheric pressure (relative to zero absolute pressure) forces Mercury up a glass tube. 14.5 psi = 29.92 in.Hg. See also Atmospheric Pressure.

Basic Fuel Quantity. In pulsed injection systems, the fuel quantity injected based only on engine rpm and load. This basic quantity is then adjusted to compensate for various operating conditions.

Basic Metering. In continuous systems, the amount of fuel delivered to the injectors by the lift of the sensor plate and control plunger by the incoming air flow. This basic fuel quantity is then adjusted by the control system to compensate for various operating conditions.

Bimetal. Usually a spring or strip made of two metals with different heat expansion rates. Heat causes the bimetal to bend or twist.

Blow-by. Unburned fuel and air, combustion by-products, and combustion gases that get past the piston rings into the crankcase.

Boost. Condition of over-pressure (above atmospheric) in the intake manifold; caused by intake air being forced in by a turbocharger or supercharger.

Bypass. A channel that permits passage (usually of air) around a closed valve such as the throttle valve.

Catalyst. Material that starts or speeds up a chemical reaction without being consumed itself. The metal coatings inside a catalytic converter.

Catalytic Converter. Device mounted in the exhaust system that converts harmful exhaust emissions into harmless gases. Works by catalytic action which promotes additional chemical reaction after combustion.

CIS. Continuous Injection System

CIS-E. Continuous Injection System-Electronic

Closed-Loop Control. A feedback system that maintains a prescribed limit in another system by monitoring the output of that system. For example, the lambda control system controls the air-fuel ratio by monitoring the oxygen content of the exhaust gas, which is the product of the air-fuel ratio.

CO. Carbon Monoxide. One of the harmful gases produced by combustion. CO in the exhaust is measured during a tune-up as an indication of combustion efficiency.

Cold-Start. Starting the engine when it is cold; when the engine has not run for several hours.

Cold-Start Injector. Solenoid-operated injector used to inject additional fuel during cold-engine starting. Mounted in the intake manifold. Also called the Cold-Start Valve.

Cold-Start Valve. (See Cold-Start Injector)

Combustion. Controlled, rapid burning of the air-fuel mixture in the engine's cylinders.

Combustion Chamber. Space left between the cylinder head and the top of the piston at TDC; where combustion of the air-fuel mixture takes place.

Combustion Knock. (See Knock)

Compression Ratio. The ratio of maximum engine cylinder volume (when the piston is at the bottom of its stroke) to minimum engine cylinder volume (with the piston at TDC). Thus, the theoretical amount that the air-fuel mixture is compressed in the cylinder.

Continuous Injection. Type of injection in which fuel flows from the injectors all the time while the engine is running. Fuel is metered at the mixture-control unit.

Continuous Injection Systems. (K-Jetronic, K-Jetronic with Lambda Control, KE-Jetronic, KE-Motronic)

Continuity. Little or no resistance in an electrical circuit to the flow of current. A solid electrical connection between two points in a circuit. The opposite of an open circuit.

Control Plunger. In continuous systems, the control plunger in the fuel distributor is lifted by the air-flow sensor plate. Plunger lift meters fuel to the injectors.

Control Pressure. Counterforce applied to top of control plunger in K-basic and K-lambda. Changes air-fuel ratio through the action of the Control-Pressure Regulator.

Control-Pressure Regulator. Device mounted on the engine that controls fuel pressure acting on the top of the control plunger. Influenced by heat. Also called the Warm-Up Regulator.

Control Unit. A transistorized device that processes electrical inputs and produces output signals to control various engine functions.

Counterforce. Fuel-pressure force applied down on the top of the control plunger to balance the force of air flow acting on the air-flow sensor plate. See also Control Pressure.

Current. Amount or intensity of flow of electricity. Measured in Amperes.

Density. The ratio of the mass of something to the volume it occupies. Air has less density when it is warm, and less density at higher altitude.

Detonation. (See Knock)

Differential Pressure. In KE-Jetronic systems, the pressure difference between system pressure and the lower-chamber pressure in the fuel distributor. See also Pressure Drop.

Differential-Pressure Regulator. (See Pressure Actuator)

Differential-Pressure Valves. Valves in the fuel distributor of continuous systems that maintain a constant pressure drop at each of the control-plunger slits, regardless of changes in the quantity of fuel flow.

Direct Injection. Injection of fuel directly into the combustion chamber, in contrast to Port Injection.

Distributor pipe. (See Fuel Rail)

Downdraft. In continuous injection systems, a configuration of the air-flow sensor in which air flows down past the sensor plate. Widely used by Mercedes. See also Updraft.

Driveability. Condition describing a car in which it starts easily and idles, accelerates, and shifts smoothly and with adequate power for varying temperatures.

Duty Cycle. In components which cycle on and off, measurement of the amount of time a component is on. The measurement is expressed in percent, with 100% the maximum.

Dwell. The amount of time that primary voltage is applied to the ignition coil to energize it. Also, a measurement of the duration of time a component is on relative to the time it is off. Dwell measurements are expressed in degrees, for example the degrees of crankshaft rotation. See also Duty Cycle.

ECU. (Electronic Control Unit — See Control Unit)

EFI. (Electronic Fuel Injection — See Pulsed Injection)

EGR. Exhaust Gas Recirculation. The process of feeding a small amount of exhaust gas back into the intake manifold to reduce combustion temperatures as a method of controlling emissions.

Electro-Hydraulic Pressure Actuator. (See Pressure Actuator)

Emissions. Unburned parts of the air-fuel mixture released in the exhaust. Refers mostly to carbon monoxide (CO), hydrocarbons (HC), and nitrous oxides (NO_x).

Engine Management System. (See Motronic)

Engine Power. Measure of the ability of the engine to move the car. See also Horsepower.

Excess-Air Factor. (See Lambda)

False Air. Air that leaks into the intake system without being measured by the fuel injection system.

Flooding. An excess of fuel in the cylinder, from an over-rich mixture, that prevents combustion.

Frequency Valve. (See Lambda Control Valve)

Fuel Distributor. Component in continuous injection systems that houses the control plunger and the differential-pressure valves. All fuel metering takes place in the fuel distributor.

Fuel Injection. Fuel delivery system that generally uses an air-flow sensing device as an input signal for precise metering of the fuel for a given air flow, injecting that fuel into the air stream at the intake ports of the engine. Replaces a carburetor or carburetors.

Fuel Metering. Control of the amount of fuel that is mixed with engine intake air to form a combustible mixture.

Fuel Rail. Pipe on pulsed injection systems delivering fuel at system pressure to the injectors. Storage volume of the fuel rail influences stability of fuel pressure in the system.

Full-Load. Load condition of engine when throttle is wide open. Can occur at any engine rpm.

Full-Load Enrichment. Additional fuel injected during acceleration to enrich the mixture as long as the throttle is wide open. Usually open-loop.

Ground. The return path for current in a circuit. Because the negative terminal of the battery is connected to the car chassis, the metal parts of the car usually serve as this path.

Hertz. Measure of frequency: cycles per second. Abbreviated as Hz.

Horsepower. The rate of doing work. A common measure of engine output also expressed in metric kilowatts (Kw).

Hot Start. Starting the engine when it is at or near normal operating temperature.

Hot-Wire Sensor. (See Air-Mass Sensor)

Ideal Air-Fuel Ratio. (See Stoichiometric Ratio)

Idle-Speed Stabilizer. Electronically-controlled air bypass around the throttle. Also called the Idle Speed Actuator or the Constant Idle System.

Ignitable Mixture. An air-fuel mixture that can be ignited by a spark. An ignitable mixture is generally one in which the excess air factor (lambda) is in the range between $\lambda = 0.7$ and $\lambda = 1.2$ (an air-fuel ratio between 10:1 and 17:1).

Ignition. The point at which the spark causes combustion to begin.

Ignition Advance/Retard. Changing the moment of combustion in relation to the point of piston travel. Ignition advance begins combustion earlier; ignition retard begins combustion later.

Ideal Air-Fuel Ratio. (See Stoichiometric Ratio)

In.Hg. (See Barometric Pressure)

Injection Valve. (See Injector)

Injector. Opens to deliver fuel to cylinder intake port. Pulsed injectors are opened by electric solenoid and closed by a spring; also called injection valves. Continuous injectors are opened by fuel pressure and closed by a spring.

Injector Fuel. Fuel flow to the injectors, through the control-plunger metering slits and the upper chambers of KE differential-pressure valves.

Intermittent Fuel Injection. (See Pulsed Injection)

K-basic. Author's term for the earliest and most basic of the Bosch continuous injection systems, K-Jetronic.

K-lambda. Author's term for Bosch K-Jetronic with Lambda Control. See also Lambda Control and Lambda Sensor.

Kilohertz (kHz). 1000 Hertz (Hz), the unit of frequency. (See Hertz)

Kilopascals (kPa). 1,000 pascals, a unit of pressure. 100 kPa = 1 bar = Atmospheric Pressure at sea level.

Knock. Sudden increase in cylinder pressure caused by pre-ignition of some of the air-fuel mixture as the flame front moves from the spark-plug ignition point. Pressure waves in the combustion chamber crash into the piston or cylinder walls. This results in the sounds known as knock or ping. Strongly influenced by fuel-octane rating, ignition timing, and compression ratio may be caused by hot carbon deposits on the piston or cylinder head.

Knock Sensor. A vibration sensor attached to the cylinder block that generates voltage when knock occurs. The voltage signals a control unit that adjusts timing (and limits boost on turbocharged cars) to stop the knock.

kPa. (See Kilopascals)

Lambda (λ). Expresses the air-fuel ratio in terms of the stoichiometric ratio compared to the oxygen content of the exhaust. At the stoichiometric ratio, when all of the fuel is burned with all of the air in the combustion chamber, the oxygen content of the exhaust is said to be at lambda = 1. If there is excess oxygen in the exhaust (lean mixture), then lambda is greater than 1 ($\lambda > 1$); if there is an excess of fuel in the exhaust (a shortage of air—a rich mixture), then lambda is less than 1 ($\lambda < 1$).

Lambda Control. Closed-loop system that adjusts air-fuel ratio to lambda = 1 based on sensing the amount of excess oxygen in the exhaust.

Lambda Control Valve. A part of K-lambda continuous injection systems that adjusts the air-fuel ratio in response to signals from the lambda sensor. Also known as the Frequency Valve or Timing Valve.

Lambda One (Lambda 1:1; Lambda = 1). (See Lambda)

Lambda Sensor. A small sensor mounted in the exhaust system that reacts to changes in the oxygen content of the exhaust gases. Voltage generated by sensor is monitored by lambda control. Also called the Oxygen Sensor.

Lean Mixture. Excess air. The air quantity drawn into the engine is greater than that required for the stoichiometric ratio, and there is oxygen left over after the fuel is burned. Lambda is greater than one, and the air-fuel ratio is greater than 14.7:1.

LED. Light Emitting Diode. A semiconductor that emits light when current is applied to it. Often used as an indicator in place of a light bulb.

Load. The amount of work the engine must do. When the car accelerates quickly from a standstill, the engine is under a heavy load.

Manifold Absolute Pressure. Manifold pressure measured on the absolute pressure scale, an indication of engine load. Abbreviation is MAP. At sea level, with the engine off, MAP equals 1 bar (14.5 psi). At a typical cruising speed MAP is about 0.7 bar (10 psi).

MAP. (See Manifold Absolute Pressure)

Map. A pictorial representation of a series of data points stored in the control unit memory of Motronic systems. The control unit refers to these maps to control as many as eight different quantities, including fuel injection and ignition timing.

Mass. The quantity of matter contained in an object. Also a measure of that object's resistance to acceleration. With normal earth gravity, it is equivalent to weight. In fuel injection, measured air volume must be corrected for temperature and density to determine its approximate mass.

Metering Slit. In continuous injection, the narrow slits in the control-plunger barrel of the fuel distributor. Fuel flows through the slits, as determined by the lift of the control plunger and the pressure drop at the slits, to the injectors.

Milliampere (mA). One-one-thousandth of one ampere. The current flow to the pressure actuator in KE systems is measured in milliamps.

Mixture. (See Air-Fuel Mixture)

Mixture-Control Unit. In continuous injection, the combination of the fuel distributor mounted on the air-flow sensor.

Motronic. Bosch term for an engine management system with a single electronic control unit that controls fuel injection and ignition timing, and in some cases also idle speed. Originally, Motronic was an extension of L-Jetronic. Now there are Motronic versions of both pulsed and continuous systems.

Multi-Port Injection. An injection system where fuel is injected into the intake manifold at each manifold port near the intake valve.

NTC. Negative Temperature Coefficient. Resistance decreases as temperature increases. See also Temperature Sensor.

Ohm. Unit of measure of resistance to flow of electrical current. The more ohms of resistance the less current flow.

Open-Loop Control. Control of an engine system based on fixed, pre-set values.

Oxygen Sensor. (See Lambda Sensor)

Part-Load. Throttle opening between idle and fully-open.

Part-Load Enrichment. Additional fuel injected during throttle opening to enrich mixture during transition. Usually closed-loop.

Pintle. The tip of the injector that opens to deliver fuel. Shape of the pintle determines the spray pattern of the atomized fuel.

Plunger. (See Control Plunger)

Port Injection. A fuel-injection system where the fuel is injected into the intake manifold by individual injectors at each cylinder intake port, upstream of the intake valve.

Post-Start. Bosch term that refers to the period of time between engine start-up and the beginning of Warm-Up. Bosch systems provide post-start enrichment based on engine temperature and the time since engine start.

Potentiometer. Variable resistance element, usually a rotary unit. In KE systems, rotation of an arm increases resistance to signal sensor-plate movement.

Power. (See Engine Power)

Pressure Actuator. In KE systems, a hydraulic valve controlled electronically to continuously adjust fuel pressure in the lower-chambers of the fuel distributor. Controls all adjustments to basic fuel metering and air-fuel ratio to compensate for changing operating conditions. Also known as Differential-Pressure Regulator, and the Electro-Hydraulic Pressure Actuator.

Pressure Drop. The difference in pressure where metering occurs. In continuous systems this is the difference between system pressure in the control-plunger slits, and upper-chamber-pressure outside the slits. In pulsed systems, this is the difference between fuel-system pressure and intake manifold pressure.

Pressure Regulator. A spring-loaded relief valve that returns excess fuel to the fuel tank to maintain system pressure.

PSI. Abbreviation for Pounds-per-Square-Inch. PSI can be a measure of air or fluid pressure. 14.7 psi = approximately 1 bar = Atmospheric Pressure at sea level.

Pulsed Injection. A system that delivers fuel in intermittent pulses by the opening and closing of solenoid-controlled injectors. Also called EFI.

Pulsed Injection Systems. (D-Jetronic, L-Jetronic, LH-Jetronic, Motronic, LH-Motronic.)

Pulse Period. The available time, dependent on the speed of crankshaft rotation, for opening of pulsed solenoid injectors.

Pulse Time. The amount of time that solenoid injectors are open to inject fuel. Also known as Pulse Width, especially when displayed on an oscilloscope as a voltage pattern.

Pulse Width. (See Pulse Time)

Relative Pressure. In pulsed injection, the difference in pressure between fuel pressure in the injector, and pressure in the intake manifold.

Residual Pressure. Fuel pressure in the fuel lines and fuel-injection system after the fuel pump shuts off.

Retard. (See Ignition Advance/Retard)

Rich Mixture. A lack of air. Less air is drawn into the engine than is required for the stoichiometric ratio. There is still fuel left after all of the oxygen has burned. Lambda is less than one, and the air-fuel mixture is less than 14.7:1.

RPM. Revolutions-Per-Minute. The speed of crankshaft rotation.

Sensor Plate. In continuous injection systems, a circular plate suspended in the intake air stream to sense the incoming air flow.

Slit. (See Metering Slit)

Solenoid. An electromagnet that moves a plunger or metal strip when current is applied.

Stoichiometric Ratio. An air-fuel ratio of 14.7:1. All of the air and all of the fuel is burned in the cylinder. The stoichiometric ratio is the best compromise between a rich air-fuel ratio for best power, and a lean air-fuel ratio for best economy. Also called the Ideal Air-Fuel Ratio. See also Lambda.

System Pressure. Fuel pressure in the fuel lines and at the pressure regulator, created by the fuel pump.

6 GLOSSARY

TBI. Throttle-Body Injection

TDC. Top Dead Center. The point where the piston is at the top of its stroke in the cylinder.

Temperature Sensor. A solid-state resistor, called a thermistor. Used to sense coolant (engine) temperature and, in some systems, air temperature. Sometimes referred to as an NTC sensor for its Negative Temperature Coefficient.

Throttle Valve. The movable plate in the intake tract controlled by the accelerator pedal. It controls the amount of air drawn into the engine.

Timing. The relation of an engine function—such as valve opening, spark-plug firing—to the position of the crankshaft and the pistons.

Timing Valve. (See Lambda Control Valve)

Torque. Twisting force that causes a shaft to turn.

Updraft. In continuous injection, a configuration of the airflow sensor in which air flows up past the sensor plate. See also Downdraft.

Vacuum. Anything less than atmospheric pressure.

Vapor Lock. A situation where fuel in the fuel system becomes so hot that it vaporizes, slowing or stopping fuel flow.

Volt. Unit of measure of electrical force. Voltage causes current (electrons) to flow in a circuit.

Warm-Up. The period of time between the end of Post-Start Enrichment and when the engine reaches operating temperature.

Warm-Up Regulator. (See Control-Pressure Regulator)

WOT. Wide-Open Throttle

Zero Absolute Pressure. A total vacuum. Zero on the absolute pressure scale.

INDEX

Subjects are indexed by chapter number in **bold**, followed by the page number(s) within the chapter where the subject can be found. Thus **3**:10 refers to page 10 of chapter 3.

Subjects are indexed by chapter number in **bold**, followed by the page number(s) within the chapter where the subject can be found. Thus **3**:10 refers to page 10 of chapter 3.

Subjects are indexed by chapter number in **bold**, followed by the page number(s) within the chapter where the subject can be found. Thus **3**:10 refers to page 10 of chapter 3.

6 INDEX

Subjects are indexed by chapter number in **bold**, followed by the page number(s) within the chapter where the subject can be found. Thus **3**:10 refers to page 10 of chapter 3.

Art courtesy of Audi of America

Chapter 6
Fig. 6-19, Fig. 6-20, Fig. 6-21, Fig. 6-22

Art courtesy of Robert Bosch Corporation

Chapter 1
Fig. 3-1, Fig. 3-2, Fig. 4-1, Fig. 4-2

Chapter 2
Fig. 2-1, Fig. 2-2, Fig. 2-3, Fig. 2-4, Fig. 2-6, Fig. 2-7, Fig. 2-8, Fig. 3-1, Fig. 3-4, Fig. 4-3, Fig. 4-5

Chapter 3
Fig. 1-1, Fig. 1-2 Fig. 2-1, Fig. 2-3, Fig. 2-4, Fig. 2-5, Fig. 2-7, Fig. 2-8, Fig. 2-9, Fig. 2-10, Fig. 2-11, Fig. 2-12, Fig. 2-13, Fig. 2-14, Fig. 2-15, Fig. 2-19, Fig. 2-20, Fig. 2-21, Fig. 2-22, Fig. 2-23, Fig. 2-24, Fig. 2-27, Fig. 2-28, Fig. 2-30, Fig. 2-32, Fig. 2-33, Fig. 2-34, Fig. 2-35, Fig. 2-36, Fig. 2-38, Fig. 2-39, Fig. 2-40, Fig. 2-41, Fig. 2-42, Fig. 2-43, Fig. 2-44, Fig. 2-45, Fig. 2-46, Fig. 2-49, Fig. 3-1, Fig. 3-2, Fig. 3-3, Fig. 3-4, Fig. 3-5, Fig. 3-6, Fig. 4-1, Fig. 4-2, Fig. 4-3, Fig. 4-4, Fig. 4-5, Fig. 4-6, Fig. 4-7, Fig. 4-8, Fig. 4-9, Fig. 4-10, Fig. 4-11, Fig. 4-12, Fig. 4-13, Fig. 4-14, Fig. 4-15, Fig. 4-16, Fig. 4-17, Fig. 4-18, Fig. 4-19, Fig. 4-20, Fig. 4-21, Fig. 4-22, Fig. 4-24, Fig. 4-25, Fig. 4-26, Fig. 4-27, Fig. 4-28, Fig. 4-29, Fig. 4-30, Fig. 5-1, Fig. 5-2, Fig. 5-3, Fig. 5-4, Fig. 5-5

Chapter 4
Fig. 1-1, Fig. 1-2, Fig. 1-4, Fig. 2-1, Fig. 2-2, Fig. 2-3, Fig. 2-4, Fig. 3-1, Fig. 3-2, Fig. 3-3, Fig. 3-4, Fig. 3-5, Fig. 3-6, Fig. 3-7, Fig. 3-9, Fig. 3-10, Fig. 3-11, Fig. 3-12, Fig. 3-13, Fig. 3-14, Fig. 4-1, Fig. 4-2, Fig. 4-3, Fig. 4-4, Fig. 5-1, Fig. 5-2, Fig. 5-3, Fig. 5-4, Fig. 5-5, Fig. 5-6, Fig. 5-7, Fig. 5-8, Fig. 6-1, Fig. 6-2, Fig. 6-3, Fig. 6-4, Fig. 6-5, Fig. 7-1, Fig. 7-2, Fig. 7-3, Fig. 7-4, Fig. 7-7, Fig. 7-8, Fig. 8-1

Chapter 5
Fig. 1-1, Fig. 1-4, Fig. 1-5, Fig. 1-7, Fig. 1-8, Fig. 2-1, Fig. 2-3, Fig. 2-4, Fig. 2-5, Fig. 2-6, Fig. 2-7, Fig. 2-8, Fig. 2-13, Fig. 2-14, Fig. 3-1, Fig. 3-2, Fig. 3-4, Fig. 3-5, Fig. 3-6, Fig. 3-7, Fig. 3-8, Fig. 4-1, Fig. 4-2, Fig. 4-3, Fig. 4-4, Fig. 4-5, Fig. 4-6, Fig. 4-7, Fig. 5-1, Fig. 5-2, Fig. 5-4, Fig. 5-5, Fig. 5-6, Fig. 5-7, Fig. 5-9, Fig. 5-10, Fig. 5-13, Fig. 5-14, Fig. 5-16, Fig. 5-17, Fig. 5-19, Fig. 6-1, Fig. 6-2, Fig. 6-3

Chapter 6
Fig. 1-1, Fig. 3-2, Fig. 3-4, Fig. 3-6, Fig. 3-8, Fig. 3-9, Fig. 3-10, Fig. 3-11, Fig. 3-12, Fig. 3-13, Fig. 3-14, Fig. 3-15, Fig. 3-16, Fig. 4-1, Fig. 4-2, Fig. 4-4, Fig. 4-6, Fig. 4-7, Fig. 4-8, Fig. 5-1, Fig. 5-2, Fig. 5-3, Fig. 5-4, Fig. 5-5, Fig. 5-6, Fig. 5-7, Fig. 5-8, Fig. 5-9, Fig. 5-10, Fig. 5-11, Fig. 5-12, Fig. 5-13, Fig. 5-14, Fig. 5-15, Fig. 5-16, Fig. 5-17, Fig. 5-18, Fig. 5-20, Fig. 6-1, Fig. 6-2, Fig. 6-3, Fig. 6-4, Fig. 6-5, Fig. 6-6, Fig. 6-7, Fig. 6-8, Fig. 6-9, Fig. 6-10, Fig. 6-11, Fig. 6-12, Fig. 6-13, Fig. 6-14, Fig. 6-15, Fig. 6-16, Fig. 6-17, Fig. 6-18, Fig. 7-1, Fig. 7-2, Fig. 7-5, Fig. 7-6, Fig. 7-7, Fig. 7-8

Chapter 7
Fig. 2-1, Fig. 2-2, Fig. 2-3, Fig. 3-1, Fig. 4-1, Fig. 4-2, Fig. 4-3, Fig. 5-4, Fig. 5-5, Fig. 5-9, Fig. 6-3, Fig. 7-1, Fig. 8-1

Art courtesy of Chevron

Chapter 4
Fig. 3-15, Fig. 3-17

Art courtesy of Cybern

Chapter 1
Fig. 2-1, Fig. 2-2, Fig. 3-3, Fig. 3-4, Fig. 3-5, Fig. 3-6, Fig. 5-1, Fig. 5-2, Fig. 6-1, Fig. 6-2, Fig. 6-3, Fig. 6-4, Fig. 6-5, Fig. 6-6, Fig. 6-7, Fig. 6-8, Fig. 6-9, Fig. 6-10, Fig. 6-11, Fig. 6-12, Fig. 6-13, Fig. 6-14, Fig. 6-15

Chapter 2
Fig. 2-5, Fig. 2-13, Fig. 2-14, Fig. 3-2, Fig. 4-2, Fig. 4-4, Fig. 5-1

Chapter 3
Fig. 2-16, Fig. 2-17, Fig. 2-18, Fig. 2-25, Fig. 2-26, Fig. 2-31, Fig. 2-37, Fig. 2-47, Fig. 2-48

Chapter 4
Fig. 1-3, Fig. 3-8

Chapter 5
Fig. 5-18, Fig. 7-3

Chapter 7
Fig. 1-2, Fig. 5-1, Fig. 5-6, Fig. 9-1, Fig. 9-2, Fig. 9-3, Fig. 9-4, Fig. 9-5, Fig. 9-6, Fig. 9-7, Fig. 9-8

Art courtesy of Dinan Engineering

Chapter 3
Fig. 2-2

Chapter 5
Fig. 1-2

Chapter 7
Fig. 1-1, Fig. 5-7

Art courtesy of Micro Dynamics

Chapter 7
Fig. 5-3, Fig. 5-8, Fig. 8-2, Fig. 10-1

Art courtesy of Mitsubishi

Chapter 2
Fig. 4-1

2 ART CREDITS

Art courtesy of Robert Bentley, Inc.

Chapter 2
Fig. 2-9, Fig. 2-10, Fig. 2-11, Fig. 2-12

Chapter 3
Fig. 2-6, Fig. 2-29

Chapter 5
Fig. 1-6, Fig. 2-12, Fig. 5-11, Fig. 5-12, Fig. 5-15, Fig. 6-6,

Chapter 6
Fig. 3-7, Fig. 5-19, Fig. 5-22, Fig. 7-3, Fig. 7-4

Chapter 7
Fig. 6-1, RBI/Tom Wilson; Fig. 6-2, RBI/Tom Wilson

Art courtesy of SAE

Chapter 4
Fig. 3-16, Fig. 3-18

Art courtesy of Volkswagen of America

Chapter 2
Fig. 4-3

Chapter 3
Fig. 4-23

Chapter 4
Fig. 7-5

Chapter 5
Fig. 1-3, Fig. 2-2, Fig. 2-9, Fig. 2-10, Fig. 2-11, Fig. 2-15, Fig. 3-3, Fig. 3-9, Fig. 3-10, Fig. 3-11, Fig. 5-3, Fig. 5-8, Fig. 6-4, Fig. 6-5, Fig. 7-1, Fig. 7-2, Fig. 7-4, Fig. 7-5

Chapter 6
Fig. 1-2, Fig. 1-3, Fig. 3-1, Fig. 3-3, Fig. 3-5, Fig. 4-3, Fig. 4-5, Fig. 5-21

Chapter 7
Fig. 5-2

Art courtesy of Wray Electronics

Chapter 7
Fig. 5-10

About the Author

Charles O. Probst received his BSE (ME–Automotive) from the University of Michigan. His career specialty is technical communication, primarily in automotive subjects. He works as an instructional-system designer, writer and filmmaker, and has been responsible for numerous video and film productions, manuals, job guides, and other systems for improving human performance.

Probst was the Senior Author for Motor's Auto Engines & Electrical Systems, and also served as Technical Editor for a 60-volume set of self-instructional auto technology courses developed by the Commercial Trades Institute. Among his many publications are articles in Automotive Engineering, Automobile Quarterly, Car Life, Consumer Digest, and Road Test.

He is active in the SAE Technical Board as Chairman of the Technician Training Committee, which is responsible for SAE Recommended Practices for technician-training systems.

In a second career, Probst served 27 years (with 8 on active duty) in the Air Force Reserve, concluding as a Colonel.